# HEAT TRANSFER

# Heat Transfer

## J. P. HOLMAN

Professor of Mechanical Engineering
Southern Methodist University

### THIRD EDITION

**McGRAW-HILL BOOK COMPANY**
New York    St. Louis    San Francisco
Düsseldorf    Johannesburg    Kuala Lumpur    London
Mexico    Montreal    New Delhi    Panama
Rio de Janeiro    Singapore
Sydney    Toronto

**HEAT TRANSFER**

*Library of Congress Catalog Card Number* 78-168452

07-029603-0

67890 KPKP 7987654


This book was set in Modern by The Maple Press Company,
and printed and bound by Kingsport Press, Inc.
The cover was designed by Edward A. Butler;
the drawings were done by Hank Iken.
The editors were B. J. Clark and J. W. Maisel.
John A. Sabella supervised production.

# CONTENTS

# PREFACE

This book presents an elementary treatment of the principles of heat transfer. As a text it contains sufficient material for a one-semester course which may be presented at the junior level, or higher, depending on individual course objectives. A background in ordinary differential equations is helpful for proper understanding of the material. Although some familiarity with fluid mechanics will aid in the convection discussions, it is not essential. The concepts of thermodynamic energy balances are also useful in the various analytical developments.

Presentation of the subject follows classical lines of separate discussions for conduction, convection, and radiation, although it is emphasized that the physical mechanism of convection heat transfer is one of conduction through the stationary fluid layer near the heat transfer surface. Throughout the book emphasis has been placed on physical understanding while, at the same time, relying on meaningful experimental data in those circumstances which do not permit a simple analytical solution.

Conduction is treated from both the analytical and numerical viewpoint so that the reader is afforded the insight which is gained from analytical solutions as well as the important tools of numerical analysis which must often be used in practice. A similar procedure is followed in the presentation of convection heat transfer. An integral analysis of both free- and forced-convection boundary layers is used to present a physical picture of the convection process. From this physical description inferences may be drawn which naturally lead to the presentation of empirical and practical relations for calculating convection heat-transfer coefficients. Because it provides an easier instruction vehicle than other methods, the radiation network method is used extensively for the analysis of radiation systems.

The log-mean-temperature-difference and effectiveness approaches are presented in heat-exchanger analysis since both are in wide use and each offers its own advantages to the designer. A brief introduction to diffusion and mass transfer is presented in order to acquaint the reader with these processes and to establish more firmly the important analogies between heat, mass, and momentum transfer.

A number of special topics are discussed in Chapter 12 which give added flavor to the basic material of the preceding chapters. Chapter 13 on environmental heat transfer is also rather specialized but contains material which is of increasing interest.

Problems are included at the end of each chapter. Some of these problems are of a routine nature to familiarize the student with the numerical manipulations and orders of magnitude of various parameters which occur in the subject of heat transfer. Other problems extend the subject matter by requiring the student to apply the basic principles to new situations and develop his own equations. Both types of problems are important.

The subject of heat transfer is not a static one. New developments occur quite regularly and better analytical solutions and empirical data are continuously made available to the professional in the field. Because of the huge amount of information which is available in the research literature, the beginning student could easily be overwhelmed if too many of the nuances of the subject were displayed and expanded before him. The book is designed to serve as an elementary text, so the author has assumed a role of interpreter of the literature with those findings and equations being presented which can be of immediate utility to the reader. It is hoped that the student's attention is called to more extensive works in a sufficient number of instances to emphasize the greater depth which is available on most of the subjects of heat transfer. For the serious student, then, the end-of-chapter references offer an open door to the literature of heat transfer which can pyramid upon further investigation.

The author has enjoyed the pleasure of teaching the material in this text to students at Southern Methodist University for more than ten years. Their patience, suggestions, and comments have contributed in a large way to the particular subject matter and presentation in the text. I hope that this new edition will reflect their interests and stimulate further lively discussions in the future.

Mention should be made of the modifications which appear in this third edition. A set of review questions has been added at the end of each chapter as well as additional problems. Some emphasis is given to SI units throughout the book by expressing answers to examples in this system in addition to the English units.

More emphasis has been given to numerical methods in the chapters on conduction with discussion of techniques most applicable to computer solution. Empirical relations for flow over bluff bodies has been expressed in a more compact form. Several new correlations for free convection systems are presented in Chapter 7 and the old conventional correlations are grouped together in a more compact and usable display. Minor modifications have been made to the radiation chapter. New information on peak heat flux in boiling has been added to Chapter 9 as well as some recent references. A discussion of the heat pipe and its applications has been added to the special topics chapter. A new chapter on environmental problems has been added to stress the variety of applications of the subject of heat transfer. It is hoped that these additions throughout the book will provide a flavor which increases the general appeal of the subject.

Many people have been generous with their comments and suggestions for improvement of the book. S. J. Kline, R. J. Schoenhals, J. V. Beck, Erich Soehngen, K. V. Prasanna, M. Fleischman, J. E. Sunderland, W. G. Wyatt, and others, have been most helpful and I am grateful for their interest. Finally, I particularly appreciate the continuing support of the editorial staff at the McGraw-Hill Book Company.

**J. P. HOLMAN**

# LIST OF SYMBOLS

$a$    Local velocity of sound

$a$    Attenuation coefficient (Chap. 13)

$A$    Area

$A$    Albedo (Chap. 13)

$A_m$    Fin profile area (Chap. 2)

$\mathbf{B}$    Magnetic field strength

$c$    Specific heat, usually Btu/lb$_m$-°F

$C$    Concentration (Chap. 11)

$C_D$    Drag coefficient, defined by Eq. (6-10)

$C_f$    Friction coefficient, defined by Eq. (5-44)

$c_p$    Specific heat at constant pressure, usually Btu/lb$_m$-°F

$c_v$    Specific heat at constant volume, usually Btu/lb$_m$-°F

$d$    Diameter

$D$    Depth or diameter

$D$    Diffusion coefficient (Chap. 11)

$D_H$    Hydraulic diameter, defined by Eq. (6-9)

$e$    Internal energy per unit mass, usually Btu/lb$_m$

$E$    Internal energy, usually Btu

$E$    Emissive power, usually Btu/hr-ft² (Chap. 8)

$E_{b_0}$    Solar constant (Chaps. 8, 13)

$E_{b\lambda}$    Blackbody emissive power per unit wavelength, defined by Eq. (8-10)

$\mathbf{E}$    Electric field vector

$f$    Friction factor, defined by Eq. (5-70)

$F$    Force, usually lb$_f$

$F_{m\text{-}n}$ or $F_{mn}$    Radiation shape factor for radiation from surface $m$ to surface $n$

$g$    Acceleration of gravity

$g_c$    Conversion factor, defined by Eq. (1-12)

$G = \dfrac{\dot{m}}{A}$    Mass velocity

$G$    Irradiation (Chap. 8)

$h$    Heat-transfer coefficient, usually Btu/hr-ft²-°F

$\bar{h}$    Average heat-transfer coefficient

$h_D$    Mass-transfer coefficient, usually ft/hr

$h_{fg}$    Enthalpy of vaporization, Btu/lb$_m$

$h_r$    Radiation heat-transfer coefficient (Chap. 8)

$\mathbf{H}$    Magnetic field intensity

$i$    Enthalpy, usually Btu/lb$_m$

$I$    Intensity of radiation

$I$    Solar insolation (Chap. 13)

$I_0$    Solar insolation at outer edge of atmosphere [Eq. (13-4b)]

$J$    Radiosity (Chap. 8)

$\mathbf{J}$    Current density

$k$    Thermal conductivity, usually Btu/hr-ft-°F

$k_e$    Effective thermal conductivity of enclosed spaces (Chap. 7). Defined by Eq. (7-35)

$k_\lambda$    Scattering coefficient (Chap. 13)

$L$    Length

$L_c$ Corrected fin length (Chap. 2)

$m$ Mass

$\dot{m}$ Mass rate of flow

$M$ Molecular weight (Chap. 11)

$n$ Molecular density

$n$ Turbidity factor, defined by Eq. (13-4)

$N$ Molal diffusion rate, moles per unit time (Chap. 11)

$p$ Pressure, usually $lb_f/ft^2$

$P$ Perimeter

$q$ Heat-transfer rate, Btu per unit time

$q''$ Heat flux, Btu per unit time, per unit area

$\dot{q}$ Heat generated per unit volume

$\bar{q}_{m,n}$ Residual of a node, used in relaxation method (Chaps. 3 and 4)

$Q$ Heat, Btu

$r$ Radius or radial distance

$r$ Recovery factor, defined by Eq. (5-78)

$R$ Fixed radius

$R$ Gas constant

$R_{th}$ Thermal resistance, usually hr-°F/Btu

$s$ A characteristic dimension (Chap. 4)

$S$ Molecular speed ratio (Chap. 12)

$S$ Conduction shape factor, usually ft

$t$ Thickness, applied to fin problems (Chap. 2)

$t, T$ Temperature

$u$ Velocity

$v$ Velocity

$v$ Specific volume, usually $ft^3/lb_m$

$V$ Velocity

$V$ Molecular volume (Chap. 11)

$W$ Weight, usually $lb_f$

$x, y, z$ Space coordinates in cartesian system

$\alpha = \dfrac{k}{\rho c}$ Thermal diffusivity, usually $ft^2/hr$

$\alpha$ Absorptivity (Chap. 8)

$\alpha$ Accommodation coefficient (Chap. 12)

$\alpha$ Solar altitude angle, degrees, (Chap. 13)

$\alpha$ Ambient atmospheric lapse rate (LR)

$\beta$ Volume coefficient of expansion, 1/°R

$\beta$ Temperature coefficient of thermal conductivity, 1/°F

$\gamma = \dfrac{c_p}{c_v}$ Isentropic exponent, dimensionless

$\Gamma$ Condensate mass flow per unit depth of plate (Chap. 9)

$\Gamma$ Dry adiabatic lapse rate (DALR) (Chap. 13)

$\delta$ Hydrodynamic boundary-layer thickness

$\delta_t$ Thermal boundary-layer thickness

$\epsilon$ Heat-exchanger effectiveness

$\epsilon$ Emissivity

$\epsilon_H, \epsilon_M$ Eddy diffusivity of heat and momentum (Chap. 5)

$\zeta = \dfrac{\delta_t}{\delta}$ Ratio of thermal boundary-layer thickness to hydrodynamic boundary-layer thickness

$\eta$ Similarity variable, defined by Eq. (B-6)

$\eta_f$ Fin efficiency, dimensionless

$\theta$ Angle in spherical or cylindrical coordinate system

$\theta$ Temperature difference, $T - T_{reference}$ The reference temperature λ is chosen differently for different systems (see Chaps. 2 to 4)

λ    Wavelength

λ    Mean-free path (Chap. 12)

$\mu$    Dynamic viscosity

$\mu$    Denotes micron unit of wavelength, $1\mu = 10^{-6}$ m (Chap. 8)

$\nu$    Kinematic viscosity

$\nu$    Frequency of radiation (Chap. 8)

$\rho$    Density, usually $lb_m/ft^3$

$\rho$    Reflectivity (Chap. 8)

$\rho_e$    Charge density

$\sigma$    Electrical conductivity

$\sigma$    Stefan-Boltzmann constant

$\sigma$    Surface tension of liquid-vapor interface (Chap. 9)

$\tau$    Time

$\tau$    Shear stress between fluid layers

$\tau$    Transmissivity (Chap. 8)

$\phi$    Angle in spherical or cylindrical coordinate system

$\psi$    Stream function

*Dimensionless Groups*

$Bi = \dfrac{hs}{k}$    Biot modulus

$Ec = \dfrac{u_\infty^2}{c_p(T_\infty - T_w)}$    Eckert number

$Fo = \dfrac{\alpha\tau}{s^2}$    Fourier modulus

$Gr = \dfrac{g\beta(T_w - T_\infty)x^3}{\nu^2}$    Grashof number

$Gr^* = GrNu$    Modified Grashof number for constant heat flux

$Gz = Re\,Pr\,\dfrac{d}{L}$    Graetz number

$Kn = \dfrac{\lambda}{L}$    Knudsen number

$Le = \dfrac{\alpha}{D}$    Lewis number (Chap. 11)

$M = \dfrac{u}{a}$    Mach number

$N = \dfrac{\sigma B_y^2 x}{\rho u_\infty}$    Magnetic influence number

$Nu = \dfrac{hx}{k}$    Nusselt number

$\overline{Nu} = \dfrac{\bar{h}x}{k}$    Average Nusselt number

$Pe = Re\,Pr$    Peclet number

$Pr = \dfrac{c_p\mu}{k}$    Prandtl number

$Re = \dfrac{\rho u x}{\mu}$    Reynolds number

$Sc = \dfrac{\nu}{D}$    Schmidt number (Chap. 11)

$Sh = \dfrac{h_D x}{D}$    Sherwood number (Chap. 11)

$St = \dfrac{h}{\rho c_p u}$    Stanton number

$\overline{St} = \dfrac{\bar{h}}{\rho c_p u}$    Average Stanton number

*Subscripts*

$aw$    Adiabatic wall conditions

$b$    Refers to blackbody conditions (Chap. 8)

$b$    Evaluated at bulk conditions [see Eq. (5-61)]

$d$    Based on diameter

$f$    Evaluated at film conditions [see Eq. (5-43)]

$g$    Saturated vapor conditions (Chap. 9)

$i$    Initial or inlet conditions

$L$    Based on length of plate

$m$    Mean flow conditions

$m, n$    Denotes nodal positions in numerical solution (see Chaps. 3 and 4)

$0$    Denotes stagnation flow conditions (Chap. 5) or some initial condition at time zero

$r$    At specified radial position

$s$      Evaluated at condition of surroundings

$x$      Denotes some local position with respect to $x$ coordinate

$w$      Evaluated at wall conditions

$*$      (Superscript) Properties evaluated at reference temperature, given by Eq. (5-82)

$\infty$      Evaluated at free-stream conditions

# CHAPTER
# 1
# INTRODUCTION

Heat transfer is that science which seeks to predict the energy transfer which may take place between material bodies as a result of a temperature difference. Thermodynamics teaches that this energy transfer is defined as heat. The science of heat transfer seeks not merely to explain how heat energy may be transferred, but also to predict the rate at which the exchange will take place under certain specified conditions. The fact that a heat-transfer *rate* is the desired objective of an analysis points out the difference between heat transfer and thermodynamics. Thermodynamics deals with systems in equilibrium; it may be used to predict the amount of energy required to change a system from one equilibrium state to another; it may not be used to predict how fast a change will take place since the system is not in equilibrium during the process. Heat transfer supplements the first and second principles of thermodynamics by providing additional experimental rules which may be used to establish energy-transfer rates. As in the science of thermodynamics, the experimental rules used as a basis of the subject of heat transfer are rather simple and easily expanded to encompass a variety of practical situations.

As an example of the different kinds of problems which are treated by thermodynamics and heat transfer, consider the cooling of a hot steel bar which is placed in a pail of water. Thermodynamics may be used to predict the final equilibrium temperature of the steel bar–water combination. Thermodynamics will not tell us how long it takes to reach this equilibrium condition or what the temperature of the bar will be after a certain length of time before the equilibrium condition is attained. Heat transfer may be used to predict the temperature of both the bar and the water as a function of time.

Most readers will be familiar with the terms used to denote the three modes of heat transfer: conduction, convection, and radiation. In this chapter we seek to explain the mechanism of these modes qualitatively so that each may be considered in its proper perspective. Subsequent chapters will treat the three types of heat transfer in detail.

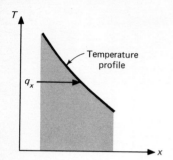

T

q_x

Temperature profile

x

**Fig. 1-1   Sketch showing direction of heat flow.**

## 1-1□CONDUCTION HEAT TRANSFER

When a temperature gradient exists in a body, experience has shown that there is an energy transfer from the high-temperature region to the low-temperature region. We say that the energy is transferred by conduction and that the heat-transfer rate per unit area is proportional to the normal temperature gradient.

$$\frac{q}{A} \sim \frac{\partial T}{\partial x}$$

When the proportionality constant is inserted,

$$q = -kA \frac{\partial T}{\partial x} \tag{1-1}$$

where $q$ is the heat-transfer rate and $\partial T/\partial x$ is the temperature gradient in the direction of the heat flow. The positive constant $k$ is called the thermal conductivity of the material, and the minus sign is inserted so that the second principle of thermodynamics will be satisfied, i.e., heat must flow downhill on the temperature scale, as indicated in the coordinate system of Fig. 1-1. Equation (1-1) is called Fourier's law of heat conduction after the French mathematical physicist Joseph Fourier, who made very significant contributions to the analytical treatment of conduction heat transfer. It is important to note that Eq. (1-1) is the defining equation for the thermal conductivity and that $k$ has the units of Btu/hr-ft-°F in a typical engineering system of units in which the heat flow is expressed in Btu per hour.

We now set ourselves the problem of determining the basic equation which governs the transfer of heat in a solid, using Eq. (1-1) as a starting point.

Consider the one-dimensional system shown in Fig. 1-2. If the system is in a steady state, i.e., if the temperature does not change with time, then

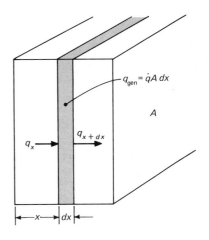

**Fig. 1-2  Elemental volume for one-dimensional heat-conduction analysis.**

the problem is a simple one and we need only integrate Eq. (1-1) and substitute the appropriate values to solve for the desired quantity. However, if the temperature of the solid is changing with time, or if there are heat sources or sinks within the solid, the situation is more complex. We consider the general case where the temperature may be changing with time and heat sources may be present within the body. For the element of thickness $dx$ the following energy balance may be made:

Energy conducted in left face + heat generated within element
        = change in internal energy + energy conducted out right face

These energy quantities are given as follows:

$$\text{Energy in left face} = q_x = -kA\frac{\partial T}{\partial x}$$

$$\text{Energy generated within element} = \dot{q}A\,dx$$

$$\text{Change in internal energy} = \rho c A\frac{\partial T}{\partial \tau}\,dx$$

$$\text{Energy out right face} = q_{x+dx} = -kA\frac{\partial T}{\partial x}\Big)_{x+dx}$$

$$= -A\left[k\frac{\partial T}{\partial x} + \frac{\partial}{\partial x}\left(k\frac{\partial T}{\partial x}\right)dx\right]$$

where $\dot{q}$ = energy generated per unit volume
        $c$ = specific heat of material
        $\rho$ = density

Combining the relations above,

$$-kA \frac{\partial T}{\partial x} + \dot{q}A \, dx = \rho cA \frac{\partial T}{\partial \tau} \, dx - A \left[ k \frac{\partial T}{\partial x} + \frac{\partial}{\partial x} \left( k \frac{\partial T}{\partial x} \right) dx \right]$$

or

$$\frac{\partial}{\partial x} \left( k \frac{\partial T}{\partial x} \right) + \dot{q} = \rho c \frac{\partial T}{\partial \tau} \tag{1-2}$$

This is the one-dimensional heat-conduction equation. To treat more than one-dimensional heat flow we need only consider the heat conducted in and out of a unit volume in all three coordinate directions, as shown in Fig. 1-3a. The energy balance yields

$$q_x + q_y + q_z + q_{\text{gen}} = q_{x+dx} + q_{y+dy} + q_{z+dz} + \frac{dE}{d\tau}$$

and the energy quantities are given by

$$q_x = -k \, dy \, dz \frac{\partial T}{\partial x}$$

$$q_{x+dx} = -\left[ k \frac{\partial T}{\partial x} + \frac{\partial}{\partial x} \left( k \frac{\partial T}{\partial x} \right) dx \right] dy \, dz$$

$$q_y = -k \, dx \, dz \frac{\partial T}{\partial y}$$

$$q_{y+dy} = -\left[ k \frac{\partial T}{\partial y} + \frac{\partial}{\partial y} \left( k \frac{\partial T}{\partial y} \right) dy \right] dx \, dz$$

$$q_z = -k \, dx \, dy \frac{\partial T}{\partial z}$$

$$q_{z+dz} = -\left[ k \frac{\partial T}{\partial z} + \frac{\partial}{\partial z} \left( k \frac{\partial T}{\partial z} \right) dz \right] dx \, dy$$

$$q_{\text{gen}} = \dot{q} \, dx \, dy \, dz$$

$$\frac{dE}{d\tau} = \rho c \, dx \, dy \, dz \frac{\partial T}{\partial \tau}$$

so that the general three-dimensional heat-conduction equation is

$$\frac{\partial}{\partial x} \left( k \frac{\partial T}{\partial x} \right) + \frac{\partial}{\partial y} \left( k \frac{\partial T}{\partial y} \right) + \frac{\partial}{\partial z} \left( k \frac{\partial T}{\partial z} \right) + \dot{q} = \rho c \frac{\partial T}{\partial \tau} \tag{1-3}$$

For constant thermal conductivity Eq. (1-3) is written

$$\frac{\partial^2 T}{\partial x^2} + \frac{\partial^2 T}{\partial y^2} + \frac{\partial^2 T}{\partial z^2} + \frac{\dot{q}}{k} = \frac{1}{\alpha} \frac{\partial T}{\partial \tau} \tag{1-3a}$$

where the quantity $\alpha = k/\rho c$ is called the thermal diffusivity of the material. The larger the value of $\alpha$, the faster will heat diffuse through the material. This may be seen by examining the quantities which make up $\alpha$.

**Fig. 1-3  Elemental volume for three-dimensional heat-conduction analysis. (a) Cartesian coordinates; (b) cylindrical coordinates; (c) spherical coordinates.**

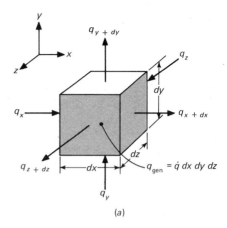

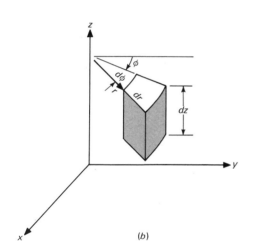

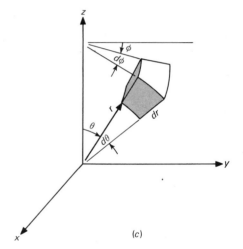

A high value of $\alpha$ could result either from a high value of thermal conductivity, which would indicate a rapid energy-transfer rate, or from a low value of the thermal heat capacity $\rho c$. A low value of the heat capacity would mean that less of the energy moving through the material would be absorbed and used to raise the temperature of the material; thus more energy would be available for further transfer.

In the derivations above, the expression for the derivatives at $x + dx$ has been written in the form of a Taylor series expansion with only the first two terms of the series employed for the development.

Equation (1-3a) may be transformed into either cylindrical or spherical coordinates by standard calculus techniques. The results are

Cylindrical coordinates:

$$\frac{\partial^2 T}{\partial r^2} + \frac{1}{r}\frac{\partial T}{\partial r} + \frac{1}{r^2}\frac{\partial^2 T}{\partial \phi^2} + \frac{\partial^2 T}{\partial z^2} + \frac{\dot{q}}{k} = \frac{1}{\alpha}\frac{\partial T}{\partial \tau} \tag{1-3b}$$

Spherical coordinates:

$$\frac{1}{r}\frac{\partial^2}{\partial r^2}(rT) + \frac{1}{r^2 \sin\theta}\frac{\partial}{\partial \theta}\left(\sin\theta\,\frac{\partial T}{\partial \theta}\right) + \frac{1}{r^2 \sin^2\theta}\frac{\partial^2 T}{\partial \phi^2} + \frac{\dot{q}}{k} = \frac{1}{\alpha}\frac{\partial T}{\partial \tau} \tag{1-3c}$$

The coordinate systems for use with Eqs. (1-3b) and (1-3c) are indicated in Figs. 1-3b and 1-3c, respectively.

Many practical problems involve only special cases of the general equations listed above. As a guide to the developments in future chapters, it is worthwhile to show the reduced form of the general equations for several cases of practical interest.

*Steady-state one-dimensional heat flow* (no heat generation):

$$\frac{d^2 T}{dx^2} = 0 \tag{1-4}$$

Note that this equation is the same as Eq. (1-1) when $q = $ const.

*Steady-state one-dimensional heat flow in cylindrical coordinates* (no heat generation):

$$\frac{d^2 T}{dr^2} + \frac{1}{r}\frac{dT}{dr} = 0 \tag{1-5}$$

*Steady-state one-dimensional heat flow with heat sources:*

$$\frac{d^2 T}{dx^2} + \frac{\dot{q}}{k} = 0 \tag{1-6}$$

*Two-dimensional steady-state conduction without heat sources:*

$$\frac{\partial^2 T}{\partial x^2} + \frac{\partial^2 T}{\partial y^2} = 0 \tag{1-7}$$

## 1-2□THERMAL CONDUCTIVITY

Equation (1-1) is the defining equation for thermal conductivity. Based on this definition, experimental measurements may be made to determine the thermal conductivity of different materials. For gases at moderately low temperatures, analytical treatments in the kinetic theory of gases may be used to predict accurately the experimentally observed values. In some cases, theories are available for the prediction of thermal conductivities in liquids and solids, but in general, there are still many open questions and concepts which need clarification where liquids and solids are concerned.

The mechanism of thermal conduction in a gas is a simple one: We identify the kinetic energy of a molecule with its temperature; thus, in a high-temperature region, the molecules have higher velocities than in some

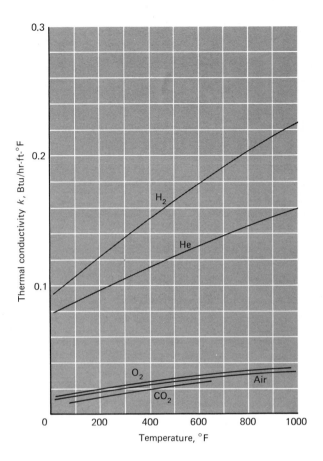

Fig. 1-4 Thermal conductivities of some typical gases.

lower-temperature region. The molecules are in continuous random motion, colliding with one another and exchanging energy and momentum. The molecules have this random motion whether or not a temperature gradient exists in the gas. If a molecule moves from a high-temperature region to a region of lower temperature, it transports kinetic energy to the lower-temperature part of the system and gives up this energy through collisions with lower-energy molecules.

We noted that thermal conductivity has the units of Btu/hr-ft-°F when the heat flow is expressed in Btu per hour. Note that a heat *rate* is involved, and the numerical value of the thermal conductivity indicates how fast heat will flow in a given material. How is the rate of energy transfer taken into account in the molecular model discussed above? Clearly, the faster the molecules move, the faster they will transport energy. Therefore the thermal conductivity of a gas should be dependent on temperature. A simplified analytical treatment shows the thermal conductivity of a gas to vary with

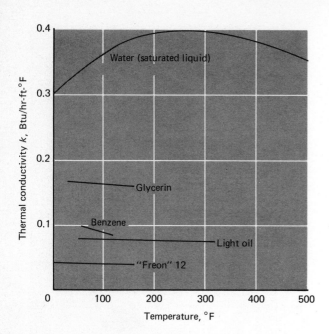

**Fig. 1-5  Thermal conductivities of some typical liquids.**

the square root of the absolute temperature. (It may be recalled that the velocity of sound in a gas varies with the square root of the absolute temperature; this velocity is approximately the mean speed of the molecules.) Thermal conductivities of some typical gases are shown in Fig. 1-4. For most gases at moderate pressures the thermal conductivity is a function of temperature alone. This means that the gaseous data for one atmosphere as given in Appendix A may be used for a rather wide range of pressures. When the pressure of the gas becomes of the order of its critical pressure, or more generally, when nonideal gas behavior is encountered, other sources must be consulted for thermal conductivity data.

The physical mechanism of thermal-energy conduction in liquids is qualitatively the same as in gases; however, the situation is considerably more complex since the molecules are more closely spaced and molecular force fields exert a strong influence on the energy exchange in the collision process. Thermal conductivities of some typical liquids are shown in Fig. 1-5.

Thermal energy may be conducted in solids by two modes: lattice vibration and transport by free electrons. In good electrical conductors a rather large number of free electrons move about in the lattice structure of the material. Just as these electrons may transport electric charge, they may also carry thermal energy from a high-temperature region to a low-temperature region, as in the case of gases. In fact, these electrons are fre-

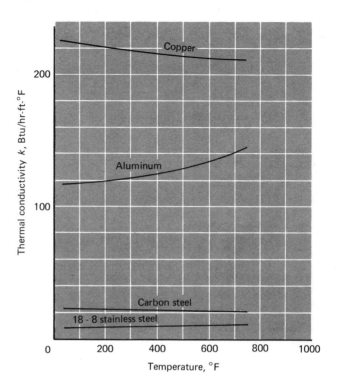

Fig. 1-6  Thermal conductivities of some typical solids.

quently referred to as the *electron gas*. Energy may also be transmitted as vibrational energy in the lattice structure of the material. In general, however, this latter mode of energy transfer is not as large as the electron transport, and it is for this reason that good electrical conductors are almost always good heat conductors, viz., copper, aluminum, and silver, and electrical insulators are usually good heat insulators. Thermal conductivities of some typical solids are shown in Fig. 1-6. Other data are given in Appendix A.

The thermal conductivities of various insulating materials are also given in Appendix A. Some typical values are 0.022 Btu/hr-ft-°F for glass wool and 0.28 Btu/hr-ft-°F for gypsum plaster. At high temperatures, the energy transfer through insulating materials may involve several modes: conduction through the fibrous or porous solid material; conduction through the air trapped in the void spaces; and at sufficiently high temperatures, radiation.

An important technical problem is the storage and transport of cryogenic liquids like liquid hydrogen over extended periods of time. Such applications have led to the development of *superinsulations* for use at these very low temperatures (down to about −430°F). The most effective of these superinsulations consists of multiple layers of highly reflective materials

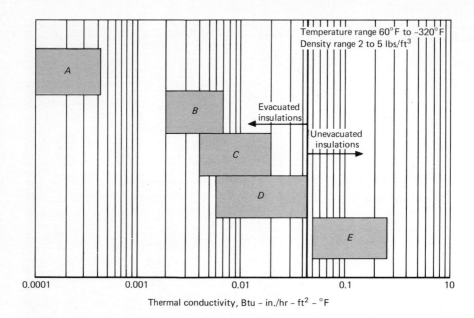

Fig. 1-7  Apparent thermal conductivities of typical cryogenic insulation material. A—multilayer insulations; B—opacified powders; C—glass fibers; D—powders; E—foams, powders, and fibers, according to Ref. 1.

separated by insulating spacers. The entire system is evacuated to minimize air conduction, and thermal conductivities as low as $2 \times 10^{-4}$ Btu/hr-ft-°F are possible. A convenient summary of the thermal conductivities of insulating materials at cryogenic temperatures is given in Fig. 1-7 according to Glaser et al. [1]. Further information on multilayer insulation is given in Ref. 3 and in the text by Barron [2].

## 1-3□CONVECTION HEAT TRANSFER

It is well known that a hot plate of metal will cool faster when placed in front of a fan than when exposed to still air. We say that the heat is convected away, and we call the process convection heat transfer. The term *convection* provides the reader with an intuitive notion concerning the heat-transfer process; however, this intuitive notion must be expanded to enable one to arrive at anything like an adequate analytical treatment of the problem. For example, we know that the velocity at which the air blows over the hot plate obviously influences the heat-transfer rate. But does it influence the cooling in a linear way; i.e., if the velocity is doubled, will the heat-transfer rate double? We should suspect that the heat-transfer rate might be different if we cooled the plate with water instead of air, but again,

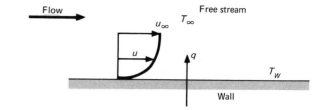

**Fig. 1-8  Sketch showing convection heat transfer from a plate.**

how much difference would there be? These questions may be answered with the aid of some rather basic analyses which will be presented in later chapters. For now, we sketch the physical mechanism of convection heat transfer and show its relation to the conduction process.

Consider the heated plate shown in Fig. 1-8. The temperature of the plate is $T_w$, and the temperature of the fluid is $T_\infty$. The velocity of the flow will appear as shown, being reduced to zero at the plate as a result of viscous action. Since the velocity of the fluid layer at the wall will be zero, the heat must be transferred only by conduction at that point. Thus we might compute the heat transfer, using Eq. (1-1), with the thermal conductivity of the fluid and the fluid temperature gradient at the wall. Why then, if the heat flows by conduction in this layer, do we speak of *convection* heat transfer and need to consider the velocity of the fluid? The answer is that the temperature gradient is dependent on the rate at which the fluid carries the heat away; a high velocity produces a large temperature gradient, and so on. Therefore the temperature gradient at the wall is dependent on the flow field, and we must develop in our later analysis an expression relating the two quantities. Nevertheless, it must be remembered that the physical mechanism of the heat transfer at the wall is a conduction process.

To express the overall effect of convection we use Newton's law of cooling:

$$q = hA(T_w - T_\infty) \tag{1-8}$$

Here the heat-transfer rate is related to the overall temperature difference between the wall and fluid and the surface area $A$. The quantity $h$ is called the convection heat-transfer coefficient, and Eq. (1-8) is the defining equation. An analytical calculation of $h$ may be made for some systems. For complex situations it must be determined experimentally. The heat-transfer coefficient is sometimes called the *film conductance* because of its relation to the conduction process in the thin stationary layer of fluid at the wall surface. From Eq. (1-8) we note that the units of $h$ are in Btu/hr-ft²-°F when the heat flow is in Btu per hour. This is the most frequently used set of units.

In view of the foregoing discussion, one may anticipate that convection heat transfer will have a dependence on the viscosity of the fluid in addition

**Table 1-1  Approximate Values of Convection Heat-transfer Coefficients**

| Mode | $h$, Btu/hr-ft²-°F | Watts/m²-°C |
|---|---|---|
| Free convection, air | 1–5 | 5–25 |
| Forced convection, air | 2–100 | 10–500 |
| Forced convection, water | 20–3,000 | 100–15,000 |
| Boiling water | 500–5,000 | 2,500–25,000 |
| Condensation of water vapor | 1,000–20,000 | 5,000–100,000 |

to its dependence on the thermal properties of the fluid (thermal conductivity, specific heat, density). This is expected because viscosity influences the velocity profile and, correspondingly, the energy-transfer rate in the region near the wall.

If a heated plate were exposed to ambient room air without an external source of motion, a movement of the air would be experienced as a result of the density gradients near the plate. We call this *natural*, or *free*, convection as opposed to *forced* convection, which is experienced in the case of the fan blowing air over a plate. Boiling and condensation phenomena are also grouped under the general subject of convection heat transfer. The approximate ranges of convection heat transfer coefficients are indicated in Table 1-1.

## 1-4□RADIATION HEAT TRANSFER

In contrast to the mechanisms of conduction and convection, where energy transfer through a material medium is involved, heat may also be transferred into regions where a perfect vacuum exists. The mechanism in this case is electromagnetic radiation. We shall limit our discussion to electromagnetic radiation which is propagated as a result of a temperature difference; this is called thermal radiation.

Thermodynamic considerations show† that an ideal radiator, or *blackbody*, will emit energy at a rate proportional to the fourth power of the absolute temperature of the body. When two bodies exchange heat by radiation, the net heat exchange is then proportional to the difference in $T^4$. Thus

$$q = \sigma A (T_1{}^4 - T_2{}^4) \tag{1-9}$$

where $\sigma$ is the proportionality constant, and is called the Stefan-Boltzmann constant with the value of $0.1714 \times 10^{-8}$ Btu/hr-ft²-°R⁴ in the engineering system of units. Equation (1-9) is called the Stefan-Boltzmann law of ther-

---

† See, for example, J. P. Holman, "Thermodynamics," p. 222, McGraw-Hill Book Company, New York, 1969.

mal radiation, and it applies only to blackbodies. It is important to note that this equation is valid only for thermal radiation; other types of electromagnetic radiation may not be treated so simply.

We have mentioned that a blackbody is a body which radiates energy according to the $T^4$ law. We call such a body black because black surfaces, such as a piece of metal covered with carbon black, approximate this type of behavior. Other types of surfaces, like a glossy painted surface or a polished metal plate, do not radiate as much energy as the blackbody; however, the total radiation emitted by these bodies still generally follows the $T^4$ proportionality. To take account of the "gray" nature of such surfaces we introduce another factor into Eq. (1-9), called the emissivity $\epsilon$, which relates the radiation of the "gray" surface to that of an ideal black surface. In addition, we must take into account the fact that not all the radiation leaving one surface will reach the other surface since electromagnetic radiation travels in straight lines and some will be lost to the surroundings. We therefore introduce two new factors in Eq. (1-9) to take into account both of these situations, so that

$$q = F_\epsilon F_G \sigma A (T_1{}^4 - T_2{}^4) \qquad (1\text{-}10)$$

where $F_\epsilon$ is the emissivity function and $F_G$ is the geometric "view factor" function. The determination of the form of these functions for specific configurations will be the subject of a subsequent chapter. It is important to alert the reader at this time, however, to the fact that these functions are usually not independent of one another as indicated in Eq. (1-10).

### 1-5□DIMENSIONS AND UNITS
In this section we shall outline the systems of units which will be used throughout the book. One must be careful not to confuse the meaning of the terms units and dimensions. A dimension is a physical variable used to specify the behavior or nature of a particular system. For example, the length of a rod is a dimension of the rod. In like manner, the temperature of a gas may be considered one of the thermodynamic dimensions of the gas. When we say the rod is so many inches long, or the gas has a temperature of so many degrees Fahrenheit, we have given the units with which we choose to measure the dimension. In our development of heat transfer we shall use the dimensions

$L$ = length
$M$ = mass
$F$ = force
$\tau$ = time
$T$ = temperature

All the physical quantities used in heat transfer may be expressed in terms of these fundamental dimensions. The units to be used for certain dimensions are selected by somewhat arbitrary definitions which usually relate to a physical phenomenon or law. For example, Newton's second law of motion may be written

$$\text{Force} \sim \text{time rate of change of momentum}$$

$$F = k \frac{d(mv)}{d\tau}$$

where $k$ is the proportionality constant. If the mass is constant,

$$F = kma \tag{1-11}$$

where the acceleration is $a = dv/d\tau$. Equation (1-11) is usually written

$$F = \frac{1}{g_c} ma \tag{1-12}$$

with $1/g_c = k$. Equation (1-12) is used to define our systems of units for mass, force, length, and time. Some typical systems of units are

1.   1 lb force will accelerate 1 lb mass 32.16 ft/sec².
2.   1 lb force will accelerate 1 slug mass 1 ft/sec².
3.   1 dyne force will accelerate 1 g mass 1 cm/sec².
4.   1 newton force will accelerate 1 kg mass 1 m/sec².

Since Eq. (1-11) must be dimensionally homogeneous, we shall have a different value of the constant $g_c$ for each of the unit systems in items 1 to 4 above. These values are

1.   $g_c = 32.16 \ \text{lb}_m/\text{lb}_f \ \text{ft/sec}^2$
2.   $g_c = 1 \ \text{slug}/\text{lb}_f \ \text{ft/sec}^2$
3.   $g_c = 1 \ \text{g-cm/dyne-sec}^2$
4.   $g_c = 1 \ \text{kg-m/newton-sec}^2$

It matters not which system of units is used so long as it is consistent with the above definitions.

Work has the dimensions of a product of force times a distance. Energy has the same dimensions. The units for work and energy may be chosen from any of the systems used above. Thus the units of work and energy for each of the above systems are

1.   $\text{lb}_f$-ft
2.   $\text{lb}_f$-ft
3.   dyne-cm = 1 erg
4.   newton-m = 1 joule

In addition, we may use the units of energy which are based on thermal phenomena:

1 Btu will raise 1 $\text{lb}_m$ of water 1°F at 68°F.

1 cal will raise 1 g of water 1°C at 20°C.
1 kcal will raise 1 kg of water 1°C at 20°C.

The conversion factors for the various units of work and energy are

1 Btu = 778.16 lb$_f$-ft
1 Btu = 1055 joules
1 kcal = 4182 joules
1 lb$_f$-ft = 1.356 joules
1 Btu = 252 cal

The weight of a body is defined as the force exerted on the body as a result of the acceleration of gravity. Thus

$$W = \frac{g}{g_c} m \qquad (1\text{-}13)$$

where $W$ is the weight and $g$ is the acceleration of gravity. Note that the weight of a body has the dimensions of a force.

For the most part, we shall use the ft-lb$_f$-lb$_m$-sec-°F-Btu system of units in this book since they are employed most widely in industry in the United States. There is increasing impetus, however, to institute the SI (Système International d'Unités) units as a worldwide standard. In this system, the fundamental units are meter, newton, kilogram mass, second, degrees Celsius, and a "thermal" energy unit is not used, i.e., the joule (newton-meter) becomes the energy unit used throughout. The watt (joule/sec) is the unit of power in this system. In the SI system, the standard units for thermal conductivity would become

k in watt/m-°C

and the convection heat-transfer coefficient would be expressed as

h in watt/m²-°C

Eventually the SI system of units may be expected to prevail, but the timing of the changeover is uncertain. In the solution of example problems, we will give the answers in both sets of units so that the reader will acquire a feel for the orders of magnitudes of various quantities in both systems. Appropriate conversion factors are given in Table A-12.

### Example 1-1

One face of a copper plate 1 in. thick is maintained at 500°F, and the other face is maintained at 100°F. How much heat is transferred through the plate? Assume $k = 215$ Btu/hr-ft-°F.
**Solution** From Fourier's law

$$\frac{q}{A} = -k \frac{dT}{dx}$$

Integrating,

$$\frac{q}{A} = -k\frac{\Delta T}{\Delta x}$$

$$= -215 \times \frac{100 - 500}{\frac{1}{12}}$$

$$= 1.032 \times 10^6 \text{ Btu/hr-ft}^2 \qquad (5.86 \times 10^6 \text{ W/m}^2\text{-°C})$$

## Example 1-2

Air at 60°F blows over a hot plate 1 by 2 ft maintained at 300°F. The heat-transfer coefficient is 5 Btu/hr-ft²-°F. Calculate the heat transfer.
**Solution**    From Newton's law of cooling

$$q = hA(T_w - T_\infty)$$
$$= (5)(2)(1)(300 - 60) = 2400 \text{ Btu/hr} \qquad (703 \text{ watts})$$

## Example 1-3

Assuming that the plate in Example 1-2 is made of steel 2 in. thick with $k = 25$ Btu/hr-ft-°F and that 1100 Btu/hr is lost from the plate surface by radiation, calculate the inside plate temperature.
**Solution**    The heat conducted through the plate must be equal to the sum of the convection and radiation losses.

$$q_{\text{cond}} = q_{\text{conv}} + q_{\text{rad}}$$

$$-kA\frac{\Delta T}{\Delta x} = 2400 + 1100 = 3500 \text{ Btu/hr}$$

$$\Delta T = \frac{-(3500)(\frac{2}{12})}{(2)(25)} = -11.7°F$$

so that the inside plate temperature is

$$T_i = 300 + 11.7 = 311.7°F$$

## 1-6□SUMMARY

We may summarize our introductory remarks very simply. Heat transfer may take place by one or more of three modes: conduction, convection, and radiation. It has been noted that the physical mechanism of convection is related to the heat conduction through the thin layer of fluid adjacent to the heat-transfer surface. In both conduction and convection Fourier's law is applicable, although fluid mechanics must be brought into play in the convection problem in order to establish the temperature gradient.

Radiation heat transfer involves a different physical mechanism—that of propagation of electromagnetic energy. To study this type of energy transfer we introduce the concept of an ideal radiator, or blackbody, which

**Fig. 1-9  Sketch illustrating combination of conduction, convection, and radiation heat transfer.**

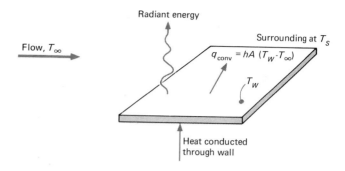

radiates energy at a rate proportional to its absolute temperature to the fourth power.

It is easy to envision cases in which all three modes of heat transfer are present, as in Fig. 1-9. In this case the heat conducted through the plate is removed from the plate surface by a combination of convection and radiation. An energy balance would give

$$-kA \frac{dT}{dy}\Big)_{\text{wall}} = hA(T_w - T_\infty) + F_\epsilon F_G \sigma A(T_w{}^4 - T_s{}^4)$$

where $T_s$ = temperature of surroundings

$T_w$ = surface temperature

$T_\infty$ = fluid temperature

To apply the science of heat transfer to practical situations, a thorough knowledge of all three modes of heat transfer must be obtained.

## PROBLEMS

1-1.   Define thermal conductivity.

1-2.   Define the convection-heat-transfer coefficient.

1-3.   Discuss the mechanism of thermal conduction in gases and solids.

1-4.   Discuss the mechanism of heat convection.

1-5.   One thousand Btu/hr is conducted through a section of insulating material 1 ft² in cross section. The thickness is 1 in., and the thermal conductivity may be taken as 0.12 Btu/hr-ft-°F. Compute the temperature difference across the material.

1-6.   Using the basic definitions of units and dimensions given in Sec. 1-5, arrive at expressions

(a)  To convert joules to Btu

(b)  To convert dyne-centimeters to joules

(c)  To convert Btu to calories

1-7.   A temperature difference of 150°F is impressed across a fiber-glass layer of 5-in. thickness. The thermal conductivity of the fiber glass is 0.02 Btu/hr-ft-°F. Compute the heat transferred through the material per hour per unit area.

1-8.    A flat wall is exposed to an environmental temperature of 100°F. The wall is covered with a layer of insulation 1 in. thick whose thermal conductivity is 0.8 Btu/hr-ft-°F, and the temperature of the wall on the inside of the insulation is 600°F. The wall loses heat to the environment by convection. Compute the value of the convection heat-transfer coefficient which must be maintained on the outer surface of the insulation in order that the outer surface temperature does not exceed 105°F.

1-9.    Two perfectly black surfaces are constructed such that all the radiant energy leaving a surface at 1500°F reaches the other surface. The temperature of the other surface is maintained at 500°F. Calculate the heat transfer between the surfaces per hour and per unit area of the surface maintained at 1500°F.

1-10.    Two very large parallel planes having surface conditions which very nearly approximate those of a blackbody are maintained at 2000 and 800°F, respectively. Calculate the heat transfer by radiation between the planes per unit time and per unit surface area.

1-11.    A $\frac{1}{4}$-in. steel plate having a thermal conductivity of 25 Btu/ hr-ft-°F is exposed to a radiant heat flux of 1500 Btu/hr-ft$^2$ in a vacuum space where the convection heat transfer is negligible. Assuming that the surface temperature of the steel exposed to the radiant energy is maintained at 100°F, what will be the other surface temperature if all the radiant energy striking the plate is transferred through the plate by conduction?

1-12.    Beginning with the three-dimensional heat-conduction equation in cartesian coordinates [Eq. (1-3a)], obtain the general heat-conduction equation in cylindrical coordinates [Eq. (1-3b)].

1-13.    Consider a wall heated by convection on one side and cooled by convection on the other side. Show that the heat-transfer rate through the wall is

$$q = \frac{T_1 - T_2}{1/h_1A + \Delta x/kA + 1/h_2A}$$

where $T_1$ and $T_2$ are the fluid temperatures on each side of the wall and $h_1$ and $h_2$ are the corresponding heat-transfer coefficients.

1-14.    A truncated cone 1 ft high is constructed of aluminum. The diameter at the top is 3 in., and the diameter at the bottom is 5 in. The lower surface is maintained at 200°F, and the upper surface at 1000°F. The other surface is insulated. Assuming one-dimensional heat flow, what is the rate of heat transfer in Btu per hour?

1-15.    The temperatures on the faces of a plane wall 6 in. thick are 700 and 200°F. The wall is constructed of a special glass with the following properties: $k = 0.45$ Btu/hr-ft-°F, $\rho = 170$ lb$_m$/ft³, $C_p = 0.2$ Btu/lb$_m$-°F. What is the heat flow through the wall at steady-state conditions?

1-16.    A housewife informs her engineer-husband that she frequently feels cooler in the summer when standing in front of an open refrigerator. Her husband tells her that she is only "imagining things" because there is no fan in the refrigerator to blow the cool air over her. A lively argument ensues. Whose side of the argument do you take? Why?

1-17.    A housewife informs her engineer-husband that "hot water will freeze faster than cold water." He calls this statement nonsense. She answers by saying that she has actually timed the freezing process for ice trays in the home refrigerator and found that hot water does indeed freeze faster. As a friend you are asked to settle the argument. Is there any logical explanation for the housewife's observations?

1-18.    An air-conditioned classroom in Texas is maintained at 72°F in the summer. The students attend classes in shorts, sandals, and skimpy shirts and are quite comfortable. In the same classroom during the winter the same students wear wool slacks, long-sleeve shirts, and sweaters and are equally comfortable with the room temperature maintained at 75°F. Assuming that humidity is not a factor, explain this apparent anomaly in "temperature comfort."

1-19.    A certain superinsulation material having a thermal conductivity of $10^{-4}$ Btu/hr-ft-°F is used to insulate a tank of liquid nitrogen that is maintained at $-320$°F. 85.9 Btu are required to vaporize each lb$_m$ of nitrogen at this temperature. Assuming that the tank is a sphere having an inside diameter of 2 ft, estimate the amount of nitrogen vaporized per day for an insulation thickness of 1.0 in. and an ambient temperature of 70°F. Assume that the outer temperature of the insulation is at 70°F.

1-20.    Calculate the radiation heat exchange in 1 day between two black planes having the area of the surface of a 2-ft-diameter sphere when the planes are maintained at $-320$°F and 70°F. What does this calculation indicate in regard to Prob. 1-19.

1-21.    Assuming that the heat transfer to the sphere in Prob. 1-19 occurs by free convection with a heat-transfer coefficient of 1.0 Btu/hr-ft²-°F, calculate the temperature difference between the outer surface of the sphere and the environment.

1-22.    Rank the following materials in order of (a) transient response and (b) steady-state conduction. Taking the material with the highest rank give the other materials as a percent of the maximum.

1.    aluminum
2.    copper
3.    silver
4.    iron
5.    lead

6.   chrome steel—18% Cr, 8% N

7.   magnesium

What do you conclude from this ranking?

## REFERENCES

1.   Glaser, P. E., I. A. Black, and P. Doherty: Multilayer Insulation, *Mech. Eng.*, p. 23, August, 1965.

2.   Barron, R.: "Cryogenic Systems," McGraw-Hill Book Company, New York, 1967.

3.   Dewitt, W. D., Gibbon, N. C., and Reid, R. L.: Multifoil Type Thermal Insulation, *IEEE Trans. Aerospace Electron. Systems*, vol. 4, no. 5, Suppl., (1968), pp. 263–271.

# CHAPTER

# 2

# STEADY-STATE
# CONDUCTION—ONE
# DIMENSION

## 2-1□INTRODUCTION

We now wish to examine the applications of Fourier's law of heat conduction to calculation of heat flow in some simple one-dimensional systems. Several different physical shapes may fall in the category of one-dimensional systems: Cylindrical and spherical systems are one-dimensional when the temperature in the body is a function only of radial distance and is independent of azimuth angle or axial distance. In some two-dimensional problems the effect of a second-space coordinate may be so small as to justify its neglect, and the multidimensional heat-flow problem may be approximated with a one-dimensional analysis. In these cases the differential equations are simplified, and we are led to a much easier solution as a result of this simplification.

## 2-2□THE PLANE WALL

First consider the plane wall where a direct application of Fourier's law [Eq. (1-1)] may be made. Integration yields

$$q = -\frac{kA}{\Delta x}(T_2 - T_1) \tag{2-1}$$

when the thermal conductivity is considered constant. The wall thickness is $\Delta x$, and $T_1$ and $T_2$ are the wall-face temperatures. If the thermal conductivity varies with temperature according to some linear relation $k = k_0(1 + \beta T)$, the resultant equation for the heat flow is

$$q = -\frac{k_0 A}{\Delta x}\left[(T_2 - T_1) + \frac{\beta}{2}(T_2{}^2 - T_1{}^2)\right] \tag{2-2}$$

If more than one material is present, as might be used in the multilayer wall

Fig. 2-1  One-dimensional heat transfer through a composite wall and electrical analog.

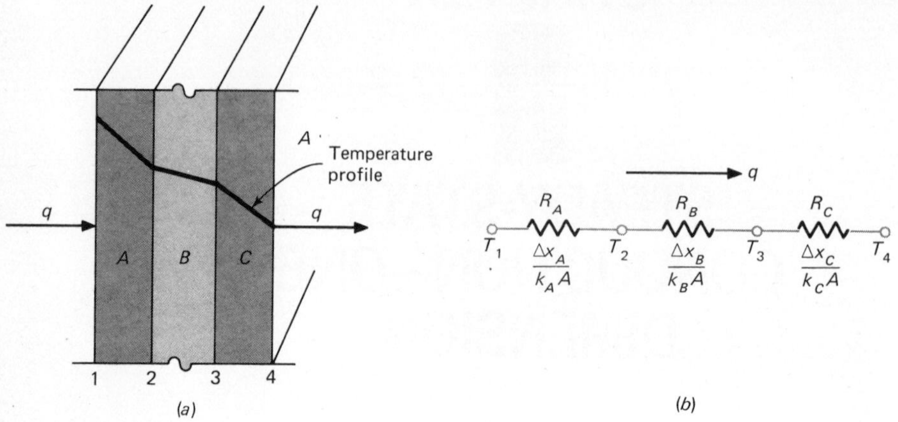

shown in Fig. 2-1, the analysis would proceed as follows:   The temperature gradients in the three materials are shown, and the heat flow may be written

$$q = -k_A A \frac{T_2 - T_1}{\Delta x_A} = -k_B A \frac{T_3 - T_2}{\Delta x_B} = -k_C A \frac{T_4 - T_3}{\Delta x_C}$$

Note that the heat flow must be the same through all sections.

Solving these three equations simultaneously, the heat flow is written

$$q = \frac{T_1 - T_4}{\Delta x_A/k_A A + \Delta x_B/k_B A + \Delta x_C/k_C A} \qquad (2\text{-}3)$$

At this point we retrace our development slightly to introduce a different conceptual viewpoint for Fourier's law. The heat-transfer rate may be considered as a flow, and the combination of thermal conductivity, thickness of material, and area, as a resistance to this flow. The temperature is the potential, or driving, function for the heat flow, and the Fourier equation may be written

$$\text{Heat flow} = \frac{\text{thermal potential difference}}{\text{thermal resistance}} \qquad (2\text{-}4)$$

a relation quite like Ohn's law in electric-circuit theory. In Eq. (2-1) the thermal resistance is $\Delta x/kA$, and in Eq. (2-3) it is the sum of the three terms in the denominator. We should expect this situation in Eq. (2-3) because the three walls side by side act as three thermal resistances in series. The equivalent electric circuit is shown in Fig. 2-1$b$.

Fig. 2-2  Series and parallel one-dimensional heat transfer through a composite wall and electric analog.

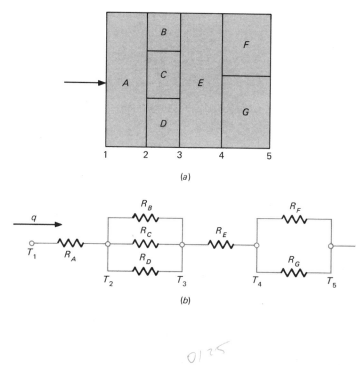

(a)

(b)

*0125*

The electrical analogy may be used to solve more complex problems involving both series and parallel thermal resistances. A typical problem and its analogous electric circuit are shown in Fig. 2-2. The one-dimensional heat-flow equation for this type of problem may be written

$$q = \frac{\Delta T_{\text{overall}}}{\Sigma R_{\text{th}}} \qquad (2\text{-}5)$$

where the $R_{\text{th}}$ are the thermal resistances of the various materials.

It is well to mention that in some systems like that in Fig. 2-2 two-dimensional heat flow may result if the thermal conductivities of materials $B$, $C$, and $D$ differ by an appreciable amount. In these cases other techniques must be employed to effect a solution.

## 2-3□RADIAL SYSTEMS—CYLINDERS

Consider a long cylinder of inside radius $r_i$, outside radius $r_o$, and length $L$, such as the one shown in Fig. 2-3. We expose this cylinder to a temperature differential $T_i - T_o$ and ask what the heat flow will be. It may be assumed that the heat flows in a radial direction so that the only space coordinate

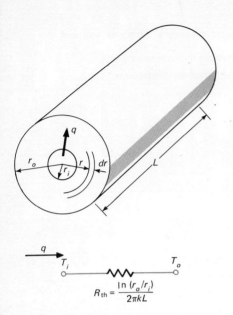

**Fig. 2-3  One-dimensional heat flow through a hollow cylinder and electrical analog.**

needed to specify the system is $r$. Again, Fourier's law is used by inserting the proper area relation. The area for heat flow in the cylindrical system is

$$A_r = 2\pi r L$$

so that Fourier's law is written

$$q_r = -kA_r \frac{dT}{dr}$$

or

$$q_r = -2\pi k r L \frac{dT}{dr} \tag{2-6}$$

with the boundary conditions

$$T = T_i \qquad \text{at } r = r_i$$
$$T = T_o \qquad \text{at } r = r_o$$

The solution to Eq. (2-6) is

$$q = \frac{2\pi k L (T_i - T_o)}{\ln (r_o/r_i)} \tag{2-7}$$

and the thermal resistance in this case is

$$R_{\text{th}} = \frac{\ln (r_o/r_i)}{2\pi k L}$$

Fig. 2-4 One-dimensional heat flow through
multiple cylindrical sections and electrical analog.

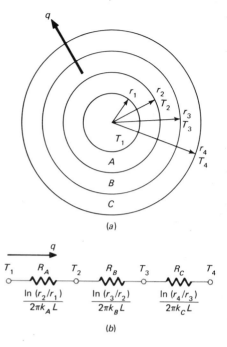

(a)

(b)

The thermal-resistance concept may be used for multiple-layer cylindrical walls just as it was used for plane walls. For the three-layer system shown in Fig. 2-4 the solution is

$$q = \frac{2\pi L(T_1 - T_4)}{\ln(r_2/r_1)/k_A + \ln(r_3/r_2)/k_B + \ln(r_4/r_3)/k_C} \qquad (2\text{-}8)$$

The thermal circuit is shown in Fig. 2-4b.

Spherical systems may also be treated as one-dimensional when the temperature is a function only of radius. The heat flow is then

$$q = \frac{4\pi k(T_i - T_o)}{1/r_i - 1/r_o} \qquad (2\text{-}9)$$

The derivation of Eq. (2-9) is left as an exercise.

## Example 2-1

A thick-walled tube of 18-8 stainless steel ($k = 12.5$) 1 in. ID and 2 in. OD is covered with a 1-in. layer of asbestos insulation ($k = 0.14$). If the inside wall temperature of the pipe is maintained at 1000°F and the

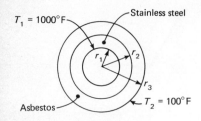

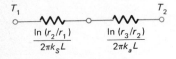

outside of the insulation at 100°F, calculate the heat loss per foot of length.
**Solution**   The accompanying figure shows the thermal network for the
problem. The heat flow is given by

$$
\begin{aligned}
\frac{q}{L} &= \frac{2\pi(T_1 - T_2)}{\ln\,(r_2/r_1)/k_s + \ln\,(r_3/r_2)/k_a} \\
&= \frac{2\pi(1000 - 100)}{\ln\,(\frac{2}{1})/12.5 + \ln\,(\frac{4}{2})/0.14} \\
&= 1130 \text{ Btu/hr-ft} \qquad (1{,}086 \text{ W/m})
\end{aligned}
$$

## 2-4□HEAT-SOURCE SYSTEMS

A number of interesting applications of the principles of heat transfer are
concerned with systems in which heat may be generated internally. Nuclear
reactors are one example; electrical conductors and chemically reacting systems are others. At this point we shall confine our discussion to one-dimensional systems, or more specifically, systems where the temperature is a
function of only one space coordinate.

## 2-5□PLANE WALL WITH HEAT SOURCES

Consider the plane wall with uniformly distributed heat sources as shown in
Fig. 2-5. The thickness of the wall in the $x$ direction is $2L$, and it is assumed
that the dimensions in the other directions are sufficiently large so that the
heat flow may be considered as one-dimensional. The heat generated per
unit volume is $\dot{q}$, and we assume that the thermal conductivity does not
vary with temperature. This situation might be produced in a practical
situation by passing a current through an electrically conducting material.

**Fig. 2-5  Sketch illustrating one-dimensional conduction problem with heat generation.**

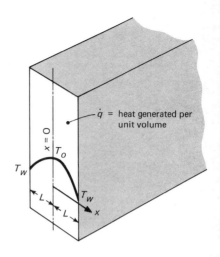

From Chap. 1, the differential equation which governs the heat flow is

$$\frac{d^2T}{dx^2} + \frac{\dot{q}}{k} = 0 \tag{2-10}$$

For the boundary conditions we specify the temperatures on either side of the wall, i.e.,

$$T = T_w \qquad \text{at } x = \pm L \tag{2-11}$$

The general solution to Eq. (2-10) is

$$T = -\frac{\dot{q}}{2k}\,x^2 + C_1 x + C_2 \tag{2-12}$$

Since the temperature must be the same on each side of the wall, $C_1$ must be zero. The temperature at the midplane is denoted by $T_0$ and from Eq. (2-12)

$$T_0 = C_2$$

The temperature distribution is therefore

$$T - T_0 = -\frac{\dot{q}}{2k}\,x^2 \tag{2-13a}$$

$$\frac{T - T_0}{T_w - T_0} = \left(\frac{x}{L}\right)^2 \tag{2-13b}$$

a parabolic distribution. An expression for the midplane temperature $T_0$ may be obtained through an energy balance. At steady-state conditions the total heat generated must equal the heat lost at the faces. Thus

$$2\left[-kA\frac{dT}{dx}\Big)_{x=L}\right] = \dot{q}A2L$$

where $A$ is the cross-sectional area of the plate. The temperature gradient at the wall is obtained by differentiating Eq. (2-13b).

$$\frac{dT}{dx}\Big)_{x=L} = (T_w - T_0)\left(\frac{2x}{L^2}\right)_{x=L} = (T_w - T_0)\frac{2}{L}.$$

Then

$$-k(T_w - T_0)\frac{2}{L} = \dot{q}L$$

and

$$T_0 = \frac{\dot{q}L^2}{2k} + T_w \tag{2-14}$$

## 2-6□CYLINDER WITH HEAT SOURCES

Consider a cylinder of radius $R$ with uniformly distributed heat sources and constant thermal conductivity. If the cylinder is sufficiently long so that the temperature may be considered a function of radius only, the appropriate differential equation may be obtained by neglecting the axial, azimuth, and time-dependent terms in Eq. (1-3b),

$$\frac{d^2T}{dr^2} + \frac{1}{r}\frac{dT}{dr} + \frac{\dot{q}}{k} = 0 \tag{2-15}$$

The boundary conditions are

$$T = T_w \qquad \text{at } r = R$$

and heat generated equals heat lost at surface, i.e.,

$$\dot{q}\pi R^2 L = -k2\pi RL\frac{dT}{dr}\Big)_{r=R}$$

Since the temperature function must be continuous at the center of the cylinder, we could specify that

$$\frac{dT}{dr} = 0 \qquad \text{at } r = 0$$

However, it will not be necessary to use this condition since it will be satisfied automatically when the two boundary conditions are satisfied.

We rewrite Eq. (2-15)

$$r\frac{d^2T}{dr^2} + \frac{dT}{dr} = \frac{-\dot{q}r}{k}$$

and note that

$$r\frac{d^2T}{dr^2} + \frac{dT}{dr} = \frac{d}{dr}\left(r\frac{dT}{dr}\right)$$

Then integration yields

$$r \frac{dT}{dr} = \frac{-\dot{q}r^2}{2k} + C_1$$

and

$$T = \frac{-\dot{q}r^2}{4k} + C_1 \ln r + C_2$$

From the second boundary condition above,

$$\frac{dT}{dr}\bigg)_{r=R} = \frac{-\dot{q}R}{2k} = \frac{-\dot{q}R}{2k} + \frac{C_1}{R}$$

Thus

$$C_1 = 0$$

From the first boundary condition,

$$T = T_w = \frac{-\dot{q}R^2}{4k} + C_2 \qquad \text{at } r = R$$

so that

$$C_2 = T_w + \frac{\dot{q}R^2}{4k}$$

The final solution for the temperature distribution is then

$$T - T_w = \frac{\dot{q}}{4k} (R^2 - r^2) \tag{2-16a}$$

or in dimensionless form,

$$\frac{T - T_w}{T_0 - T_w} = 1 - \left(\frac{r}{R}\right)^2 \tag{2-16b}$$

where $T_0$ is the temperature at $r = 0$ and is given by

$$T_0 = \frac{\dot{q}R^2}{4k} + T_w \tag{2-17}$$

It is left as an exercise to show that the temperature gradient at $r = 0$ is zero.

## Example 2-2

A current of 200 amp is passed through a stainless-steel wire 0.1 in. in diameter. The resistivity of the steel may be taken at 70 $\mu$ohm-cm, and the length of the wire is 3 ft. If the outer surface temperature of the wire is maintained at 350°F, calculate the center temperature. Assume $k = 13$ Btu/hr-ft-°F for stainless steel.

**Solution**  Equation (2-17) may be used to calculate the center temperature of the wire when the heat generated per unit volume $\dot{q}$ is known. $\dot{q}$ may be calculated by making the energy balance

$$\text{Power input} = I^2 R = \dot{q}\pi r_o^2 L$$

where $R$ = resistance of wire
$\quad\;\; I$ = electric current
$\quad\; r_o$ = outside radius of wire
The resistance is calculated from

$$R = \rho \frac{L}{A} = \frac{(70 \times 10^{-6})(36)(2.54)}{\pi(0.05)^2(2.54)^2} = 0.126 \text{ ohm}$$

where $\rho$ is the resistivity of the wire. We may now calculate $\dot{q}$.

$$(200)^2(0.126) = \dot{q}\pi\left(\frac{0.05}{12}\right)^2 (3)$$
$$\dot{q} = 3.07 \times 10^7 \text{ watts/ft}^3 = 1.047 \times 10^8 \text{ Btu/hr-ft}^3$$

From Eq. (2-17)

$$T_0 = \frac{\dot{q}r_o^2}{4k} + T_W = \frac{(1.047 \times 10^8)(0.05/12)^2}{(4)(13)} + 350$$
$$= 34.9 + 350 = 385°\text{F}$$

## 2-7□CONDUCTION - CONVECTION SYSTEMS

The heat which is conducted through a body must frequently be removed (or delivered) by some convection process. For example, the heat lost by conduction through a furnace wall must be dissipated to the surroundings through convection. In heat-exchanger applications a finned-tube arrangement might be used to remove heat from a hot liquid. The heat transfer from the liquid to the finned tube is by convection. The heat is conducted through the material and finally dissipated to the surroundings by convection. Obviously, an analysis of combined conduction-convection systems is very important from a practical standpoint.

We shall defer part of our analysis of conduction-convection systems to Chap. 10 on Heat Exchangers. For the present we wish to examine some simple extended surface problems. Consider the one-dimensional fin exposed to a surrounding fluid at a temperature $T_\infty$ as shown in Fig. 2-6. The temperature of the base of the fin is $T_0$. We approach the problem by making an energy balance on an element of the fin of thickness $dx$ as shown in the figure. Thus

Energy in left face = energy out right face + energy lost by convection

The defining equation for the convection heat-transfer coefficient is recalled as

$$q = hA(T_w - T_\infty) \tag{2-18}$$

where the area in this equation is the surface area for convection. Let the cross-sectional area of the fin be $A$ and the perimeter be $P$. Then the energy

Fig. 2-6 Sketch illustrating one-
dimensional conduction and convection through a rectangular fin.

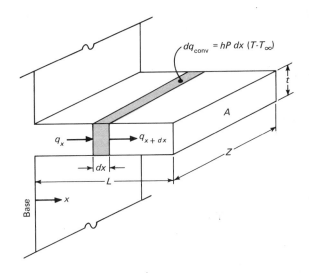

Fig. 2-6 Sketch illustrating one-
dimensional conduction and convection through a rectangular fin.

quantities are

$$\text{Energy in left face} = q_x = -kA\,\frac{dT}{dx}$$

$$\text{Energy out right face} = q_{x+dx} = -kA\,\frac{dT}{dx}\bigg)_{x+dx}$$

$$= -kA\left(\frac{dT}{dx} + \frac{d^2T}{dx^2}\,dx\right)$$

$$\text{Energy lost by convection} = hP\,dx\,(T - T_\infty)$$

Here it is noted that the differential surface area for convection is the product of the perimeter of the fin and the differential length $dx$. Combining the quantities, the energy balance yields

$$\frac{d^2T}{dx^2} - \frac{hP}{kA}(T - T_\infty) = 0 \qquad (2\text{-}19a)$$

Let $\theta = T - T_\infty$. Then Eq. (2-19a) becomes

$$\frac{d^2\theta}{dx^2} - \frac{hP}{kA}\,\theta = 0 \qquad (2\text{-}19b)$$

One boundary condition is

$$\theta = \theta_0 = T_0 - T_\infty \quad \text{at } x = 0$$

The other boundary condition depends on the physical situation. Several cases may be considered.

Case 1: The fin is very long and the temperature at the end of the fin is essentially that of the surrounding fluid.

Case 2: The fin is of finite length and loses heat by convection from its end.

Case 3: The end of the fin is insulated so that $dT/dx = 0$.

If we let $m^2 = hP/kA$, the general solution for Eq. (2-19b) may be written

$$\theta = C_1 e^{-mx} + C_2 e^{mx} \tag{2-20}$$

For case 1 the boundary conditions are

$$\theta = \theta_0 \qquad \text{at } x = 0$$
$$\theta = 0 \qquad \text{at } x = \infty$$

and the solution becomes

$$\frac{\theta}{\theta_0} = \frac{T - T_\infty}{T_0 - T_\infty} = e^{-mx} \tag{2-21}$$

For case 3 the boundary conditions are

$$\theta = \theta_0 \qquad \text{at } x = 0$$
$$\frac{d\theta}{dx} = 0 \qquad \text{at } x = L$$

Thus
$$\theta_0 = C_1 + C_2$$
$$0 = m(-C_1 e^{-mL} + C_2 e^{mL})$$

Solving for the constants $C_1$ and $C_2$, the solution is obtained as

$$\frac{\theta}{\theta_0} = \frac{e^{-mx}}{1 + e^{-2mL}} + \frac{e^{mx}}{1 + e^{2mL}} \tag{2-22a}$$

$$\frac{\theta}{\theta_0} = \frac{\cosh [m(L - x)]}{\cosh mL} \tag{2-22b}$$

The solution for case 2 is a little more involved algebraically, and is left as an exercise for the reader.

Using the equations for the temperature distributions, the heat lost by the fin may be computed rather easily since

$$q = -kA \left. \frac{dT}{dx} \right)_{x=0}$$

An alternative method of integrating the convection heat loss could be used.

$$q = \int_0^L hP(T - T_\infty) \, dx = \int_0^L hP\theta \, dx$$

In most cases, however, the first equation is easier to apply. For case 1 above,

$$q = -kA(-m\theta_0 e^{-m(0)}) = \sqrt{hPkA}\ \theta_0 \qquad (2\text{-}23)$$

For case 3,

$$q = -kA\theta_0 m \left( \frac{1}{1 + e^{-2mL}} - \frac{1}{1 + e^{+2mL}} \right)$$
$$= \sqrt{hPkA}\ \theta_0 \tanh mL \qquad (2\text{-}24)$$

In the above development it has been assumed that the substantial temperature gradients occur only in the $x$ direction. This assumption will be satisfied if the fin is sufficiently thin. For most fins of practical interest the error introduced by this assumption is less than 1 percent. The overall accuracy of practical fin calculations will usually be limited by uncertainties in values of the convection coefficient $h$. It is worthwhile to note that the convection coefficient is seldom uniform over the entire surface as has been assumed above. If severe nonuniform behavior is encountered numerical finite-difference techniques must be employed to solve the problem. Such techniques are discussed in Chap. 3.

## 2-8□FINS

In the foregoing development we derived relations for the heat transfer from a rod or fin of uniform cross-sectional area protruding from a flat wall. In practical applications fins may have varying cross-sectional areas and may be attached to circular surfaces. In either of these cases the area must be considered as a variable in the derivation, and solution of the basic differential equation and the mathematical techniques become more tedious. We shall present only the results for these more complex situations. The reader is referred to Ref. 1 for details on the mathematical methods used to obtain the solutions.

To indicate the effectiveness of a fin in transferring a given quantity of heat, a new parameter called fin efficiency is defined by

$$\text{Fin efficiency} = \frac{\text{actual heat transferred}}{\begin{array}{c}\text{heat which would be transferred}\\ \text{if entire fin area were at base}\\ \text{temperature}\end{array}} = \eta_f$$

For case 3 above the fin efficiency becomes

$$\eta_f = \frac{\sqrt{hPkA}\ \theta_0 \tanh mL}{hPL\theta_0} = \frac{\tanh mL}{mL} \qquad (2\text{-}25)$$

The fins discussed above were assumed to be sufficiently deep so that the heat flow could be considered one-dimensional. The expression for $mL$

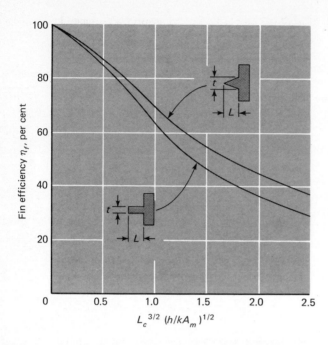

**Fig. 2-7   Efficiencies of rectangular and triangular fins:**

$$L_c = \begin{cases} L + \dfrac{t}{2} & \text{rectangular fin} \\ \\ L & \text{triangular fin} \end{cases}$$

$$A_m = \begin{cases} t L_c & \text{rectangular fin} \\ \\ \dfrac{t}{2} L & \text{triangular fin} \end{cases}$$

may be written

$$mL = \sqrt{\frac{hP}{kA}}\, L = \sqrt{\frac{h(2z + 2t)}{kzt}}\, L$$

where $z$ is the depth of the fin and $t$ is the thickness. Now, if the fin is sufficiently deep, the term $2z$ will be large compared with $2t$. Accordingly,

$$mL = \sqrt{\frac{2hz}{ktz}}\, L = \sqrt{\frac{2h}{kt}}\, L$$

Multiplying numerator and denominator by $L^{\frac{1}{2}}$,

$$mL = \sqrt{\frac{2h}{kLt}}\, L^{\frac{3}{2}}$$

$Lt$ is the profile area of the fin, which we define as

$$A_m = Lt$$

so that

$$mL = \sqrt{\frac{2h}{kA_m}}\, L^{\frac{3}{2}} \tag{2-26}$$

We may therefore use the expression in Eq. (2-26) to compute the efficiency of a fin with insulated tip as given by Eq. (2-25).

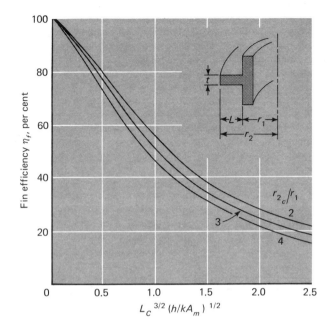

**Fig. 2-8  Efficiencies of circumferential fins of rectangular profile:**

$$L_e = L + \frac{t}{2}$$

$$r_{2_e} = r_1 + L_e$$

$$A_m = t(r_{2_e} - r_1)$$

Harper and Brown [2] have shown that the solution in case 2 above may be expressed in the same form as Eq. (2-25) when the length of the fin is extended by one-half the thickness of the fin. A corrected length $L_c$ is then used in all the equations which apply for the case of the fin with an insulated tip. Thus

$$L_c = L + \frac{t}{2} \qquad (2\text{-}27)$$

The error which results from this approximation will be less than 8 percent when

$$\left(\frac{ht}{2k}\right)^{\frac{1}{2}} = \frac{1}{2}$$

Figure 2-7 presents a comparison of the efficiencies of a triangular fin and a straight rectangular fin corresponding to case 2 above. Figure 2-8 shows the efficiencies of circumferential fins of rectangular cross-sectional area [3]. Notice that the corrected fin lengths $L_c$ and profile area $A_m$ have been used in Figs. 2-7 and 2-8.

It is interesting to note that the fin efficiency reaches its maximum value for the trivial case of $L = 0$, or no fin at all. Therefore, we should not expect to be able to maximize fin performance with respect to fin length. It is possible, however, to maximize the efficiency with respect to the quantity of fin material (mass, volume, or cost), and such a maximization process has rather obvious economic significance. We have not discussed the subject of radiation heat transfer from fins. The radiant transfer is an important consideration in a number of applications, and the interested reader should consult the text by Sparrow and Cess [9] for information on this subject.

In some cases a valid method of evaluating fin performance is to compare the heat transfer with the fin to that which would be obtained without the fin. The ratio of these quantities is

$$\frac{q \text{ with fin}}{q \text{ without fin}} = \frac{\eta_f A_f h \theta_0}{h A_b \theta_0}$$

where $A_f$ is the total surface area of the fin and $A_b$ is the base area. For the insulated tip fin described by Eq. (2-25),

$$A_f = PL$$
$$A_b = A$$

and the heat ratio would become

$$\frac{q \text{ with fin}}{q \text{ without fin}} = \frac{\tanh mL}{\sqrt{hA/kP}}$$

Still another method of evaluating fin performance is discussed in Prob. 2-25. Kern and Kraus [8] give a very complete discussion of extended surface heat transfer. Some photographs of different fin shapes used in electronic cooling applications are shown in Fig. 2-9.

### Example 2-3

A long steel rod ($k = 26$) 1 in. in diameter protrudes from a wall which is maintained at 400°F. The rod is surrounded by a fluid which removes heat from it, $h = 5$ Btu/hr-ft²-°F. The fluid temperature is 100°F. Calculate the heat loss from the rod.

**Solution**    This situation corresponds to case 1 above. We may calculate the heat loss immediately, using Eq. (2-23).

$$q = \sqrt{hPkA}\ \theta_0 = \left[ (5)\pi(\tfrac{1}{12})(26)\ \frac{\pi}{4}\ (\tfrac{1}{12})^2 \right]^{\frac{1}{2}} (400 - 100)$$
$$= 129 \text{ Btu/hr} \qquad (37.8 \text{ W})$$

### Example 2-4

An aluminum fin ($k = 124$) $\tfrac{1}{8}$ in. thick and 3 in. long protrudes from a wall as shown in Fig. 2-6. The base is maintained at 600°F, and the ambient temperature is 125°F. $h = 2$ Btu/hr-ft²-°F. Calculate the heat loss from the

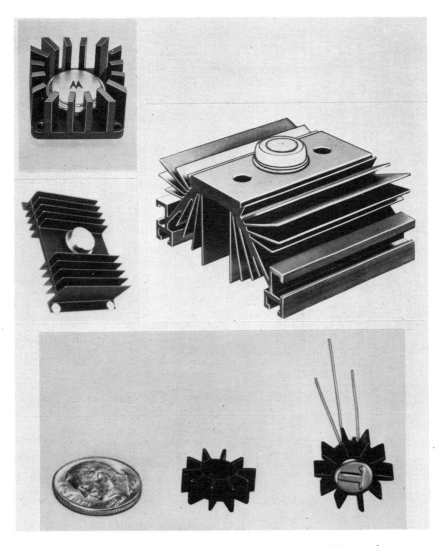

Fig. 2-9  Some fin arrangements used in electronic cooling applications. (Photographs courtesy of Wakefield Engineering, Inc., Wakefield, Mass.)

fin per unit depth of material.

**Solution**  This situation corresponds to case 2 above. We may use the approximate method of solution by extending the rod by a fictitious length $t/2$ and then computing the heat flow from a fin with insulated tip as given by Eq. (2-24).

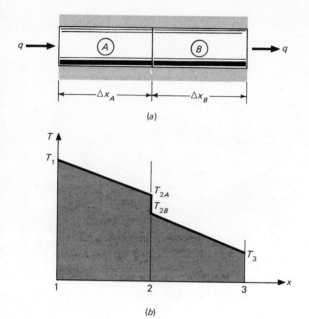

**Fig. 2-10** Illustrations of thermal-contract-resistance effect. (a) Physical situation; (b) temperature profile.

$$L_c = L + \frac{t}{2} = 3 + \tfrac{1}{16} = 3.062 \text{ in.}$$

$$m = \sqrt{\frac{hP}{kA}} = \sqrt{\frac{h(2z + 2t)}{ktz}} \approx \sqrt{\frac{2h}{kt}}$$

$$= \sqrt{\frac{(2)(2)(12)}{(124)(0.125)}} = 1.76$$

$$q = \tanh{(mL_c)}\sqrt{hPkA}\ \theta_0$$
$$= mkA\ \theta_0 \tanh{mL_c}$$

For a 1-ft depth

$$A = (1)(\tfrac{1}{8})(\tfrac{1}{12}) = 0.0104 \text{ ft}^2$$

and $\quad q = (1.76)(124)(0.0104)(600 - 125) \tanh\left(1.76 \times \dfrac{3.062}{12}\right)$

$$= 453 \text{ Btu/hr-ft} \qquad (435 \text{ W/m})$$

### Example 2-5

Aluminum fins $\frac{1}{2}$ in. wide and $\frac{1}{16}$ in. thick are placed on a 1-in.-diameter tube to dissipate the heat. The tube surface temperature is 340°F, and the ambient fluid temperature is 70°F. Calculate the heat loss per fin when $h = 25$ Btu/hr-ft²-°F. Assume $k = 124$ Btu/hr-ft-°F for aluminum.

**Solution**   We may compute the heat transfer using the fin efficiency given in Fig. 2-8. The parameters needed are calculated as

$$L_c = L + \frac{t}{2} = \frac{1}{2} + \frac{1}{32} = 0.531 \text{ in.} = 0.0443 \text{ ft}$$

$$r_{2c} = r_1 + L_c = \frac{1}{2} + 0.531 = 1.031 \text{ in.}$$

$$\frac{r_{2c}}{r_1} = \frac{1.031}{0.5} = 2.062$$

$$A_m = t(r_{2c} - r_1) = \frac{(\frac{1}{16})(0.531)}{144} = 2.30 \times 10^{-4} \text{ ft}^2$$

$$L_c^{\frac{3}{2}} \left(\frac{h}{kA_m}\right)^{\frac{1}{2}} = (0.0443)^{\frac{3}{2}} \left[\frac{25}{(124)(2.30 \times 10^{-4})}\right]^{\frac{1}{2}} = 0.277$$

From Fig. 2-8, $\eta_f = 90$ percent. The heat which would be transferred if the entire fin were at the base temperature is

$$q_{max} = 2h\pi(r_{2c}^2 - r_1^2)(340 - 70)$$

$$= \frac{(2)(25)\pi}{144}[(1.031)^2 - (0.5)^2](340 - 70) = 238 \text{ Btu/hr} \qquad (69.7 \text{ W})$$

The actual heat transfer is then the product of this heat flow and the fin efficiency.

$$q = (0.90)(238) = 214 \text{ Btu/hr} \qquad (62.7 \text{ W})$$

## 2-9□THERMAL CONTACT RESISTANCE

Imagine two solid bars brought into contact as indicated in Fig. 2-10, with the sides of the bars insulated so that heat flows only in the axial direction. The materials may have different thermal conductivities, but if the sides are insulated, the heat flux must be the same through both materials under steady-state conditions. Experience shows that the actual temperature profile through the two materials varies approximately as shown in Fig. 2-10b. The temperature drop at plane 2, the contact plane between the two materials, is said to be the result of a *thermal contact resistance*. Performing an energy balance on the two materials, we obtain

$$q = k_A A \frac{T_1 - T_{2A}}{\Delta x_A} = \frac{T_{2A} - T_{2B}}{1/h_c A} = k_B A \frac{T_{2B} - T_3}{\Delta x_B}$$

or

$$q = \frac{T_1 - T_3}{\Delta x_A/k_A A + 1/h_c A + \Delta x_B/k_B A} \qquad (2-28)$$

where the quantity $1/h_c A$ is called the thermal contact resistance and $h_c$ is called the contact coefficient. This factor can be extremely important in a number of applications because of the many heat-transfer situations which involve mechanical joining of two materials.

The physical mechanism of contact resistance may be better understood by examining a joint in more detail, as shown in Fig. 2-11. The actual

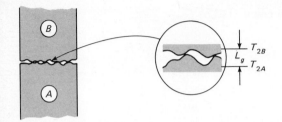

**Fig. 2-11** Joint-roughness model for analysis of thermal contact resistance.

surface roughness is exaggerated to implement the discussion. No real surface is perfectly smooth, and the actual surface roughness is believed to play a central role in determining the contact resistance. There are two principal contributions to the heat transfer at the joint:

1. The solid-to-solid conduction at the spots of contact
2. The conduction through entrapped gases in the void spaces created by the contact

It is the second factor that is believed to represent the major resistance to heat flow since the thermal conductivity of the gas is quite small in comparison with that of the solids.

Designating the contact area by $A_c$ and the void area by $A_v$, we may write for the heat flow across the joint

$$q = \frac{T_{2A} - T_{2B}}{L_g/2k_A A_c + L_g/2k_B A_c} + k_f A_v \frac{T_{2A} - T_{2B}}{L_g} = \frac{T_{2A} - T_{2B}}{1/h_c A}$$

where $L_g$ is the thickness of the void space and $k_f$ is the thermal conductivity of the fluid which fills the void space. $A$ is the *total* cross-sectional area of the bars. Solving for $h_c$, the contact coefficient, we obtain

$$h_c = \frac{1}{L_g}\left(\frac{A_c}{A}\frac{2k_A k_B}{k_A + k_B} + \frac{A_v}{A} k_f\right) \tag{2-29}$$

In most instances air is the fluid filling the void space and $k_f$ is small compared with $k_A$ and $k_B$. If the contact area is small, the major thermal resistance results from the void space. The main problem with this simple theory is that it is extremely difficult to determine effective values of $A_c$, $A_v$, and $L_g$ for surfaces in contact.

From the above physical model, we may tentatively conclude:

1. The contact resistance should increase with a decrease in the ambient gas pressure when the pressure is decreased below the value where the mean-free path of the molecules is large compared with

a characteristic dimension of the void space, since the effective thermal conductance of the entrapped gas will be decreased for this condition.

2.  The contact resistance should be decreased for an increase in the joint pressure since this results in a deformation of the high spots of the contact surfaces, thereby creating a greater contact area between the solids.

A very complete survey of the contact-resistance problem is presented in Refs. 4, 6, and 7. Unfortunately, there is no satisfactory theory which will predict thermal contact resistance for all types of engineering materials, nor have experimental studies yielded completely reliable empirical correlations. This is understandable because of the many complex surface conditions which may be encountered in practice.

Radiation heat transfer across the joint can also be important when high temperatures are encountered. This energy transfer may be calculated by the methods to be discussed in Chap. 8.

## REVIEW QUESTIONS

1.  What is meant by the term one dimensional when applied to conduction problems?

2.  What is meant by thermal resistance?

3.  Why is the one-dimensional heat-flow assumption important in the analysis of fins?

4.  Define fin efficiency.

5.  Why is the insulated-tip solution important for the fin problems.

6.  What is meant by thermal-contact resistance? Upon what parameters does this resistance depend?

## PROBLEMS

2-1.   A wall of 0.8 ft thickness is to be constructed from material which has an average thermal conductivity of 0.75 Btu/hr-ft-°F. The wall is to be insulated with material having an average thermal conductivity of 0.2 Btu/hr-ft-°F so that the heat loss per square foot will not exceed 580 Btu/hr. Assuming that the inner and outer surface temperatures of the insulated wall are 2400 and 80°F, calculate the thickness of insulation required.

2-2.   A certain material 1 in. thick, having a cross-sectional area of 1 ft², has one side maintained at 100°F and the other side at 200°F. The temperature at the center plane of the material is 140°F, and the heat flow through the material is 3500 Btu/hr. Obtain an expression for the thermal conductivity of the material as a function of temperature.

2-3.   A composite wall is formed of a 1-in. copper plate, a $\frac{1}{8}$-in. layer of asbestos, and a 2-in. layer of fiber glass. The wall is subjected to an overall temperature difference of 1000°F. Calculate the heat flow per unit area through the composite structure.

2-4.    Find the heat transfer through the composite wall below per unit area. Assume one-dimensional heat flow.

2-5.    One side of a copper block 2 in. thick is maintained at 500°F. The other side is covered with a layer of fiber glass 1 in. thick. The outside of the fiber glass is maintained at 100°F, and the total heat flow through the copper–fiber-glass combination is 150,000 Btu/hr. What is the area of the slab?

2-6.    A steel pipe with a 2 in. OD is covered with a $\frac{1}{4}$-in. layer of asbestos insulation ($k = 0.096$ Btu/hr-ft-°F) followed by a 1-in. layer of fiber glass insulation ($k = 0.028$ Btu/hr-ft-°F). The pipe wall temperature is 600°F and the outside insulation temperature is 100°F. Calculate the interface temperature between the asbestos and fiber glass.

2-7.    Derive an expression for the thermal resistance through a hollow spherical shell of inside radius $r_i$ and outside radius $r_o$ having a thermal conductivity $k$.

2-8.    Derive an expression for the temperature distribution in a plane wall having uniformly distributed heat sources and one face maintained at a temperature $T_1$ while the other face is maintained at a temperature $T_2$. The thickness of the wall may be taken as $2L$.

2-9.    Derive an expression for the temperature distribution in a plane

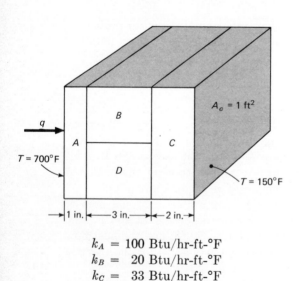

$$k_A = 100 \text{ Btu/hr-ft-°F}$$
$$k_B = 20 \text{ Btu/hr-ft-°F}$$
$$k_C = 33 \text{ Btu/hr-ft-°F}$$
$$k_D = 45 \text{ Btu/hr-ft-°F}$$
$$A_B = A_D$$

wall in which there are uniformly distributed heat sources which vary according to the linear relation

$$\dot{q} = \dot{q}_w[1 + \beta(T - T_w)]$$

where $\dot{q}_w$ is a constant and equal to the heat generated per unit volume at the wall temperature $T_w$. Both sides of the plate are maintained at $T_w$, and the plate thickness is $2L$.

2-10.   A $\frac{1}{8}$-in.-diameter stainless-steel wire 1 ft long has a voltage of 10 volts impressed on it. The outer surface temperature of the wire is maintained at 200°F. Calculate the center temperature of the wire. Take the resistivity of the wire as 70 $\mu$ohm-cm, and the thermal conductivity as 13 Btu/hr-ft-°F.

2-11.   The heater wire of Example 2-2 is submerged in a fluid maintained at 100°F. The convection heat-transfer coefficient is 1000 Btu/hr-ft²-°F. Calculate the center temperature of the wire.

2-12.   A plane wall 3 in. thick generates heat internally at the rate of $10^4$ Btu/hr-ft³. One side of the wall is insulated, and the other side is exposed to an environment at 100°F. The convection heat-transfer coefficient between the wall and the environment is 100 Btu/hr-ft²-°F. The thermal conductivity of the wall is 12 Btu/hr-ft-°F. Calculate the maximum temperature in the wall.

2-13.   Consider a shielding wall for a nuclear reactor. The wall receives a gamma-ray flux such that heat is generated within the wall according to the relation

$$\dot{q} = \dot{q}_0 e^{-ax}$$

where $\dot{q}_0$ is the heat generation at the inner face of the wall exposed to the gamma-ray flux and $a$ is a constant. Using this relation for heat generation, derive an expression for the temperature distribution in a wall of thickness $L$, where the inside and outside temperatures are maintained at $T_i$ and $T_o$, respectively. Also obtain an expression for the maximum temperature in the wall.

2-14.   Repeat Prob. 2-13, assuming that the outer surface is adiabatic while the inner surface temperature is maintained at $T_i$.

2-15.   Show that the temperature distribution for case 2 in Sec. 2-7 is

$$\frac{T - T_\infty}{T_0 - T_\infty} = \frac{\cosh m(L - x) + (h/km) \sinh m(L - x)}{\cosh mL + (h/mk) \sinh mL}$$

Subsequently show that the heat transfer is

$$q = \sqrt{hPkA}\,(T_0 - T_\infty)\frac{\sinh mL + (h/mk) \cosh mL}{\cosh mL + (h/mk) \sinh mL}$$

2-16.    A thin rod of length $L$ has its two ends connected to two walls which are maintained at temperatures $T_1$ and $T_2$, respectively. The rod loses heat to the environment at $T_\infty$ by convection. Derive an expression (a) for the temperature distribution in the rod, and (b) for the total heat lost by the rod.

2-17.    A rod of length $L$ has one end maintained at temperature $T_0$ and is exposed to an environment of temperature $T_\infty$. An electrical-heating element is placed in the rod so that heat is generated uniformly along the length at a rate of $\dot{q}$ Btu/ft³-hr. Derive an expression (a) for the temperature distribution in the rod and (b) for the total heat transferred to the environment. Obtain an expression for the value of $\dot{q}$ which will make the heat transfer zero at the end which is maintained at $T_0$.

2-18.    An aluminum rod 1 in. in diameter and 6 in. long protrudes from a wall which is maintained at 500°F. The rod is exposed to an environment at 60°F. The convection heat-transfer coefficient is 2.6 Btu/hr-ft²-°F. Calculate the heat lost by the rod.

2-19.    Derive Eq. (2-23) by integrating the convection heat loss from the rod of case 1 in Sec. 2-7.

2-20.    Derive Eq. (2-24) by integrating the convection heat loss from the rod of case 3 in Sec. 2-7.

2-21.    One end of a copper rod 1 ft long is firmly connected to a wall which is maintained at 400°F. The other end is firmly connected to a wall which is maintained at 200°F. Air is blown across the rod so that a heat-transfer coefficient of 3 Btu/hr-ft²-°F is maintained. The diameter of the rod is $\frac{1}{2}$ in. The temperature of the air is 100°F. What is the net heat lost to the air in Btu per hour?

2-22.    A long thin copper rod $\frac{1}{4}$ in. in diameter is exposed to an environment at 70°F. The base temperature of the rod is 300°F. The heat-transfer coefficient between the rod and the environment is 4.26 Btu/hr-ft²-°F. Calculate the heat given up by the rod.

2-23.    A very long copper rod ($k = 215$ Btu/hr-ft-°F) 1 in. in diameter has one end maintained at 200°F. The rod is exposed to a fluid whose temperature is 100°F. The heat-transfer coefficient is 2 Btu/hr-ft²-°F. How much heat is lost by the rod?

2-24.    An aluminum fin $\frac{1}{16}$ in. thick is placed on a circular tube with 1-in. OD. The fin is $\frac{1}{4}$ in. long. The tube wall is maintained at 300°F, the environment temperature is 60°F, and the convection heat-transfer coefficient is 4 Btu/hr-ft²-°F. Calculate the heat lost by the fin.

2-25.    The *total* efficiency for a finned surface may be defined as the ratio of the *total* heat transfer of the combined area of the surface and fins to the heat which would be transferred if this *total* area were maintained at the base temperature $T_0$. Show that this efficiency can be calculated from

$$\eta_t = 1 - \frac{A_f}{A}(1 - \eta_f)$$

where $\eta_t$ = total efficiency

$A_f$ = surface area of all fins

$A$ = total heat-transfer area, including fins and exposed tube or other surface

$\eta_f$ = fin efficiency

2-26.    A triangular fin of 18-8 stainless steel is attached to a plane wall maintained at 860°F. The fin thickness is $\frac{1}{4}$ in., and the length is 1 in. The environment is at 200°F, and the convection heat-transfer coefficient is 5 Btu/hr-ft²-°F. Calculate the heat lost from the fin.

2-27.    A 1-in.-diameter tube has circumferential fins of rectangular profile spaced at $\frac{3}{8}$-in. increments along its length. The fins are constructed of aluminum and are $\frac{1}{32}$ in. thick and $\frac{1}{2}$ in. long. The tube wall temperature is maintained at 400°F, and the environment temperature is 100°F. The heat-transfer coefficient is 20 Btu/hr-ft²-°F. Calculate the heat loss from the tube per foot of length.

2-28.    A circumferential fin of rectangular cross section surrounds a 1-in.-diameter tube. The length of the fin is $\frac{1}{4}$ in., and the thickness is $\frac{1}{8}$ in. The fin is constructed of mild steel. If air blows over the fin so that a heat-transfer coefficient of 5 Btu/hr-ft²-°F is experienced and the temperatures of the base and air are 500 and 100°F, respectively, calculate the heat transfer from the fin.

2-29.    A straight rectangular fin 1 in. thick and 6 in. long is constructed of steel and placed on the outside of a wall maintained at 400°F. The environment temperature is 60°F, and the heat-transfer coefficient for convection is 3 Btu/hr-ft²-°F. Calculate the heat lost from the fin per unit depth.

2-30.    An aluminum fin $\frac{1}{16}$ in. thick surrounds a tube 1 in. in diameter. The length of the fin is $\frac{1}{2}$ in. The tube wall temperature is 400°F, and the environment temperature is 70°F. The heat-transfer coefficient is 10 Btu/hr-ft²-°F. What is the heat lost by the fin?

2-31.    Rework Prob. 2-9 assuming that the plate is subjected to a convection environment on both sides of temperature $T_\infty$ with a heat-transfer coefficient $h$. $T_w$ is now some reference temperature *not* necessarily the same as the surface temperature.

2-32.    An electric current is used to heat a tube through which flows a suitable cooling fluid. The outside of the tube is covered with insulation to minimize heat loss to the surroundings, and thermocouples are attached to the outer surface of the tube to measure the temperature. Assuming uniform heat generation in the tube, derive an expression for the convection heat-transfer coefficient on the inside of the tube in terms of the measured variables: voltage $E$, current $I$, outside tube wall temperature $T_o$, inside and outside radii $r_i$ and $r_o$, tube length $L$, and fluid temperature $T_f$.

2-33.  Obtain an expression for the optimum thickness of a straight rectangular fin for a given profile area. Use the simplified insulated-tip solution.

2-34.  Derive a differential equation (do not solve) for the temperature distribution in a straight triangular fin. For convenience take the coordinate axis as shown and assume one-dimensional heat flow.

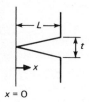

2-35.  An outside wall for a building consists of a 4-in. layer of common brick and a 1-in. layer of fiber glass ($k = 0.03$). Calculate the heat flow through the wall for an 80°F temperature differential.

2-36.  Heat is generated in a 1-in.-square copper rod at the rate of $10^6$ Btu/hr-ft³. The rod is exposed to a convection environment at 70°F, and the heat-transfer coefficient is 700 Btu/hr-ft²-°F. Calculate the surface temperature of the rod.

2-37.  A plane wall of thickness $2L$ has an internal heat generation which varies according to

$$\dot{q} = \dot{q}_0 \cos ax$$

where $\dot{q}_0$ is the heat generated per unit volume at the center of the wall ($x = 0$) and $a$ is a constant. If both sides of the wall are maintained at a constant temperature of $T_w$, derive an expression for the total heat loss from the wall per unit surface area.

2-38.  One side of a copper block $1\frac{1}{2}$ in. thick is maintained at 400°F. The other side is covered with a layer of fiber glass 1 in. thick. The outside of the fiber glass is maintained at 100°F, and the total heat flow through the composite slab is 1000 Btu/hr. What is the area of the slab?

2-39.  A long stainless-steel rod ($k = 9.4$) has a square cross section of $\frac{1}{2}$ by $\frac{1}{2}$ in. and has one end maintained at 400°F. The rod is exposed to a convection environment at 100°F. The heat-transfer coefficient is 7 Btu/hr-ft²-°F. Calculate the heat lost by the rod.

2-40.  Two 1-in.-diameter bars of stainless steel ($k = 10$) are brought into end-to-end contact so that only 0.1 percent of the cross-section area is in contact at the joint. The bars are 3 in. long and subjected to an axial temperature difference of 500°F. The roughness depth in each bar $(L_g/2)$ is estimated to be 50 $\mu$in. The surrounding fluid is air, whose thermal conductivity may be taken as 0.02 Btu/hr-ft-°F for this problem. Estimate the value of the contact resistance and the axial heat flow. What would the heat flow be for a continuous 6-in. stainless-steel bar?

2-41. When the "joint pressure" for two surfaces in contact is increased, the high spots of the surfaces are deformed so that the contact area $A_c$ is increased and the roughness depth $L_g$ is decreased. Discuss this effect in the light of the presentation of Sec. 2-9. (Experimental work shows that joint conductance varies almost directly with pressure.)

2-42. A pipe of outside radius $r_1$ is covered with an insulating material having a thermal conductivity $k$ and subjected to an environment with a convection heat-transfer coefficient $h$. Derive an expression for the thickness of insulation that will produce the maximum heat transfer. Discuss the result.

2-43. A straight fin of rectangular profile is constructed of duralumin (94 % Al, 3 %Cu) with a thickness of $\frac{3}{32}$ in. The length of the fin is 0.75 in. and it is subjected to a convection environment with $h = 15$ Btu/hr-ft²-°F. If the base temperature is 200°F and the environment is at 80°F, calculate the heat transfer per unit length of fin.

2-44. A certain semiconductor material has a conductivity of 0.0124 watt/cm-°C. A rectangular bar of the material has a cross sectional area of 1 cm² and a length of 3 cm. One end is maintained at 300°C and the other end at 100°C and the bar carries a current of 50 amp. Assuming the longitudinal surface is insulated, calculate the midpoint temperature in the bar.

2-45. A plane wall is constructed of a material having a thermal conductivity that varies as the square of temperature according to the relation

$$k = k_o(1 + \beta T^2)$$

Derive an expression for the heat transfer in such a wall.

2-46. The temperature distribution in a certain plane wall is

$$\frac{T - T_1}{T_2 - T_1} = C_1 + C_2 x^2 + C_3 x^3$$

where $T_1$ and $T_2$ are the temperatures on each side of the wall. If the thermal conductivity of the wall is constant and the wall thickness is $L$, derive an expression for the heat generation per unit volume as a function of $x$, the distance from the plane where $T = T_1$. Let the heat generation rate be $\dot{q}_o$ at $x = 0$.

2-47. A certain internal combustion engine is air-cooled and has a cylinder constructed of cast iron ($k = 35$ Btu/hr-ft-°F). The fins on the cylinder have a length of $\frac{5}{8}$ in. and thickness of $\frac{1}{8}$ in. The convection coefficient is 12 Btu/hr-ft²-°F. The cylinder diameter is 4 in. Calculate the heat loss per fin for a base temperature of 450°F and environment temperature of 100°F.

2-48.    Derive an expression for the temperature distribution in a sphere of radius $r$ with uniform heat generation $\dot{q}$ and constant surface temperature $T_w$.

2-49.    A $\frac{1}{16}$-in.-diameter stainless steel rod ($k = 12.5$ Btu/hr-ft-°F) protrudes from a wall maintained at 120°F. The rod is 0.500 in. long and the convection coefficient is 100 Btu/hr-ft²-°F. The environment temperature is 80°F. Calculate the temperature of the tip of the rod. Repeat the calculation for $h = 30$ and 200 Btu/hr-ft²-°F.

## REFERENCES

1.  Schneider, P. J.: "Conduction Heat Transfer," Addison-Wesley Publishing Company, Inc., Reading, Mass., 1955.
2.  Harper, W. B., and D. R. Brown: Mathematical Equations for Heat Conduction in the Fins of Air-cooled Engines, *NACA Rept.* 158, 1922.
3.  Gardner, K. A.: Efficiency of Extended Surfaces, *Trans. ASME*, vol. 67, pp. 621–631, 1945.
4.  Moore, C. J.: Heat Transfer across Surfaces in Contact: Studies of Transients in One-dimensional Composite Systems, *Southern Methodist Univ., Thermal/Fluid Sci. Ctr. Res. Rept.* 67-2, Dallas, Tex., March, 1967.
5.  Ybarrondo, L. J., and J. E. Sunderland: Heat Transfer from Extended Surfaces, *Bull. Mech. Engr. Educ.*, vol. 5, pp. 229–234, 1966.
6.  Moore, C. J., Jr., H. A. Blum, and H. Atkins: Subject Classification Bibliography for Thermal Contact Resistance Studies, *ASME Paper* 68-WA/HT-18, December, 1968.
7.  Clausing, A. M.: Heat Transfer at the Interface of Dissimilar Metals—The Influence of Thermal Strain, *Intern. J. Heat and Mass Transfer*, vol. 9, p. 791, 1966.
8.  Kern, D. Q., and A. D. Kraus: "Extended Surface Heat Transfer," McGraw-Hill Book Company, New York, 1972.
9.  Sparrow, E. M. and R. D. Cess: "Radiation Heat Transfer," Wadsworth Publishing Company, 1966.

# CHAPTER
# 3

# STEADY-STATE CONDUCTION—TWO DIMENSIONS

## 3-1□INTRODUCTION

In the preceding chapter steady-state heat transfer was calculated in systems in which the temperature gradient and area could be expressed in terms of one space coordinate. We now wish to analyze the more general case of two-dimensional heat flow. For steady state, the Laplace equation applies:

$$\frac{\partial^2 T}{\partial x^2} + \frac{\partial^2 T}{\partial y^2} = 0 \tag{3-1}$$

assuming constant thermal conductivity. The solution to this equation may be obtained by either analytical, numerical, or graphical techniques.

The objective of any heat-transfer analysis is usually to predict heat flow or the temperature which results from a certain heat flow. The solution to Eq. (3-1) will give the temperature in a two-dimensional body as a function of the two independent space coordinates $x$ and $y$. Then the heat flow in the $x$ and $y$ directions may be calculated from the Fourier equations

$$q_x = -kA_x \frac{\partial T}{\partial x} \tag{3-2}$$

$$q_y = -kA_y \frac{\partial T}{\partial y} \tag{3-3}$$

These heat-flow quantities are directed either in the $x$ direction or in the $y$ direction. The total heat flow at any point in the material is the resultant of the $q_x$ and $q_y$ at that point. Thus the total heat-flow vector is directed so that it is perpendicular to the lines of constant temperature in the material, as shown in Fig. 3-1. So if the temperature distribution in the material is known, we may easily establish the heat flow.

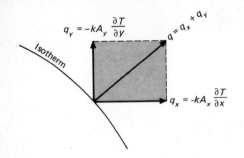

**Fig. 3-1  Sketch showing heat flow in two dimensions.**

## 3-2□MATHEMATICAL ANALYSIS OF TWO - DIMENSIONAL HEAT CONDUCTION

We shall first consider an analytical approach to a two-dimensional problem and then indicate the numerical and graphical methods which may be used to advantage in many other problems. It is worthwhile to mention here that analytical solutions are not always possible to obtain, and indeed, in many instances they are very cumbersome and difficult to use. In these cases numerical techniques are frequently used to advantage. For a more extensive treatment of the analytical methods used in conduction problems the reader is referred to the books by Carslaw and Jaeger [1], Schneider [2], Arpaci [13] and Ozisik [12].

Consider the rectangular plate shown in Fig. 3-2. Three sides of the plate are maintained at the constant temperature $T_1$, and the upper side

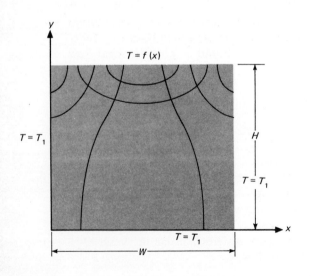

**Fig. 3-2  Isotherms and heat-flow lines in a rectangular plate.**

has some temperature distribution impressed upon it. This distribution could be simply a constant temperature or something more complex, such as a sine-wave distribution. We shall consider both cases.

To solve Eq. (3-1), the separation-of-variables method is used. The essential point of this method is that the solution to the differential equation is assumed to take a product form

$$T = XY \tag{3-4}$$

where

$$X = X(x)$$

and

$$Y = Y(y)$$

The boundary conditions are then applied to determine the form of the functions $X$ and $Y$. The basic assumption as given by Eq. (3-4) can be justified *only* if it is possible to find a solution of this form which satisfies the boundary conditions.

First consider the boundary conditions with a sine-wave temperature distribution impressed on the upper edge of the plate. Thus

$$
\begin{array}{ll}
T = T_1 & \text{at } y = 0 \\
T = T_1 & \text{at } x = 0 \\
T = T_1 & \text{at } x = W \\
T = T_m \sin \dfrac{\pi x}{W} + T_1 & \text{at } y = H
\end{array}
\tag{3-5}
$$

where $T_m$ is the amplitude of the sine function. Substituting Eq. (3-4) in (3-1),

$$-\frac{1}{X}\frac{d^2X}{dx^2} = \frac{1}{Y}\frac{d^2Y}{dy^2} \tag{3-6}$$

Observe that each side of Eq. (3-6) is independent of the other since $x$ and $y$ are independent variables. This requires that each side be equal to some constant. We may thus obtain two ordinary differential equations in terms of this constant.

$$\frac{d^2X}{dx^2} + \lambda^2 X = 0 \tag{3-7}$$

$$\frac{d^2Y}{dy^2} - \lambda^2 Y = 0 \tag{3-8}$$

where $\lambda^2$ is called the separation constant and its value must be determined from the boundary conditions. Note that the form of the solution to Eqs. (3-7) and (3-8) will depend on the sign of $\lambda^2$; a different form would also result if $\lambda^2$ were zero. The only way that the correct form can be determined is through an application of the boundary conditions of the problem. So we shall first write down all possible solutions and then see which one fits the problem under consideration.

For $\lambda^2 = 0$:
$$X = C_1 + C_2 x$$
$$Y = C_3 + C_4 y \tag{3-9}$$
$$T = (C_1 + C_2 x)(C_3 + C_4 y)$$

This function cannot fit the sine-function boundary condition, so that the $\lambda^2 = 0$ solution may be excluded.

For $\lambda^2 < 0$:
$$X = C_5 e^{-\lambda x} + C_6 e^{\lambda x}$$
$$Y = C_7 \cos \lambda y + C_8 \sin \lambda y \tag{3-10}$$
$$T = (C_5 e^{-\lambda x} + C_6 e^{\lambda x})(C_7 \cos \lambda y + C_8 \sin \lambda y)$$

Again, the sine-function boundary condition cannot be satisfied, so that this solution is excluded also.

For $\lambda^2 > 0$:
$$X = C_9 \cos \lambda x + C_{10} \sin \lambda x$$
$$Y = C_{11} e^{-\lambda y} + C_{12} e^{\lambda y} \tag{3-11}$$
$$T = (C_9 \cos \lambda x + C_{10} \sin \lambda x)(C_{11} e^{-\lambda y} + C_{12} e^{\lambda y})$$

Now, it is possible to satisfy the sine-function boundary condition; so we shall attempt to satisfy the other conditions. The algebra is somewhat easier to handle when the substitution

$$\theta = T - T_1$$

is made. The differential equation and the solution then retain the same form in the new variable $\theta$, and we need only transform the boundary conditions. Thus

$$\begin{aligned} \theta &= 0 && \text{at } y = 0 \\ \theta &= 0 && \text{at } x = 0 \\ \theta &= 0 && \text{at } x = W \\ \theta &= T_m \sin \frac{\pi x}{W} && \text{at } y = H \end{aligned} \tag{3-12}$$

Applying these conditions,

$$0 = (C_9 \cos \lambda x + C_{10} \sin \lambda x)(C_{11} + C_{12}) \tag{a}$$
$$0 = C_9 (C_{11} e^{-\lambda y} + C_{12} e^{\lambda y}) \tag{b}$$
$$0 = (C_9 \cos \lambda W + C_{10} \sin \lambda W)(C_{11} e^{-\lambda y} + C_{12} e^{\lambda y}) \tag{c}$$
$$T_m \sin \frac{\pi x}{W} = (C_9 \cos \lambda x + C_{10} \sin \lambda x)(C_{11} e^{-\lambda H} + C_{12} e^{\lambda H}) \tag{d}$$

Accordingly,
$$C_{11} = -C_{12}$$
$$C_9 = 0$$

and from (c),
$$0 = C_{10} C_{12} \sin \lambda W (e^{\lambda y} - e^{-\lambda y})$$

This requires that
$$\sin \lambda W = 0 \tag{3-13}$$

It will be recalled that $\lambda$ was an undetermined separation constant. There are several values which will satisfy Eq. (3-13), and these may be written

$$\lambda = \frac{n\pi}{W} \qquad (3\text{-}14)$$

where $n$ is an integer. The solution to the differential equation may thus be written as a sum of the solutions for each value of $n$. This is an infinite sum, so that the final solution is the infinite series

$$\theta = T - T_1 = \sum_{n=1}^{\infty} C_n \sin \frac{n\pi x}{W} \sinh \frac{n\pi y}{W} \qquad (3\text{-}15)$$

where the constants have been combined and the exponential terms converted to the hyperbolic function. The final boundary condition may now be applied.

$$T_m \sin \frac{\pi x}{W} = \sum_{n=1}^{\infty} C_n \sin \frac{n\pi x}{W} \sinh \frac{n\pi H}{W}$$

which requires that $C_n = 0$ for $n > 1$. The final solution is therefore

$$T = T_m \frac{\sinh (\pi y/W)}{\sinh (\pi H/W)} \sin \left(\frac{\pi x}{W}\right) + T_1 \qquad (3\text{-}16)$$

The temperature field for this problem is shown in Fig. 3-2. Note that the heat-flow lines are perpendicular to the isotherms.

We now consider the set of boundary conditions

$$\begin{aligned}
T &= T_1 & \text{at } y &= 0 \\
T &= T_1 & \text{at } x &= 0 \\
T &= T_1 & \text{at } x &= W \\
T &= T_2 & \text{at } y &= H
\end{aligned}$$

Using the first three boundary conditions, the solution is obtained in the form of Eq. (3-15).

$$T - T_1 = \sum_{n=1}^{\infty} C_n \sin \frac{n\pi x}{W} \sinh \frac{n\pi y}{W} \qquad (3\text{-}17)$$

Applying the fourth boundary condition,

$$T_2 - T_1 = \sum_{n=1}^{\infty} C_n \sin \frac{n\pi x}{W} \sinh \frac{n\pi H}{W} \qquad (3\text{-}18)$$

This is a Fourier sine series, and the values of the $C_n$ may be determined by expanding the constant temperature difference $T_2 - T_1$ in a Fourier series over the interval $0 < x < W$. This series is

$$T_2 - T_1 = (T_2 - T_1) \frac{2}{\pi} \sum_{n=1}^{\infty} \frac{(-1)^{n+1} + 1}{n} \sin \frac{n\pi x}{W} \qquad (3\text{-}19)$$

Upon comparison of Eq. (3-18) with Eq. (3-19), we find that

$$C_n = \frac{2}{\pi} (T_2 - T_1) \frac{1}{\sinh (n\pi H/W)} \frac{(-1)^{n+1} + 1}{n}$$

and the final solution is expressed as

$$\frac{T - T_1}{T_2 - T_1} = \frac{2}{\pi} \sum_{n=1}^{\infty} \frac{(-1)^{n+1} + 1}{n} \sin \frac{n\pi x}{W} \frac{\sinh (n\pi y/W)}{\sinh (n\pi H/W)} \qquad (3\text{-}20)$$

An extensive study of analytical techniques used in conduction heat transfer requires a background in the theory of orthogonal functions. Fourier series are one example of orthogonal functions, as are Bessel functions and other special functions applicable to different geometries and boundary conditions. The interested reader may consult one or more of the conduction heat transfer texts listed in the references for further information on the subject.

### 3-3□GRAPHICAL ANALYSIS

Consider the two-dimensional system shown in Fig. 3-3. The inside surface is maintained at some temperature $T_1$, and the outer surface is maintained at $T_2$. We wish to calculate the heat transfer. Isotherms and heat-flow lines have been sketched to aid in this calculation. The isotherms and heat-flow lines form groupings of curvilinear figures like that shown in Fig. 3-3b. The heat flow across this curvilinear section is given by Fourier's law, assuming unit depth of material.

$$q = -k \, \Delta x \, (1) \frac{\Delta T}{\Delta y} \qquad (3\text{-}21)$$

This heat flow will be the same through each section within this heat-flow lane, and the total heat flow will be the sum of the heat flows through all the lanes. If the sketch is drawn so that $\Delta x = \Delta y$, the heat flow is proportional to the $\Delta T$ across the element and, since this heat flow is constant, the $\Delta T$ across each element must be the same within the same heat-flow lane. Thus the $\Delta T$ across an element is given by

$$\Delta T = \frac{\Delta T_{\text{overall}}}{N}$$

where $N$ is the number of temperature increments between the inner and

**Fig. 3-3   Sketch showing element used for curvilinear-square analysis of two-dimensional heat flow.**

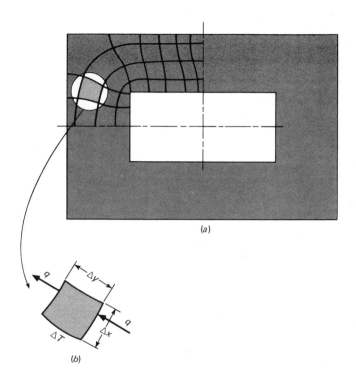

(a)

(b)

outer surface. Furthermore, the heat flow through each lane is the same since it is independent of the dimensions $\Delta x$ and $\Delta y$ when they are constructed equal. Thus we write for the total heat transfer

$$q = \frac{M}{N} k \, \Delta T_{\text{overall}} = \frac{M}{N} k(T_2 - T_1) \qquad (3\text{-}22)$$

where $M$ is the number of heat-flow lanes. So, to calculate the heat transfer, we need only construct these curvilinear square plots and count the number of temperature increments and heat-flow lanes. Care must be taken to construct the plot so that $\Delta x \approx \Delta y$ and the lines are perpendicular.

The accuracy of this method is dependent entirely on the skill of the person sketching the curvilinear squares. Even a crude sketch, however, can frequently help to give fairly good estimates of the temperatures that will occur in a body; and these estimates may then be refined with numerical techniques discussed in Sec. 3-5. An electrical analogy may be employed to sketch the curvilinear squares as discussed in Sec. 3-6.

**Fig. 3-4  Sketch illustrating dimensions for use in calculating three-dimensional shape factors.**

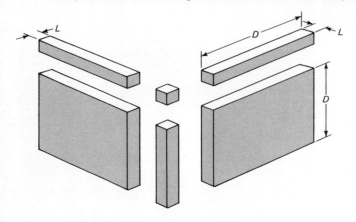

### 3-4□THE CONDUCTION SHAPE FACTOR

In a two-dimensional system where only two temperature limits are involved, we may define a conduction shape factor $S$ such that

$$q = kS \, \Delta T_{\text{overall}} \qquad (3\text{-}23)$$

The values of $S$ have been worked out for several geometries and are summarized in Table 3-1.

For a three-dimensional wall, as in a furnace, separate shape factors are used to calculate the heat flow through the edge and corner sections. When all the interior dimensions are greater than one-fifth of the wall thickness,

$$S_{\text{wall}} = \frac{A}{L}$$
$$S_{\text{edge}} = 0.54D$$
$$S_{\text{corner}} = 0.15L$$

where $A$ = area of wall
$L$ = wall thickness
$D$ = length of edge

These dimensions are illustrated in Fig. 3-4. Note that the shape factor per unit depth is given by the ratio $M/N$ when the curvilinear-squares method is used for calculations. The use of the shape factor for calculation purposes is illustrated in Examples 3-1 and 3-2.

### Example 3-1

A horizontal pipe 6 in. in diameter and 6 ft long is buried in the earth at a depth of 6 in. The pipe wall temperature is 150°F, and the earth surface temperature is 42°F. Assuming that the thermal conductivity of the earth is 0.5 Btu/hr-ft-°F, calculate the heat lost by the pipe.

**Solution**  We may calculate the shape factor for this situation using the equation given in Table 3-1. Since $D < 3r$,

$$S = \frac{2\pi L}{\cosh^{-1}(D/r)} = \frac{2\pi(6)}{\cosh^{-1}\left(\frac{6}{3}\right)} = 28.5$$

The heat flow is calculated from

$$q = kS\,\Delta T = (0.5)(28.5)(150 - 42) = 1540 \text{ Btu/hr} \qquad (451 \text{ W})$$

## Example 3-2

A small cubical furnace 1 by 1 by 1 ft on the inside is constructed of fireclay brick ($k = 0.6$ Btu/hr-ft-°F) with a wall thickness of 4 in. The inside of the furnace is maintained at 800°F, and the outside is maintained at 100°F. Calculate the heat lost through the walls.

**Solution**  We compute the total shape factor by adding the shape factors for the walls, edges, and corners.

Walls:       $S = \dfrac{A}{L} = \dfrac{(1)(1)}{\frac{4}{12}} = 3$

Edges:      $S = 0.54D = 0.54(1) = 0.54$

Corners:    $S = 0.15L = 0.15 \times \frac{4}{12} = 0.05$

There are six wall sections, twelve edges, and eight corners, so that the total shape factor is

$$S = (6)(3) + (12)(0.54) + (8)(0.05) = 24.88$$

The heat flow is calculated from

$$q = kS\,\Delta T = (0.6)(24.88)(800 - 100) = 10,450 \text{ Btu/hr} \qquad (3,062 \text{ W})$$

## 3-5□NUMERICAL METHOD OF ANALYSIS

An immense number of analytical solutions for conduction heat-transfer problems has been accumulated in the literature over the past 100 years. Even so, there are many practical situations where the geometry or boundary conditions are such that an analytical solution has not been obtained at all, or, if the solution has been developed, it involves such a complex series solution that numerical evaluation becomes exceedingly difficult. For such situations the most fruitful approach to the problem is one based on finite-difference techniques, the basic principles of which we shall outline in this section. Of course, the rapid development of high-speed digital computers has enabled the practicing heat-transfer specialist to obtain numerical

**Table 3-1** Conduction Shape Factors†

| Physical system | Schematic | Shape factor | Restrictions |
|---|---|---|---|
| Isothermal cylinder of radius $r$ buried in semi-infinite medium having isothermal surface | 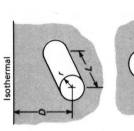 | $\dfrac{2\pi L}{\cosh^{-1}(D/r)}$ | $L \gg r$ |
| | | $\dfrac{2\pi L}{\ln(2D/r)}$ | $L \gg r$ $D > 3r$ |
| | | $\dfrac{2\pi L}{\ln(L/r)\left[1 - \dfrac{\ln(L/2D)}{\ln(L/r)}\right]}$ | $D \gg r$ $L \gg D$ |
| Isothermal sphere of radius $r$ buried in infinite medium | | $4\pi r$ | |
| Isothermal sphere of radius $r$ buried in semi-infinite medium having isothermal surface | | $\dfrac{4\pi r}{1 - r/2D}$ | |
| Conduction between two isothermal cylinders buried in infinite medium | | $\dfrac{2\pi L}{\cosh^{-1}\left(\dfrac{D^2 - r_1^2 - r_2^2}{2r_1 r_2}\right)}$ | $L \gg r$ $L \gg D$ |

| | | | |
|---|---|---|---|
| Isothermal cylinder of radius $r$ placed in semi-infinite medium as shown | Isothermal | $\dfrac{2\pi L}{\ln(2L/r)}$ | $L \gg 2r$ |
| Isothermal rectangular parallelepiped buried in semi-infinite medium having isothermal surface | Isothermal | $1.685L\left[\log\left(1+\dfrac{b}{a}\right)\right]^{-0.59}\left(\dfrac{b}{c}\right)^{-0.078}$ (see Ref. 7) | |
| Plane wall | | $\dfrac{A}{L}$ | One-dimensional heat flow |
| Hollow cylinder, length $L$ | | $\dfrac{2\pi L}{\ln(r_o/r_i)}$ | $L \gg r$ |
| Hollow sphere | | $\dfrac{4\pi r_o r_i}{r_o - r_i}$ | |

† Summarized according to Refs. 6 and 7.

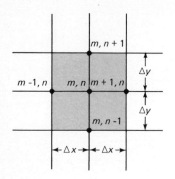

**Fig. 3-5  Sketch illustrating nomenclature used in two-dimensional numerical analysis of heat conduction.**

solutions to many problems heretofore believed to be impossible.

Consider a two-dimensional body which is to be divided into equal increments in both the $x$ and $y$ directions as shown in Fig. 3-5. The nodal points are designated as shown, the $m$ locations indicating the $x$ increment and the $n$ locations indicating the $y$ increment. We wish to establish the temperatures at any of these nodal points within the body, using Eq. (3-1) as a governing condition. Finite differences are used to approximate differential increments in the temperature and space coordinates; and the smaller we choose these finite increments, the more closely the true temperature distribution will be approximated.

The temperature gradients may be written as follows:

$$\frac{\partial T}{\partial x}\bigg)_{m+\frac{1}{2},n} \approx \frac{T_{m+1,n} - T_{m,n}}{\Delta x}$$

$$\frac{\partial T}{\partial x}\bigg)_{m-\frac{1}{2},n} \approx \frac{T_{m,n} - T_{m-1,n}}{\Delta x}$$

$$\frac{\partial T}{\partial y}\bigg)_{m,n+\frac{1}{2}} \approx \frac{T_{m,n+1} - T_{m,n}}{\Delta y}$$

$$\frac{\partial T}{\partial y}\bigg)_{m,n-\frac{1}{2}} \approx \frac{T_{m,n} - T_{m,n-1}}{\Delta y}$$

$$\frac{\partial^2 T}{\partial x^2}\bigg)_{m,n} \approx \frac{\dfrac{\partial T}{\partial x}\bigg)_{m+\frac{1}{2},n} - \dfrac{\partial T}{\partial x}\bigg)_{m-\frac{1}{2},n}}{\Delta x} = \frac{T_{m+1,n} + T_{m-1,n} - 2T_{m,n}}{(\Delta x)^2}$$

$$\frac{\partial^2 T}{\partial y^2}\bigg)_{m,n} \approx \frac{\dfrac{\partial T}{\partial y}\bigg)_{m,n+\frac{1}{2}} - \dfrac{\partial T}{\partial y}\bigg)_{m,n-\frac{1}{2}}}{\Delta y} = \frac{T_{m,n+1} + T_{m,n-1} - 2T_{m,n}}{(\Delta y)^2}$$

Thus the finite-difference approximation for Eq. (3-1) becomes

$$\frac{T_{m+1,n} + T_{m-1,n} - 2T_{m,n}}{(\Delta x)^2} + \frac{T_{m,n+1} + T_{m,n-1} - 2T_{m,n}}{(\Delta y)^2} = 0$$

If $\Delta x = \Delta y$, then

Fig. 3-6

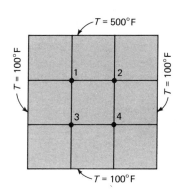

$$T_{m+1,n} + T_{m-1,n} + T_{m,n+1} + T_{m,n-1} - 4T_{m,n} = 0 \qquad (3\text{-}24)$$

Since we are considering the case of constant thermal conductivity, the heat flows may all be expressed in terms of temperature differentials. Equation (3-24) states very simply that the net heat flow into any node is zero at steady-state conditions. In effect, the numerical finite-difference approach replaces the continuous temperature distribution by fictitious heat-conducting rods connected between small nodal points which do not generate heat.

We can also devise a finite-difference scheme to take heat generation into account. We merely add the term $\dot{q}/k$ into the general equation and obtain

$$\frac{T_{m+1,n} + T_{m-1,n} - 2T_{m,n}}{(\Delta x)^2} + \frac{T_{m,n+1} + T_{m,n-1} - 2T_{m,n}}{(\Delta y)^2} + \frac{\dot{q}}{k} = 0$$

Then for a square grid in which $\Delta x = \Delta y$,

$$T_{m+1,n} + T_{m-1,n} + T_{m,n+1} + T_{m,n-1} + \frac{\dot{q}(\Delta x)^2}{k} - 4T_{m,n} = 0 \quad (3\text{-}24a)$$

To utilize the numerical method, Eq. (3-24) must be written for each node within the material and the resultant system of equations must be solved for the temperatures at the various nodes. If small subdivisions are used, the number of nodes may be very large indeed, the equations may become unwieldy, and a simultaneous solution might be very time-consuming if carried out by hand. For a modest number of nodes a hand calculation may be reasonably expedient, and we now consider the so-called relaxation technique for solving the system of equations in these circumstances. In this method the right side of Eq. (3-24) is set equal to some residual $\bar{q}_{m,n}$, which we wish to "relax" to zero. The relaxation process is carried out according to the following steps:

## Table 3-2  Relaxation Table for System of Fig. 3-6

| $T_1$ | $\bar{q}_1$ | $T_2$ | $\bar{q}_2$ | $T_3$ | $\bar{q}_3$ | $T_4$ | $\bar{q}_4$ |
|---|---|---|---|---|---|---|---|
| 300 | $-100$ | 300 | $-100$ | 200 | $-100$ | 200 | $-100$ |
| 275 | 0 | | $-125$ | | $-125$ | | |
| | $-30$ | 270 | $-5$ | | | | $-130$ |
| | | | $-45$ | | $-165$ | 160 | 30 |
| | $-70$ | | | 160 | $-5$ | | $-10$ |
| 255 | 10 | | $-65$ | | $-25$ | | |
| | 0 | 260 | $-25$ | | | | $-20$ |
| | $-5$ | | | 155 | $-5$ | | $-25$ |
| | $-15$ | 250 | 15 | | | | $-35$ |
| | | | 5 | | $-15$ | 150 | 5 |
| | $-20$ | | | 150 | 5 | | 0 |
| 250 | 0 | | 0 | | 0 | | |

1. Assume values for temperatures at the various nodes. Attempt to guess these values as near to their true values as possible. The curvilinear square plot may be a useful way to obtain the first guess.
2. Calculate the residuals at each node, using the assumed temperatures.
3. Relax the largest residual to zero by changing the corresponding nodal temperature an appropriate amount. For the two-dimensional system it will be necessary to change the nodal temperature by one-fourth of the change effected in the residual, and in the opposite sense. For example, if the residual to point $(m, n)$ is $+100$, then to relax it to zero the temperature of node $(m, n)$ must be *increased* by 25.
4. Change the residuals of the surrounding nodes to correspond with the temperature change in step 3.
5. Continue to relax residuals until all are as close to zero as desired. Note that a residual of 4 means an error in temperature of only 1°F. This accuracy is sufficient for most purposes.

An example of the relaxation technique is shown in Fig. 3-6 and Table 3-2. Notice that the residuals may be overrelaxed or underrelaxed to make the calculation easier.

Once the temperatures are determined, the heat flow may be calculated from

$$q = \sum k \, \Delta x \frac{\Delta T}{\Delta y}$$

where the $\Delta T$ is taken at the boundaries. In the example the heat flow may be calculated at either the 500°F face or the three 100°F faces. If a suf-

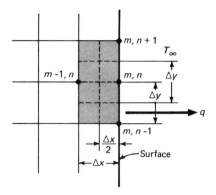

ficiently fine relaxation grid is used, the two values should be very nearly the same. As a matter of general practice it is usually best to take the arithmetic average of the two values for use in calculations. In the example the two calculations yield

500°F face:

$$q = -k\frac{\Delta y}{\Delta x}[(250 - 500) + (250 - 500)] = 500k$$

100°F face:

$$q = -k\frac{\Delta y}{\Delta x}[(250 - 100) + (150 - 100) + (150 - 100) + (150 - 100)$$
$$+ (150 - 100) + (250 - 100)] = -500k$$

and the two values agree in this case. The calculation of the heat flow in cases in which curved boundaries or complicated shapes are involved is treated by Dusinberre [3], Schneider [2], and Myers [15].

In the above example only four nodes were employed, and the equations corresponding to Eq. (3-24) could easily have been solved simultaneously; however, this simple example shows the general technique of the relaxation method that may be applied to more complex problems.

When the solid is exposed to some convection boundary condition, the temperatures at the surface must be computed differently from the method given above. Consider the boundary shown in Fig. 3-7. The energy balance on node $(m, n)$ is

$$-k\,\Delta y\,\frac{T_{m,n} - T_{m-1,n}}{\Delta x} - k\frac{\Delta x}{2}\frac{T_{m,n} - T_{m,n+1}}{\Delta y} - k\frac{\Delta x}{2}\frac{T_{m,n} - T_{m,n-1}}{\Delta y}$$
$$= h\,\Delta y\,(T_{m,n} - T_\infty)$$

If $\Delta x = \Delta y$, the boundary temperature is expressed in the equation

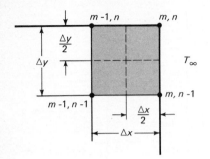

$$T_{m,n}\left(\frac{h\,\Delta x}{k} + 2\right) - \frac{h\,\Delta x}{k}\,T_\infty - \tfrac{1}{2}(2T_{m-1,n} + T_{m,n+1} + T_{m,n-1}) = 0 \quad (3\text{-}25)$$

An equation of this type must be written for each node along the surface shown in Fig. 3-7. So when a convection boundary condition is present, an equation like (3-25) is used at the boundary and an equation like (3-24) is used for the interior points. The solution would still proceed in the same way by setting Eq. (3-25) equal to some residual and applying the relaxation technique; however, the more complicated nature of the residual equation at the boundary brings about more involved arithmetical calculations in the relaxation solution.

Equation (3-25) applies to a plane surface exposed to a convection boundary condition. It will not apply for other situations, such as an insulated wall or a corner exposed to a convection boundary condition. Consider the corner section shown in Fig. 3-8. The energy balance for the corner section is

$$-k\frac{\Delta y}{2}\frac{T_{m,n} - T_{m-1,n}}{\Delta x} - k\frac{\Delta x}{2}\frac{T_{m,n} - T_{m,n-1}}{\Delta y}$$
$$= h\frac{\Delta x}{2}(T_{m,n} - T_\infty) + h\frac{\Delta y}{2}(T_{m,n} - T_\infty)$$

If $\Delta x = \Delta y$,

$$2T_{m,n}\left(\frac{h\,\Delta x}{k} + 1\right) - 2\frac{h\,\Delta x}{k}\,T_\infty - (T_{m-1,n} + T_{m,n-1}) = 0 \quad (3\text{-}26)$$

Other boundary conditions may be treated in a similar fashion.

### Example 3-3

Consider the square of the previous example. The left face is maintained at 100°F and the top face at 500°F while the other two faces are exposed to an environment at 100°F.

**Fig. 3-9   Nomenclature for Example 3-3.**

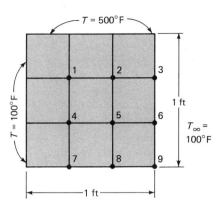

$$h = 10 \text{ Btu/hr-ft}^2\text{-}°F \quad \text{and} \quad k = 10 \text{ Btu/hr-ft-}°F$$

The block is 1 ft square. Compute the temperatures of the various nodes as indicated in Fig. 3-9.

**Solution**   The residual equation for nodes 1, 2, 4, and 5 is

$$T_{m+1,n} + T_{m-1,n} + T_{m,n+1} + T_{m,n-1} - 4T_{m,n} = \bar{q}_{m,n}$$

The residual for nodes 3, 6, 7, and 8 is given by Eq. (3-25), and the residual for node 9 is given by Eq. (3-26):

$$\frac{h\,\Delta x}{k} = \frac{(10)(1)}{(3)(10)} = \frac{1}{3}$$

The residual equation for nodes 3 and 6 are thus written

$$2T_2 + T_6 + 567 - 4.67T_3 = \bar{q}_3$$
$$2T_5 + T_3 + T_9 + 67 - 4.67T_6 = \bar{q}_6$$

The residuals for nodes 7 and 8 are given by

$$2T_4 + T_8 + 167 - 4.67T_7 = \bar{q}_7$$
$$2T_5 + T_7 + T_9 + 67 - 4.67T_8 = \bar{q}_8$$

and the residual equation for node 9 is

$$T_6 + T_8 + 67 - 2.67T_9 = \bar{q}_9$$

Note that the effect of a change in a boundary temperature on the residual at that point is not the same as for an interior point. The relaxation process for this problem is shown in Table 3-3. Notice again that the residuals may be overrelaxed or underrelaxed to speed the solution. The relaxation process has been stopped while some of the residuals still have nonzero values; however, an inspection of the residual equations shows that all temperatures should be within 1°F of their correct value with the residuals shown. This accuracy is acceptable.

A very nice feature of the relaxation method is that calculation errors need not be corrected directly. If a check of the residuals is made at a half-way point in the calculation and a discrepancy is discovered, we need not go back and find our error. We simply recalculate the residuals corresponding to the temperatures at that point in the calculations and then proceed with the computation. For relaxation problems involving a large number of nodes it is good practice to recalculate the residuals periodically in the computation in order to discover any numerical errors. The relaxation process may then be continued on the basis of the corrected values of the residuals if an error is discovered.

A convenient summary of residual equations is given in Table 3-4 for different geometrical and boundary situations. The situations $(f)$ and $(g)$ are of particular interest since they provide the calculation equations which may be employed with curved boundaries, while still using uniform increments in $\Delta x$ and $\Delta y$.

From the foregoing discussion it should be obvious that the relaxation technique is merely a method for obtaining solutions to a set of simultaneous equations. Insofar as the numerical finite difference approximation is concerned, these solutions may be obtained by any other convenient means. For many people the "other means" will be through the use of a digital computer. Most large computers have standard subroutines for obtaining the solutions to a set of simultaneous equations, and the programming is quite straightforward. For very complicated problems, involving a large number of nodes, or problems which will be solved over and over again, with only different boundary conditions, the computer is the obvious tool. For problems of modest scope, such as the examples above, it is hardly worth the time and effort to program the solution. A hand calculation will give the answer in much less time and at less expense.

In order to speed the calculations in a relaxation solution it is frequently advantageous to construct a table showing the effects of changes in nodal temperatures on the residuals of surrounding nodes. Such a group of numbers for the equations of Example 3-3 is shown in Table 3-5. Utilizing this table, we see that an increase in temperature of node 3 by one degree results in a reduction of the residual of that node by 4.67 and an increase in the residuals of adjoining nodes 2 and 6 by 1.0.

For those problems which require a computer solution there are a number of important references available to the reader. The relaxation method outlined above almost certainly would *not* be employed in such

**Table 3-3  Relaxation Method for Convection Boundary of Fig. 3-9**

| | $T_1$ | $\bar{q}_1$ | $T_2$ | $\bar{q}_2$ | $T_3$ | $\bar{q}_3$ | $T_4$ | $\bar{q}_4$ | $T_5$ | $\bar{q}_5$ | $T_6$ | $\bar{q}_6$ | $T_7$ | $\bar{q}_7$ | $T_8$ | $\bar{q}_8$ | $T_9$ | $\bar{q}_9$ |
|---|---|---|---|---|---|---|---|---|---|---|---|---|---|---|---|---|---|---|
| | 300 | −100 | 300 | 0 | 200 | 383 | 200 | −50 | 200 | 0 | 150 | 117 | 150 | 17 | 150 | 67 | 150 | −33 |
| | | | | 100 | 300 | −84 | | | | 50 | 200 | 217 | | | | | | 17 |
| | | | | | | −34 | | −75 | | | | −16 | | | | | | |
| | 275 | 0 | 325 | 75 | | 16 | 180 | 5 | | 75 | | 24 | | 27 | 160 | 20 | 170 | 27 |
| | | 25 | | −25 | | | | 25 | 220 | 85 | | 44 | | −13 | | 60 | | |
| | | 5 | | −5 | | | 190 | −15 | | 65 | 210 | −3 | | 7 | 170 | 80 | | −27 |
| | | 15 | | | | 26 | | | | −15 | | 7 | 155 | 17 | | 33 | | −17 |
| | | | | 5 | 310 | −21 | | −11 | 224 | −5 | | 15 | | −6 | | 41 | | −7 |
| | | | | 9 | 305 | 2 | | −6 | | 5 | 212 | 10 | 154 | −1 | 180 | 46 | | −5 |
| | 279 | −1 | 327 | 4 | 307 | 6 | | −2 | 227 | 15 | 214 | 1 | 156 | 9 | 182 | 45 | 173 | 5 |
| | | 1 | | 8 | | 8 | | −3 | | −1 | | 3 | | 0 | | −2 | | 7 |
| | | | | 0 | | −1 | | 0 | | 1 | | 9 | | 2 | | 4 | | −1 |
| | | | | 2 | | 1 | | 2 | | 3 | | 0 | | | | 6 | | 1 |
| | | | | 5 | | | | | | 13 | | 3 | | | | 9 | | |
| | | | | | | | | | | 1 | | | | | | 0 | | |
| | | | | | | | | | | 3 | | | | | | | | |
| | | | | | | | | | | 5 | | | | | | | | |
| Final values | 279 | 1 | 327 | 5 | 307 | 1 | 190 | 2 | 227 | 5 | 214 | 3 | 156 | 2 | 182 | 0 | 173 | 1 |

**Table 3-4  Summary of Residual Formulas for Finite-difference Calculations**

| Physical situation | Nodal equation for equal increments in $x$ and $y$ |
|---|---|
| (a) Interior node | $\bar{q}_{m,n} = T_{m+1,n} + T_{m,n+1} + T_{m-1,n} + T_{m,n-1} - 4T_{m,n}$ |

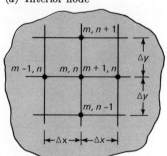

(b) Convection boundary node

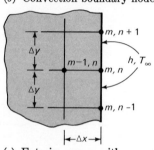

$$\bar{q}_{m,n} = \frac{h\,\Delta x}{k}\,T_\infty + \frac{1}{2}\,(2T_{m-1,n} + T_{m,n+1} + T_{m,n-1})$$
$$- \left(\frac{h\,\Delta x}{k} + 2\right)T_{m,n}$$

(c) Exterior corner with convection boundary

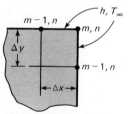

$$\bar{q}_{m,n} = 2\,\frac{h\,\Delta x}{k}\,T_\infty + (T_{m-1,n} + T_{m,n-1})$$
$$- 2\left(\frac{h\,\Delta x}{k} + 1\right)T_{m,n}$$

(d) Interior corner with convection boundary

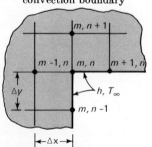

$$\bar{q}_{m,n} = 2\,\frac{h\,\Delta x}{k}\,T_\infty + 2T_{m-1,n} + 2T_{m,n+1} + T_{m+1,n}$$
$$+ T_{m,n-1} - 2\left(3 + \frac{h\,\Delta x}{k}\right)T_{m,n}$$

| Physical situation | Nodal equation for equal increments in $x$ and $y$ |
|---|---|
| (e) Insulated boundary | $\bar{q}_{m,n} = T_{m,n+1} + T_{m,n-1} + 2T_{m-1,n} - 4T_{m,n}$ |

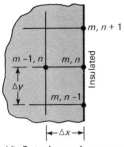

(f) Interior node near curved boundary

$$\bar{q}_{m,n} = \frac{2}{b(b+1)} T_2 + \frac{2}{a+1} T_{m+1,n}$$

$$+ \frac{2}{b+1} T_{m,n-1} + \frac{2}{a(a+1)} T_1 - 2\left(\frac{1}{a} + \frac{1}{b}\right) T_{m,n}$$

(g) Boundary node with convection along curved boundary—node 2 for (f) above

$$\bar{q}_2 = \frac{b}{\sqrt{a^2 + b^2}} T_1 + \frac{b}{\sqrt{c^2 + 1}} T_3$$

$$+ \frac{a+1}{b} T_{m,n} + \frac{h\,\Delta x}{k} (\sqrt{c^2 + 1} + \sqrt{a^2 + b^2}) T_\infty$$

$$- \left[ \frac{b}{\sqrt{a^2 + b^2}} + \frac{b}{\sqrt{c^2 + 1}} + \frac{a+1}{b} \right.$$

$$\left. + (\sqrt{c^2 + 1} + \sqrt{a^2 + b^2}) \frac{h\,\Delta x}{k} \right] T_2$$

---

solutions. The approach selected may well be a transient analysis (see Chap. 4) carried through to the steady-state limit, or direct elimination (Gauss elimination [9]) or iterative techniques [14] may be employed to solve the set of simultaneous equations. A brief summary of available numerical techniques is given by Ozisik [12].

The accuracy of a numerical solution obviously depends on the size of the divisions chosen in the space coordinates and the care that is exerted to solve the set of simultaneous equations. When a computer solution is employed, the relaxation method is seldom used. Instead, the equations are

**Table 3-5    Effects of Changes in Nodal Temperatures of 1°F on Surrounding Residuals for Equations of Example 3-3**

|            | $T_1$ | $T_2$ | $T_3$   | $T_4$ | $T_5$ | $T_6$   | $T_7$   | $T_8$   | $T_9$   |
|------------|-------|-------|---------|-------|-------|---------|---------|---------|---------|
| $\bar{q}_1$ | $-4$  | $+1$  |         | $+1$  |       |         |         |         |         |
| $\bar{q}_2$ | $+1$  | $-4$  | $+1$    |       | $+1$  |         |         |         |         |
| $\bar{q}_3$ | $0$   | $+2$  | $-4.67$ |       |       | $+1$    |         |         |         |
| $\bar{q}_4$ | $+1$  |       |         | $-4$  | $+1$  |         | $+1$    |         |         |
| $\bar{q}_5$ | $0$   | $+1$  |         | $+1$  | $-4$  | $+1$    |         | $+1$    |         |
| $\bar{q}_6$ | $0$   |       | $+1$    |       | $+2$  | $-4.67$ |         |         | $+1$    |
| $\bar{q}_7$ | $0$   |       |         | $+2$  |       |         | $-4.67$ | $+1$    |         |
| $\bar{q}_8$ | $0$   |       |         |       | $+2$  |         | $+1$    | $-4.67$ | $+1$    |
| $\bar{q}_9$ | $0$   |       |         |       |       | $+1$    |         | $+1$    | $-2.67$ |

expressed in matrix form and standard routines used to effect the inversion. Basically, the set of temperature equations takes the form

$$\begin{aligned}
a_{11}T_1 + a_{12}T_2 + \cdots + a_{1n}T_n &= C_1 \\
a_{21}T_1 + a_{22}T_2 + \cdots &= C_2 \\
a_{31}T_1 + \qquad\qquad &= C_3 \\
\cdots\cdots\cdots\cdots\cdots\cdots\cdots\cdots \\
a_{n1}T_1 + a_{n2}T_2 + \cdots + a_{nn}T_n &= C_n
\end{aligned} \tag{3-27}$$

where $T_1$, $T_2$, ..., $T_n$ are the unknown nodal temperatures. Using the matrix notation,

$$[A] = \begin{bmatrix} a_{11}a_{12} & \cdots & a_{1n} \\ a_{21}a_{22} & \cdots & \\ a_{31} & \cdots & \\ \cdots\cdots\cdots\cdots \\ a_{n1}a_{n2} & \cdots & a_{nn} \end{bmatrix}$$

$$[C] = \begin{bmatrix} C_1 \\ C_2 \\ \cdot \\ \cdot \\ \cdot \\ C_n \end{bmatrix} \qquad [T] = \begin{bmatrix} T_1 \\ T_2 \\ \cdot \\ \cdot \\ \cdot \\ T_n \end{bmatrix}$$

Eq. (3-27) can be expressed as

$$[A][T] = [C] \tag{3-28}$$

and the problem is to find the inverse of $[A]$ such that

$$[T] = [A]^{-1}[C] \tag{3-29}$$

Designating $[A]^{-1}$ by

$$[A]^{-1} = \begin{bmatrix} b_{11}b_{12} & \cdots & b_{1n} \\ b_{21} & \cdots & \\ \cdots & \cdots & \cdots \\ b_{n1}b_{n2} & \cdots & b_{nn} \end{bmatrix}$$

the final solutions for the unknown temperatures are written in expanded form as

$$\begin{aligned} T_1 &= b_{11}C_1 + b_{12}C_2 + \cdots + b_{1n}C_n \\ T_2 &= b_{21}C_1 + \cdots \\ &\cdots\cdots\cdots\cdots\cdots\cdots\cdots\cdots\cdots\cdots \\ T_n &= b_{n1}C_1 + b_{n2}C_2 + \cdots + b_{nn}C_n \end{aligned} \qquad (3\text{-}30)$$

Clearly, the larger the number of nodes the more complex and time-consuming the solution, even with a high-speed computer. For most conduction problems the matrix contains a large number of zero elements so that some simplification in the procedure is afforded. For example, the matrix notation for the system of Example 3-3 would be

$$\begin{bmatrix} -4 & 1 & 0 & 1 & 0 & 0 & 0 & 0 & 0 \\ 1 & -4 & 1 & 0 & 1 & 0 & 0 & 0 & 0 \\ 0 & 2 & -4.67 & 0 & 0 & 1 & 0 & 0 & 0 \\ 1 & 0 & 0 & -4 & 1 & 0 & 1 & 0 & 0 \\ 0 & 1 & 0 & 1 & -4 & 1 & 0 & 1 & 0 \\ 0 & 0 & 1 & 0 & 2 & -4.67 & 0 & 0 & 1 \\ 0 & 0 & 0 & 2 & 0 & 0 & -4.67 & 1 & 0 \\ 0 & 0 & 0 & 0 & 2 & 0 & 1 & -4.67 & 1 \\ 0 & 0 & 0 & 0 & 0 & 0 & 0 & 1 & -2.67 \end{bmatrix} \begin{bmatrix} T_1 \\ T_2 \\ T_3 \\ T_4 \\ T_5 \\ T_6 \\ T_7 \\ T_8 \\ T_9 \end{bmatrix} = \begin{bmatrix} -600 \\ -500 \\ -567 \\ -100 \\ 0 \\ -67 \\ -167 \\ -67 \\ -67 \end{bmatrix}$$

## Gauss-Seidel Iteration

When the number of nodes is very large, an iterative technique may frequently yield a more efficient solution to the nodal equations. From our previous nodal equations we note that it is possible to solve for the temperatures $T_{m,n}$ in terms of temperatures of adjoining nodes and certain connecting resistances. Designating the central nodal temperature by $T_i$ and the adjoining temperatures by $T_j$, we can show that the nodal equations can be expressed in the form

$$T_i = \frac{\sum\limits_j T_j/R_{ij}}{\sum\limits_j 1/R_{ij}} \qquad (3\text{-}31)$$

where $R_{ij}$ is the thermal resistance between nodes $i$ and $j$. If an additional heat $q_i$ is delivered to node $i$ (by heat generation, radiation, etc.), Eq. (3-31) would take the form

$$T_i = \frac{q_i + \sum_j T_j/R_{ij}}{\sum_j 1/R_{ij}} \tag{3-32}$$

The Gauss-Seidel iteration procedure makes use of the difference equations expressed in the form of (3-32) through the following procedure:

1. An initial set of values for the $T_i$ is assumed. This initial assumption can be obtained through any expedient method, including a rough relaxation or matrix inversion calculation.
2. Next, new values of the nodal temperatures are calculated according to Eq. (3-32).
3. The process is repeated until successive calculations differ by a sufficiently small amount. In terms of a computer program this means that a test will be inserted to stop the calculations when

$$T_{i_{n+1}} - T_{i_n} \leq \delta \qquad \text{for all } T_i$$

where $\delta$ is some selected constant and $n$ is the number of iterations. Obviously, the smaller the value of $\delta$, the greater the calculation time required to obtain the desired result. The reader should note, however, that the *accuracy* of the solution to the physical problem is not dependent on the value of $\delta$ alone. This constant governs the accuracy of the solution to the set of difference equations. The solution to the physical problem also depends on the selection of the increment $\Delta x$.

By now the reader should have realized the potential power of numerical methods as applied to conduction problems, particularly when coupled with efficient computer techniques. For further information Refs. 3, 9, 10, and 12 should be consulted.

## 3-6□ELECTRICAL ANALOGY FOR TWO - DIMENSIONAL CONDUCTION

Steady-state electric conduction in a homogeneous material of constant resistivity is analogous to steady-state heat conduction in a body of similar geometric shape. For two-dimensional electric conduction the Laplace equation applies.

$$\frac{\partial^2 E}{\partial x^2} + \frac{\partial^2 E}{\partial y^2} = 0$$

where $E$ is the electric potential. A very simple way of solving a two-dimensional heat-conduction problem is to construct an electrical analog and experimentally determine the geometric shape factors for use in Eq. (3-23).

One way to accomplish this is to use a commercially available paper which is coated with a thin conductive film. This paper may be cut to an exact geometric model of the two-dimensional heat-conduction system. At the appropriate edges of the paper good electrical conductors are attached to simulate the temperature-boundary conditions on the problem. An electric-potential difference is then impressed on the model. It may be noted that the paper has a very high resistance in comparison with the conductors attached to the edges, so that a constant potential condition can be maintained at the region of contact.

Once the electric potential is impressed on the paper, an ordinary voltmeter may be used to plot lines of constant electric potential. With these constant-potential lines available, the flux lines may be easily constructed since they are orthogonal to the potential lines. These equipotential and flux lines have precisely the same arrangement as the isotherms and heat-flux lines in the corresponding heat-conduction problem. The shape factor is calculated immediately, using the method which was applied to the curvilinear squares.

It may be noted that the conducting-sheet analogy is not applicable to problems where heat generation is present; however, by addition of appropriate resistances, convection boundary conditions may be handled with little trouble. Schneider [2] and Ozisik [12] discuss the conducting-sheet method, as well as other analogies for treating conduction heat-transfer problems, and Kayan [4, 5] gives a detailed discussion of the conducting-sheet method.

## REVIEW QUESTIONS

1.  What is the main assumption in the separation of variables method for solving Laplace's equation?

2.  Define the conduction shape factor.

3.  What is the basic procedure in setting up a numerical solution to a two-dimensional conduction problem?

4.  Once finite-difference equations are obtained for a conduction problem, what methods are available to effect a solution. What are the advantages and disadvantages of each method, and when would each technique be applied?

5.  Investigate the computer routines that are available at your computer center for solution of conduction heat-transfer problems.

## PROBLEMS

3-1.  Beginning with the separation of variables solutions for $\lambda^2 = 0$ and $\lambda^2 < 0$ [Eqs. (3-9) and (3-10)], show that it is not possible to satisfy the boundary conditions for the constant temperature at $y = H$ with either of these two forms of solution. That is, show that in order to satisfy the boundary conditions

$$T = T_1 \quad \text{at } y = 0$$
$$T = T_1 \quad \text{at } x = 0$$
$$T = T_1 \quad \text{at } x = W$$
$$T = T_2 \quad \text{at } y = H$$

either a trivial or physically unreasonable solution results when either Eq. (3-9) or (3-10) is used.

3-2.  Write out the first four terms of the series solution given in Eq. (3-20). What percentage error results from using only these first four terms at $y = H$ and $x = W/2$?

3-3.  A 3-in.-diameter pipe whose surface temperature is maintained at 400°F passes through the center of a concrete slab 18 in. thick. The outer surface temperatures of the slab are maintained at 60°F. Using the flux plot, estimate the heat loss from the pipe per unit length.

3-4.  A furnace of 1- by 2- by 3-ft inside dimensions is constructed of a material having a thermal conductivity of 0.5 Btu/hr-ft-°F. The wall thickness is 6 in. The inner and outer surface temperatures are 1000 and 200°F, respectively. Calculate the heat loss through the furnace wall.

3-5.  A cube 14 in. on each external side is constructed of fireclay brick. The wall thickness is 2 in. The inner surface temperature is 1000°F, and the outer surface temperature is 200°F. Compute the heat flow in Btu per hour.

3-6.  A heavy wall tube of Monel 1-in. ID and 2-in. OD is covered with a 1-in. layer of glass wool. The inside tube temperature is 500°F, and the temperature at the outside of the insulation is 100°F. How much heat is lost per foot of length?

3-7.  A symmetrical furnace wall has the dimensions shown. Using the flux plot, obtain the shape factor for this wall.

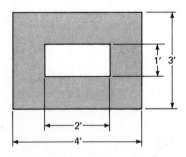

3-8.  Derive an equation equivalent to Eq. (3-24) for an interior node in a three-dimensional heat-flow problem.

3-9.  Derive an equation equivalent to Eq. (3-24) for an interior node in a one-dimensional heat-flow problem.

3-10.   Derive an equation equivalent to Eq. (3-25) for a one-dimensional convection boundary condition.

3-11.   Considering the one-dimensional fin problems of Chap. 2, show that a residual equation for nodes along the fin shown.in the accompanying figure may be expressed as

$$T_m \left[ \frac{hP(\Delta x)^2}{kA} + 2 \right] - \frac{hP(\Delta x)^2}{kA} T_\infty - (T_{m-1} + T_{m+1}) = \bar{q}_m$$

3-12.   An aluminum rod 1 in. in diameter and 6 in. long protrudes from a wall maintained at 600°F. The environment temperature is 100°F. The heat-transfer coefficient is 3 Btu/hr-ft²-°F. Using a numerical technique in accordance with the result of Prob. 3-11, obtain values for the temperatures along the rod. Subsequently obtain the heat flow from the wall at $x = 0$. HINT: The boundary condition at the end of the rod may be expressed by

$$T_m \left[ \frac{h\,\Delta x}{k} + \frac{hP(\Delta x)^2}{2kA} + 1 \right] - T_\infty \left[ \frac{h\,\Delta x}{k} + \frac{hP(\Delta x)^2}{2kA} \right] - T_{m-1} = \bar{q}_m$$

where $m$ denotes the node at the tip of the fin. The heat flow at the base is

$$q_{x=0} = \frac{-kA}{\Delta x} (T_{m+1} - T_m)$$

where $T_m$ is the base temperature and $T_{m+1}$ is the temperature at the first increment.

3-13.   Repeat Prob. 3-12 using a linear variation of heat-transfer coefficient between base temperature and the tip of the fin. Assume $h = 5$ Btu/hr-ft²-°F at the base and $h = 2$ Btu/hr-ft²-°F at the tip.

3-14.   For the wall in Prob. 3-7 a material with $k = 0.8$ Btu/hr-ft-°F is used. The inner and outer wall temperatures are 1200 and 300°F, respectively. Using the relaxation technique, calculate the heat flow through the wall.

3-15.   Repeat Prob. 3-14 assuming that the outer wall is exposed to an environment at 100°F and that the convection heat-transfer coefficient is 3 Btu/hr-ft²-°F. Assume that the inner surface temperature is maintained at 1200°F.

3-16.   Repeat Prob. 3-3 using the relaxation technique.

3-17.   Show that the residual equation corresponding to an insulated wall shown in the accompanying figure is

$$T_{m,n+1} + T_{m,n-1} + 2T_{m-1,n} - 4T_{m,n} = \bar{q}_{m,n}$$

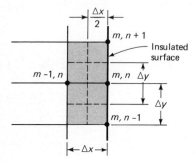

3-18.   In the section given below, the surface 1-4-7 is insulated. The convection heat-transfer coefficient at surface 1-2-3 is 5 Btu/hr-ft²-°F. The thermal conductivity of the solid material is 2 Btu/hr-ft-°F. Using the relaxation technique, compute the temperatures at nodes 1, 2, 4, and 5.

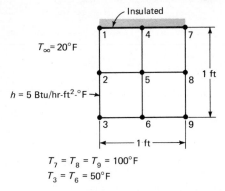

3-19.   A 3- by 12- by 12-in. glass plate ($k = 0.4$) is oriented with the 12 by 12 face in a vertical position. One face loses heat by convection to the surroundings at 70°F. The other vertical face is placed in contact with a constant-temperature block at 400°F. The other four faces are insulated. The convection heat-transfer coefficient varies approximately as

$$h_x = 0.22(T_s - T_\infty)^{\frac{1}{4}} x^{-\frac{1}{4}} \qquad \text{Btu/hr-ft}^2\text{-}°F$$

where $T_s$ is the local surface temperature and $x$ is the vertical distance from the bottom of the plate. Determine the convection heat loss from the plate, using an appropriate numerical analysis. Employ a digital computer for the solution if such facilities are available.

3-20.   Calculate the temperatures at points 1, 2, 3, and 4, using the relaxation method.

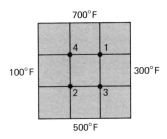

3-21.   For the insulated corner section shown, derive an expression for the residual of node $(m, n)$ under steady-state conditions.

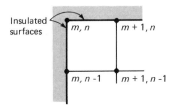

3-22.   Two long cylinders 3 in. and 1 in. in diameter are completely surrounded by a medium with $k = 0.8$ Btu/hr-ft-°F. The distance between centers is 4 in., and the cylinders are maintained at 400 and 100°F. Calculate the heat-transfer rate per unit length.

3-23.   A 3-ft-diameter sphere is maintained at 100°F and is buried in the earth at a place where $k = 1.0$ Btu/hr-ft-°F. The depth to the centerline is 8 ft, and the surface temperature is 40°F. Calculate the heat lost by the sphere.

3-24.   For the block shown, calculate the steady-state temperature distribution at appropriate nodal locations using the relaxation method. Employ a digital computer for the solution if possible, and take advantage of library subroutines available in the computer center.

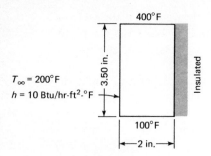

**3-25.** Derive the nodal equation given in Table 3-4($f$).

**3-26.** Derive an expression for the residual of a boundary node subjected to a constant heat flux from the environment. Use the nomenclature of Fig. 3-7.

**3-27.** An 8-in.-diameter sphere is totally enclosed by a large mass of glass wool. A heater inside the sphere maintains its outer surface temperature at 350°F while the temperature at the outer edge of the glass wool is 70°F. How much power must be supplied to the heater to maintain equilibrium conditions?

**3-28.** Rework the example of Fig. 3-6 using the Gauss-Seidel iteration method.

**3-29.** Rework Example 3-3 using the Gauss-Seidel iteration method.

**3-30.** Rework Prob. 3-18 using the Gauss-Seidel iteration method.

**3-31.** A scheme is devised to measure the thermal conductivity of soil by immersing a long electrically heated rod in the ground. For design purposes, the rod is taken as 1 in. diameter with a length of 3 ft. To avoid improper alteration of the soil, the maximum surface temperature of the rod is 135°F while the soil temperature is 55°F. Assuming a soil conductivity of 1.0 Btu/hr-ft-°F, what are the power requirements of the electric heater in watts?

## REFERENCES

1. Carslaw, H. S., and J. C. Jaeger: "Conduction of Heat in Solids," 2d ed. Oxford University Press, Fair Lawn, N.J., 1959.
2. Schneider, P. J.: "Conduction Heat Transfer," Addison-Wesley Publishing Company, Inc., Reading, Mass., 1955.
3. Dusinberre, G. M.: "Heat Transfer Calculations by Finite Differences," International Textbook Company, Scranton, Pa., 1961.
4. Kayan, C. F.: An Electrical Geometrical Analogue for Complex Heat Flow, *Trans. ASME*, vol. 67, p. 713, 1945.
5. Kayan, C. F.: Heat Transfer Temperature Patterns of a Multicomponent Structure by Comparative Methods, *Trans. ASME*, vol. 71, p. 9, 1949.
6. Rudenberg, R.: Die Ausbreitung der Luft-und Erdfelder um Hochspannungs-

leitungen, besonders bei Erd- und Kurzschlussen, *Elecktrotech. Z.*, vol. 46, p. 1342, 1925.

7. Andrews, R. V.: Solving Conductive Heat Transfer Problems with Electrical-analogue Shape Factors, *Chem. Eng. Progr.*, vol. 51, no. 2, p. 67, 1955.

8. Sunderland, J. E., and K. R. Johnson: Shape Factors for Heat Conduction through Bodies with Isothermal or Convective Boundary Conditions, *Trans. ASHAE*, vol. 70, pp. 237–241, 1964.

9. Richtmeyer, R. D.: "Difference Methods for Initial Value Problems," Interscience Publishers, Inc., New York, 1957.

10. Cranck, J., and P. Nicolson: A Practical Method for Numerical Evaluation of Solutions of P.D.E. of the Heat Conduction Type, *Proc. Cambridge Phil. Soc.*, vol. 43, p. 50, 1947.

11. Barakat, H. Z., and J. A. Clark: On the Solution of Diffusion Equation by Numerical Methods, *J. Heat Transfer*, p. 421, November, 1966.

12. Ozisik, M. N.: "Boundary Value Problems of Heat Conduction," International Textbook Company, Scranton, Pa., 1968.

13. Arpaci, V. S.: "Conduction Heat Transfer," Addison-Wesley Publishing Company, Inc., Reading, Mass., 1966.

14. Ames, W. F.: "Nonlinear Partial Differential Equations in Engineering," Academic Press, Inc., New York, 1965.

15. Myers, R. F.: "Conduction Heat Transfer," McGraw-Hill Book Company, New York, 1972.

# CHAPTER

# 4

# UNSTEADY-STATE HEAT CONDUCTION

## 4-1□INTRODUCTION

If a solid body is suddenly subjected to a change in environment, some time must elapse before an equilibrium temperature condition will prevail in the body. We refer to the equilibrium condition as the steady state and calculate the temperature distribution and heat transfer by methods described in Chaps. 2 and 3. In the transient heating or cooling process which takes place in the interim period before equilibrium is established, the analysis must be modified to take into account the change in internal energy of the body with time, and the boundary conditions must be adjusted to match the physical situation which is apparent in the unsteady-state heat-transfer problem. Unsteady-state heat-transfer analysis is obviously of significant practical interest because of the large number of heating and cooling processes which must be calculated in industrial applications.

To analyze a transient heat-transfer problem, we could proceed by solving the general heat-conduction equation by the separation-of-variables method, similar to the analytical treatment used for the two-dimensional steady-state problem discussed in Sec. 3-2. We shall give one illustration of this method of solution for a case of simple geometry and then refer the reader to the references for analysis of more complicated cases. Consider the infinite plate of thickness $2L$ shown in Fig. 4-1. Initially the plate is at a uniform temperature $T_i$, and at time zero the surfaces are suddenly lowered to $T = T_1$. The differential equation is

$$\frac{\partial^2 T}{\partial x^2} = \frac{1}{\alpha} \frac{\partial T}{\partial \tau} \tag{4-1}$$

The equation may be arranged in a more convenient form by introduction of the variable $\theta = T - T_1$. Then

$$\frac{\partial^2 \theta}{\partial x^2} = \frac{1}{\alpha} \frac{\partial \theta}{\partial \tau} \tag{4-2}$$

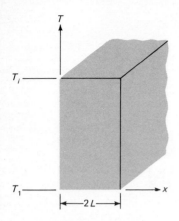

**Fig. 4-1   Infinite plate subjected to sudden cooling of surfaces.**

with the initial and boundary conditions

$$\theta = \theta_i = T_i - T_1 \qquad \text{at } \tau = 0, 0 \leq x \leq 2L \qquad (a)$$
$$\theta = 0 \qquad\qquad\qquad \text{at } x = 0, \tau > 0 \qquad\qquad (b)$$
$$\theta = 0 \qquad\qquad\qquad \text{at } x = 2L, \tau > 0 \qquad\quad (c)$$

Assuming a product solution $\theta(x,\tau) = X(x)\mathcal{H}(\tau)$ produces the two ordinary differential equations

$$\frac{d^2 X}{dx^2} + \lambda^2 X = 0$$

$$\frac{d\mathcal{H}}{d\tau} - \lambda^2 \mathcal{H} = 0$$

where $\lambda^2$ is the separation constant. In order to satisfy the boundary conditions it is necessary that $\lambda^2 > 0$ so that the form of the solution becomes

$$\theta = (C_1 \cos \lambda x + C_2 \sin \lambda x)e^{-\lambda^2 \alpha \tau}$$

From boundary condition (b), $C_1 = 0$ for $\tau > 0$. Because $C_2$ cannot also be zero, we find from boundary condition (c) that $\sin 2L\lambda = 0$, or

$$\lambda = \frac{n\pi}{2L} \qquad n = 1, 2, 3, \ldots$$

The final series form of the solution is therefore

$$\theta = \sum_{n=1}^{\infty} C_n \exp\left[-(n\pi/2L)^2 \alpha \tau\right] \sin \frac{n\pi x}{2L}$$

This equation may be recognized as a Fourier sine expansion with the constants $C_n$ determined from the initial condition (a) and the following equation

$$C_n = \frac{1}{L} \int_0^{2L} \theta_i \sin \frac{n\pi x}{2L} \, dx$$

$$= \frac{4}{n\pi} \theta_i \qquad\qquad n = 1, 3, 5, \ldots$$

The final series solution is therefore

$$\frac{\theta}{\theta_i} = \frac{T - T_1}{T_i - T_1} = \frac{4}{\pi} \sum_{n=1}^{\infty} \frac{1}{n} e^{-(n\pi/2L)^2 \alpha\tau} \sin \frac{n\pi x}{2L} \qquad n = 1, 3, 5, \ldots \quad (4\text{-}3)$$

In Sec. 4-4, this solution will be presented in graphical form for calculation purposes. For now, our purpose has been to show how the unsteady heat conduction equation can be solved, for at least one case, with the separation of variables method. Further information on analytical methods in unsteady state problems is given in the references.

## 4-2□LUMPED - HEAT - CAPACITY SYSTEM

We begin our discussion of transient heat conduction by analyzing systems which may be considered uniform in temperature. This type of analysis is called the lumped-heat-capacity method. Such systems are obviously idealized since a temperature gradient must exist in a material if heat is to be conducted into or out of the material. In general, the smaller the physical size of the body, the more realistic the assumption of a uniform temperature throughout; in the limit a differential volume could be employed as in the derivation of the general heat-conduction equation.

If a hot steel ball were immersed in a cool pan of water, the lumped-heat-capacity method of analysis might be used if we could justify an assumption of uniform ball temperature during the cooling process. Clearly, the temperature distribution in the ball would depend on the thermal conductivity of the ball material and the heat-transfer conditions from the surface of the ball to the surrounding fluid, i.e., the surface-convection heat-transfer coefficient. We should obtain a reasonably uniform temperature distribution in the ball if the resistance to heat transfer by conduction were small compared with the convection resistance at the surface so that the major temperature gradient would occur through the fluid layer at the surface. The lumped-heat-capacity analysis, then, is one which assumes that the internal resistance of the body is negligible in comparison with the external resistance.

The convection heat loss from the body is evidenced as a decrease in the internal energy of the body, as shown in Fig. 4-2. Thus

$$q = hA(T - T_\infty) = -C\rho V \frac{dT}{d\tau} \qquad\qquad (4\text{-}4)$$

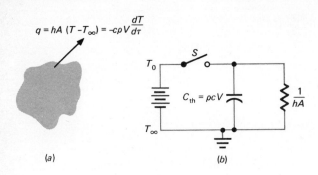

$$q = hA\,(T - T_\infty) = -c\rho V \frac{dT}{d\tau}$$

**Fig. 4-2 Nomenclature for single-lump heat-capacity analysis.**

$T_0$

$S$

$C_{th} = \rho c V$

$\dfrac{1}{hA}$

$T_\infty$

(a)                    (b)

where $A$ is the surface area for convection and $V$ is the volume. The initial condition is written

$$T = T_0 \qquad \text{at } \tau = 0$$

so that the solution to Eq. (4-4) is

$$\frac{T - T_\infty}{T_0 - T_\infty} = e^{-(hA/\rho CV)\tau} \tag{4-5}$$

The thermal network for the single-capacity system is shown in Fig. 4-2b. In this network we notice that the thermal capacity of the system is "charged" initially at the potential $T_0$ by closing the switch $S$. Then, when the switch is opened, the energy stored in the thermal capacitance is dissipated through the resistance $1/hA$. The analogy between this thermal system and an electric system is apparent, and we could easily construct an electric system which would behave exactly like the thermal system as long as we made the ratio

$$\frac{hA}{\rho CV} = \frac{1}{R_{th}C_{th}} \qquad R_{th} = \frac{1}{hA} \qquad C_{th} = \rho CV$$

equal to $1/R_eC_e$, where $R_e$ and $C_e$ are the electric resistance and capacitance, respectively. In the thermal system we store energy, while in the electric system we store electric charge. The flow of energy in the thermal system is called heat, and the flow of charge is called electric current. The quantity $C\rho V/hA$ is called the time constant of the system since it has the dimensions of time. When

$$\tau = \frac{C\rho V}{hA}$$

it is noted that the temperature difference $T - T_\infty$ has a value of 36.8 percent of the initial difference $T_0 - T_\infty$.

### Example 4-1

A steel ball 2 in. in diameter, initially at a uniform temperature of

850°F, is suddenly placed in a controlled environment in which the temperature is maintained at 200°F. The convection heat-transfer coefficient is 2 Btu/hr-ft²-°F. Calculate the time required for the ball to attain a temperature of 300°F using the lumped-capacity method of analysis.

**Solution**    Equation (4-5) applies for this problem. The properties and quantities for use in this equation are

$$T = 300°F \qquad\qquad h = 2 \text{ Btu/hr-ft}^2\text{-°F}$$
$$T_\infty = 200°F \qquad\qquad \rho = 490 \text{ lb}_m/\text{ft}^3$$
$$T_0 = 850°F \qquad\qquad C = 0.11 \text{ Btu/lb}_m\text{-°F}$$

$$\frac{hA}{\rho CV} = \frac{(2)(4\pi)(\frac{1}{12})^2}{(490)(0.11)(4\pi/3)(\frac{1}{12})^3}$$
$$= 1.34 \text{ hr}^{-1}$$

$$\frac{T - T_\infty}{T_0 - T_\infty} = e^{-(hA/\rho CV)\tau}$$

$$\frac{300 - 200}{850 - 200} = e^{-1.34\tau}$$

$$\tau = 1.39 \text{ hr}$$

Multiple-lumped-capacity systems may be treated in a manner similar to that analyzed above. These systems may be encountered when composite materials are subjected to environmental change. If the size and type of material are known, it is easy to determine the thermal capacity of each "lump" of the system; but for multiple systems the thermal resistance connecting the lumps must also be ascertained. If we were considering a two-lump system consisting of two solid materials bonded together, the connecting resistance would be a conduction resistance determined from the conductivity and type of each material. On the other hand, if the two-lump system consisted of a quantity of water in a thick-walled metal container, the connecting resistance would be that due to convection between the water and the container. For each particular problem a different method may be necessary for determining the connecting resistance between lumps. For purposes of illustration a two-lump system such as the container of water will be analyzed to show the method of attack.

Suppose a container of water (or other liquid) is available as shown in Fig. 4-3 The heat-transfer coefficient between the water and container is $h_1$, and that between the container and the environment is $h_2$. The water and container are initially at some temperature $T_0$ and are allowed to cool in the presence of the environment temperature $T_\infty$. An energy balance is made on each of the lumps.

Body 1: 
$$h_1 A_1 (T_1 - T_2) = -\rho_1 C_1 V_1 \frac{dT_1}{d\tau} \qquad (4\text{-}6)$$

Body 2:    
$$h_1 A_1 (T_2 - T_1) + h_2 A_2 (T_2 - T_\infty) = -\rho_2 C_2 V_2 \frac{dT_2}{d\tau} \qquad (4\text{-}7)$$

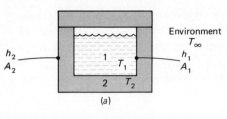

Fig. 4-3  Nomenclature for two-lump heat-capacity analysis.

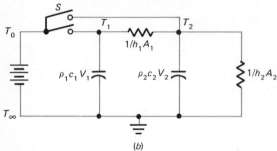

We thus have two simultaneous linear differential equations which may be solved for the temperature history in each of the bodies. The boundary conditions for use with these equations are

$$T_1 = T_2 = T_0 \quad \text{at } \tau = 0$$

which implies $dT_1/d\tau = 0$ at $\tau = 0$ from Eq. (4-3).

Equations (4-3) and (4-4) may be rewritten in operator form as

$$\left(D + \frac{h_1 A_1}{\rho_1 C_1 V_1}\right) T_1 - \frac{h_1 A_1}{\rho_1 C_1 V_1} T_2 = 0 \qquad (4\text{-}6a)$$

$$-\frac{h_1 A_1}{\rho_2 C_2 V_2} T_1 + \left(D + \frac{h_1 A_1 + h_2 A_2}{\rho_2 C_2 V_2}\right) T_2 = \frac{h_2 A_2}{\rho_2 C_2 V_2} T_\infty \qquad (4\text{-}7a)$$

where the symbol $D$ denotes differentiation with respect to time. For convenience let

$$K_1 = \frac{h_1 A_1}{\rho_1 C_1 V_1}$$

$$K_2 = \frac{h_1 A_1}{\rho_2 C_2 V_2}$$

$$K_3 = \frac{h_2 A_2}{\rho_2 C_2 V_2}$$

Then

$$(D + K_1) T_1 - K_1 T_2 = 0$$
$$-K_2 T_1 + (D + K_2 + K_3) T_2 = K_3 T_\infty$$

Solving the equations simultaneously, we get for the differential equation involving only $T_1$

$$[D^2 + (K_1 + K_2 + K_3)D + K_1K_3]T_1 = K_1K_3T_\infty \tag{4-8}$$

whose general solution is

$$T = T_\infty + Me^{m_1\tau} + Ne^{m_2\tau} \tag{4-9}$$

where $m_1$ and $m_2$ are given by

$$m_1 = \frac{-(K_1 + K_2 + K_3) + [(K_1 + K_2 + K_3)^2 - 4K_1K_3]^{\frac{1}{2}}}{2} \tag{4-10}$$

$$m_2 = \frac{-(K_1 + K_2 + K_3) - [(K_1 + K_2 + K_3)^2 - 4K_1K_3]^{\frac{1}{2}}}{2} \tag{4-11}$$

The arbitrary constants $M$ and $N$ may be obtained by applying the initial conditions

$$T_1 = T_0 \qquad \text{at } \tau = 0$$

and

$$\frac{dT_1}{d\tau} = 0 \qquad \text{at } \tau = 0$$

so that

$$T_0 = T_\infty + M + N$$
$$0 = m_1M + m_2N$$

The final solution is

$$\frac{T_1 - T_\infty}{T_0 - T_\infty} = \frac{m_2}{m_2 - m_1} e^{m_1\tau} - \frac{m_1}{m_2 - m_1} e^{m_2\tau} \tag{4-12}$$

The solution for $T_2$ may be obtained by substituting the relation for $T_1$ from Eq. (4-12) into Eq. (4-6).

When a three-lump system is involved, three simultaneous differential equations must be written; four equations are necessary for a four-lump system, and so on. In addition, time-varying boundary conditions may be applied. In these more involved cases the Laplace transform method is often used to advantage in obtaining the solutions to the differential equations.

The network analogy for the two-lump system is shown in Fig. 4-3b. When the switch $S$ is closed, the two thermal capacitances are charged to the potential $T_0$. At time zero the switch is opened and these capacitances discharge through the thermal resistances shown.

To indicate the application of lumped-capacity analysis to solids consider the system shown in Fig. 4-4. To show how the system is analyzed we are using only two nodes to indicate the connecting resistance. In a practical problem there could be additional connecting nodes and additional resistances. As in the examples above, a differential equation is written for each node, and the resulting set of equations is solved to obtain the tran-

Fig. 4-4  Lumped capacities for solids in contact.

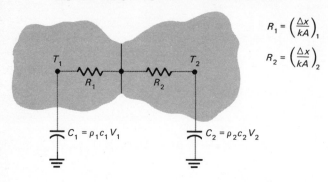

$$R_1 = \left(\frac{\Delta x}{kA}\right)_1$$

$$R_2 = \left(\frac{\Delta x}{kA}\right)_2$$

sient temperature response. If the number of nodes is very large, it is usually best to resort to a numerical solution, as indicated in later sections of this chapter.

## Applicability of Lumped-capacity Analysis

We have already noted that the lumped-capacity type of analysis assumes a uniform temperature distribution throughout the solid body and that the assumption is equivalent to saying that the surface-convection resistance is large compared with the internal-conduction resistance. Such an analysis may be expected to yield reasonable estimates when the following condition is met:

$$\frac{h(V/A)}{k} < 0.1$$

where $k$ is the thermal conductivity of the solid. In sections which follow we shall examine those situations for which this condition does not apply.

## 4-3□TRANSIENT HEAT FLOW IN A SEMI-INFINITE SOLID

Consider the semi-infinite solid shown in Fig. 4-5 maintained at some initial temperature $T_i$. The surface temperature is suddenly lowered and maintained at a temperature $T_0$, and we seek an expression for the temperature distribution in the plate as a function of time. This temperature distribution may subsequently be used to calculate heat flow at any $x$ position in the solid as a function of time. Assuming constant properties, the differential equation for the temperature distribution $T(x, \tau)$ is

$$\frac{\partial^2 T}{\partial x^2} = \frac{1}{\alpha} \frac{\partial T}{\partial \tau} \tag{4-13}$$

Fig. 4-5  Nomenclature for transient heat flow in a
semi-infinite solid.

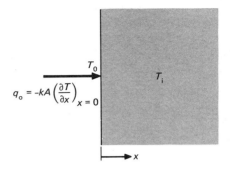

Fig. 4-5  Nomenclature for transient heat flow in a
semi-infinite solid.

The boundary and initial conditions are

$$T(x, 0) = T_i$$
$$T(0, \tau) = T_0 \quad \text{for } \tau > 0$$

This is a problem which may be solved by the Laplace transform technique. The solution is given in Ref. 1 as

$$\frac{T(x, \tau) - T_0}{T_i - T_0} = \text{erf} \frac{x}{2\sqrt{\alpha\tau}} \tag{4-14}$$

where the gauss error function is defined as

$$\text{erf} \frac{x}{2\sqrt{\alpha\tau}} = \frac{2}{\sqrt{\pi}} \int_0^{x/2\sqrt{\alpha\tau}} e^{-\eta^2} \, d\eta \tag{4-15}$$

It will be noted that in this definition $\eta$ is a dummy variable and the integral is a function of its upper limit. Inserting the definition of the error function in Eq. (4-14), the expression for the temperature distribution becomes

$$\frac{T(x, \tau) - T_0}{T_i - T_0} = \frac{2}{\sqrt{\pi}} \int_0^{x/2\sqrt{\alpha\tau}} e^{-\eta^2} \, d\eta \tag{4-16}$$

The heat flow at any $x$ position may be obtained from

$$q_x = -kA \frac{\partial T}{\partial x}$$

Performing the partial differentiation of Eq. (4-16),

$$\frac{\partial T}{\partial x} = (T_i - T_0) \frac{2}{\sqrt{\pi}} e^{-x^2/4\alpha\tau} \frac{\partial}{\partial x} \left( \frac{x}{2\sqrt{\alpha\tau}} \right)$$

$$= \frac{T_i - T_0}{\sqrt{\pi\alpha\tau}} e^{-x^2/4\alpha\tau} \tag{4-17}$$

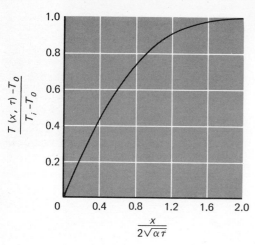

**Fig. 4-6  Temperature distribution in the semi-infinite solid.**

At the surface the heat flow is

$$q_0 = \frac{kA(T_0 - T_i)}{\sqrt{\pi\alpha\tau}} \qquad (4\text{-}18)$$

The surface heat flux is determined by evaluating the temperature gradient at $x = 0$ from Eq. (4-17). A plot of the temperature distribution for the semi-infinite solid is given in Fig. 4-6. Values of the error function are tabulated in Ref. 3, and an abbreviated tabulation is given in App. A.

### Constant Heat Flux

For the same uniform initial temperature distribution we could suddenly expose the surface to a constant surface heat flux $q_0/A$. The initial and boundary conditions on Eq. (4-13) would then become

$$T(x,0) = T_i$$
$$\frac{q_0}{A} = -k\left.\frac{\partial T}{\partial x}\right|_{x=0} \qquad \text{for } \tau > 0$$

The solution for this case is

$$T - T_i = \frac{2q_0\sqrt{\alpha\tau/\pi}}{kA}\exp\left(\frac{-x^2}{4\alpha\tau}\right) - \frac{q_0 x}{kA}\left[1 - \operatorname{erf}\left(\frac{x}{2\sqrt{\alpha\tau}}\right)\right] \quad (4\text{-}19)$$

### Example 4-2

A large slab of aluminum at a uniform temperature of 450°F suddenly has its surface temperature lowered to 150°F. Calculate the time required for the temperature to reach 250°F at a depth of $1\frac{1}{2}$ in. What is the total

energy removed from the slab per unit surface area during this time?
**Solution**   We may use either Eq. (4-14) or Fig. 4-6 to compute the time, and Eq. (4-18) to calculate the total heat flow. The properties of aluminum are

$$\alpha = 3.66 \text{ ft}^2/\text{hr}$$
$$k = 120 \text{ Btu/hr-ft-}°F$$
$$T_i = 450°F$$
$$T_0 = 150°F$$
$$T(x, \tau) = 250°F$$

$$\frac{T(x, \tau) - T_0}{T_i - T_0} = \frac{250 - 150}{450 - 150} = 0.333 = \text{erf } \frac{x}{2\sqrt{\alpha\tau}}$$

From Fig. 4-6, or App. A,

$$\frac{x}{2\sqrt{\alpha\tau}} = 0.302$$
$$\tau = 0.0117 \text{ hr} = 0.702 \text{ min}$$
$$Q_0 = \int_0^\tau q_0 \, d\tau = \frac{2kA(T_0 - T_i)}{\sqrt{\pi\alpha/\tau}} = \frac{(2)(120)(150 - 450)}{(\pi 3.66/0.0117)^{\frac{1}{2}}}$$
$$= 2290 \text{ Btu} = 2.416 \times 10^6 \text{ joules}$$

## 4-4☐CONVECTION BOUNDARY CONDITIONS

In many practical situations the transient heat-conduction problem is connected with a convection boundary condition at the surface of the solid. Naturally, the boundary conditions for the differential equation must be modified to take into account this convection heat transfer at the surface. For the semi-infinite solid problem above this would be expressed by

$$hA(T_\infty - T)_{x=0} = -kA \frac{\partial T}{\partial x}\bigg)_{x=0} \tag{4-20}$$

The solution for this problem is rather involved, and is worked out in detail by Schneider [1]. The result is

$$\frac{T - T_i}{T_\infty - T_i} = 1 - \text{erf } X - \left[ \exp\left(\frac{hx}{k} + \frac{h^2\alpha\tau}{k^2}\right) \right]\left[ 1 - \text{erf }\left(X + \frac{h\sqrt{\alpha\tau}}{k}\right) \right] \tag{4-21}$$

where $X = x/(2\sqrt{\alpha\tau})$
  $T_i$ = initial temperature of solid
  $T_\infty$ = environment temperature
This solution is presented in graphical form in Fig. 4-7.

Solutions have been worked out for other geometries. The most important cases are those dealing with (1) plates whose thickness is small in relation to the other dimensions, (2) long cylinders, and (3) spheres which

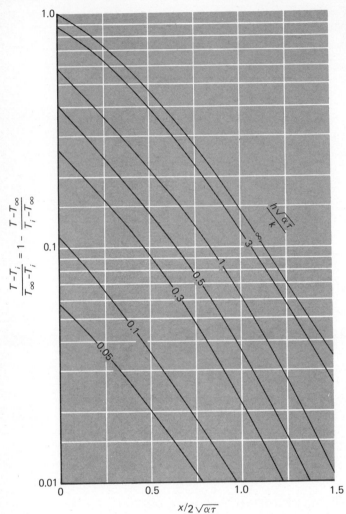

are initially at a uniform temperature $T_i$ and suddenly immersed in a fluid at temperature $T_\infty$ with a convection heat-transfer coefficient $h$. The results of these analyses have been presented in graphical form by Heisler [2] and are given in Figs. 4-8 to 4-14. The heat losses for these three cases are given in Figs. 4-15 to 4-17. In these figures $Q_0$ represents the initial internal energy content of the body in reference to the environment temperature. Thus

$$Q_0 = \rho C V (T_i - T_\infty) = \rho C V \theta_i \qquad (4\text{-}22)$$

Examples 4-3 to 4-5 illustrate the use of the various charts for the solution of problems.

**Fig. 4-8** Midplane temperature for an infinite plate of thickness 2L, from Heisler [2].

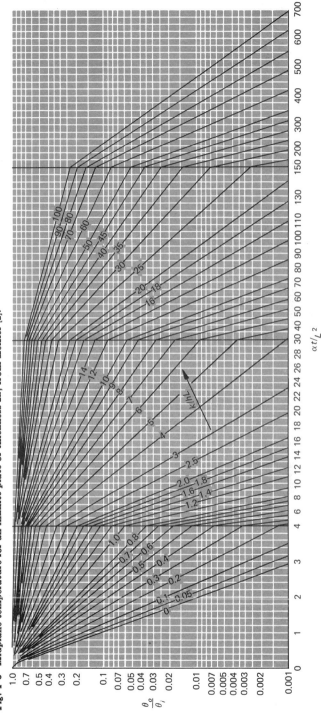

**Fig. 4-9** Axis temperature for an infinite cylinder of radius $r_0$, from Heisler [2].

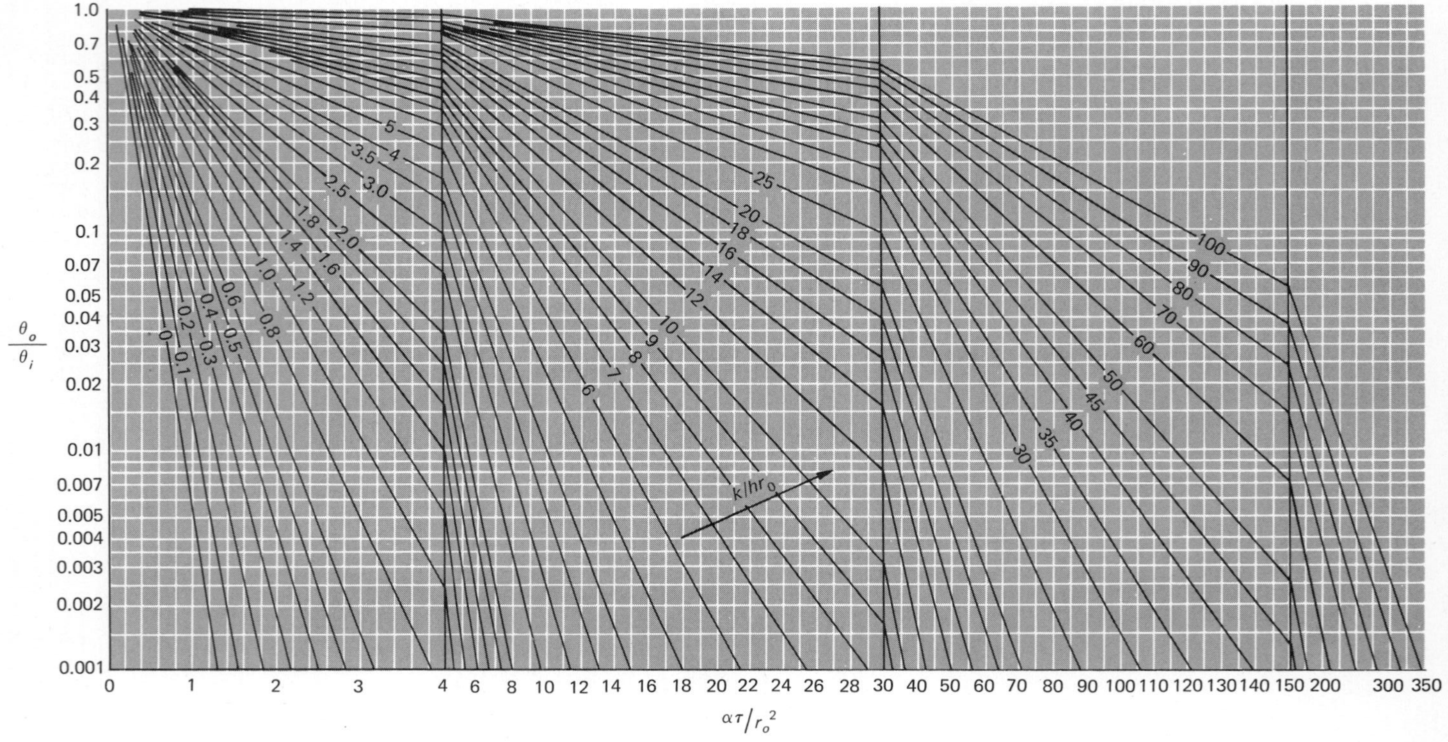

**Fig. 4-10** Center temperature for a sphere of radius $r_o$, from Heisler [2].

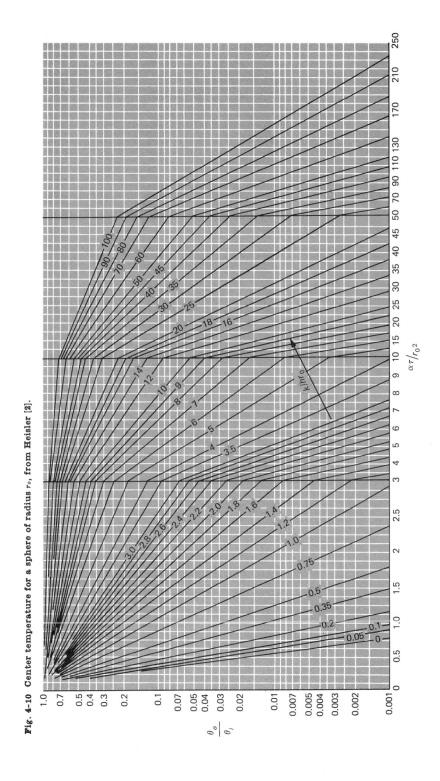

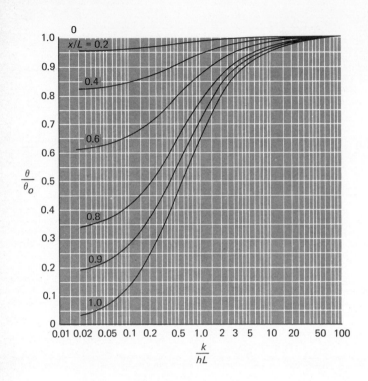

**Fig. 4-11** Temperature as a function of center temperature in an infinite plate of thickness $2L$, from Heisler [2].

Obviously, there are many other practical heating and cooling problems of interest. The solutions for a large number of cases are presented in graphical form by Schneider [7] and readers interested in such calculations will find this reference to be of great utility.

### The Biot and Fourier Moduli

A quick inspection of Figs. 4-7 to 4-17 indicates that the dimensionless temperature profiles and heat flows may all be expressed in terms of two dimensionless parameters called the Biot and Fourier moduli:

$$\text{Biot modulus} = \text{Bi} = \frac{hs}{k}$$

$$\text{Fourier modulus} = \text{Fo} = \frac{\alpha\tau}{s^2} = \frac{k\tau}{\rho c s^2}$$

In these parameters $s$ designates some characteristic dimension of the body; for the plate it is the half-thickness, whereas for the cylinder and sphere it is the radius. The Biot modulus compares the relative magnitudes of surface convection and internal conduction resistances to heat transfer. The Fourier modulus compares a characteristic body dimension with an approximate temperature wave penetration depth for a given time $\tau$.

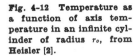

**Fig. 4-12** Temperature as a function of axis temperature in an infinite cylinder of radius $r_0$, from Heisler [2].

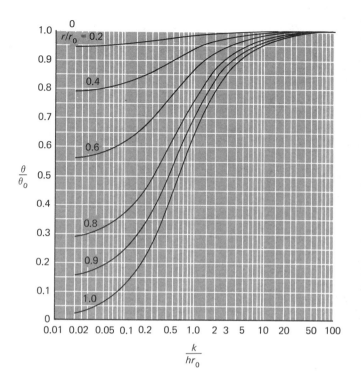

A very low value of the Biot modulus means that internal-conduction resistance is negligible in comparison with surface-convection resistance. This in turn implies that the temperature will be nearly uniform throughout the solid and its behavior may be approximated by the lumped-capacity method of analysis. It is interesting to note that the exponent of Eq. (4-5) may be expressed in terms of the Biot and Fourier moduli if one takes the ratio $V/A$ as the characteristic dimension $s$. Then,

$$\frac{hA}{\rho c V}\tau = \frac{h\tau}{\rho cs} = \frac{hs}{k}\frac{k\tau}{\rho cs^2} = \text{BiFo}$$

## Applicability of the Heisler Charts

The Heisler charts are restricted to values of the Fourier modulus $(\alpha\tau/s^2)$ greater than 0.2. For smaller values of this parameter the reader should consult the solutions and charts given in the various references at the end of the chapter.

## Example 4-3

The slab of Example 4-2 is suddenly exposed to a convection surface environment at 150°F. The heat-transfer coefficient is 100 Btu/hr-ft²-°F.

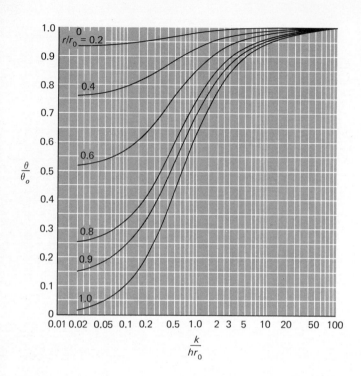

**Fig. 4-13** Temperature as a function of center temperature for a sphere of radius $r_o$, from Heisler [2].

Calculate the time required for the temperature to reach 250°F at a depth of $1\frac{1}{2}$ in.

**Solution**   We may use either Eq. (4-21) or Fig. 4-7 for the solution of this problem. Figure 4-7 is somewhat easier to apply; so it will be used. Note that a trial-and-error solution is required because the time appears in both of the variables $(h\sqrt{\alpha\tau})/k$ and $x/(2\sqrt{\alpha\tau})$. We therefore try values for $\tau$ until we obtain

$$\frac{T - T_i}{T_\infty - T_i} = \frac{250 - 450}{150 - 450} = 0.667$$

The trials are listed below:

| $\tau$, hr | $\dfrac{h\sqrt{\alpha\tau}}{k}$ | $\dfrac{x}{2\sqrt{\alpha\tau}}$ | $\dfrac{T - T_i}{T_\infty - T_i}$ (from Fig. 4-7) |
|---|---|---|---|
| 0.3 | 0.871 | 0.0597 | 0.475 |
| 1.0 | 1.59 | 0.0345 | 0.66 |

Consequently, the time required is approximately 1 hr.

**Fig. 4-14**   Center temperatures for plates, cylinders, and spheres for small values of $h$, according to Heisler [2].

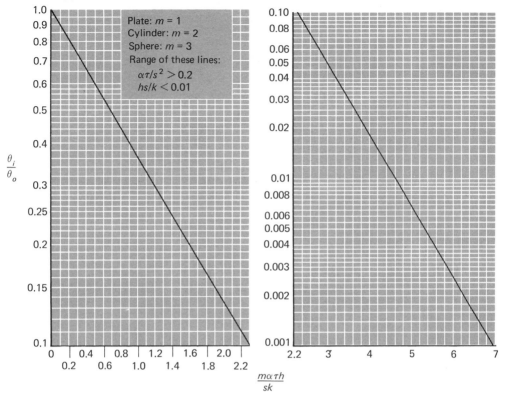

$$\frac{m\alpha\tau h}{sk}$$

## Example 4-4

A large slab of aluminum 2 in. thick and initially at 450°F is suddenly exposed to the convection environment of Example 4-3. Calculate the temperature $\frac{1}{2}$ in. from one of the faces 1 min after the slab has been exposed to the environment. How much energy has been removed per unit area from the slab in this time?

**Solution**   The Heisler charts of Figs. 4-8 and 4-11 may be used for the solution of this problem. We first calculate the center temperature of the slab using Fig. 4-8, and then use Fig. 4-11 to calculate the temperature at the specified $x$ position. From the conditions of the problem we have

$\theta_i = T_i - T_\infty = 450 - 150 = 300$

$\alpha = 3.66 \text{ ft}^2/\text{hr}$

$L = 1 \text{ in.} = 0.0833 \text{ ft}$

$\tau = 1 \text{ min} = 0.0167 \text{ hr}$

$k = 120 \text{ Btu/hr-ft-°F}$

$h = 100 \text{ Btu/hr-ft}^2\text{-°F}$

**Fig. 4-15**  Dimensionless heat loss $Q/Q_0$ of an infinite plate of thickness $2L$ with time, from Gröber [6].

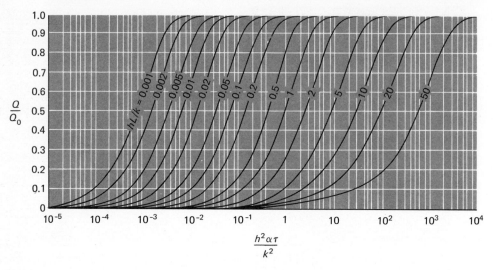

$$\frac{\alpha\tau}{L^2} = \frac{(3.66)(0.0167)}{(0.0833)^2} = 8.79$$

$$\frac{k}{hL} = \frac{120}{(100)(0.0833)} = 14.4$$

$$\frac{x}{L} = \frac{\frac{1}{2}}{1} = 0.5$$

From Fig. 4-8

$$\frac{\theta_0}{\theta_i} = 0.6$$

$$\theta_0 = T_0 - T_\infty = (0.6)(300) = 180$$

From Fig. 4-11, at $x/L = 0.5$,

$$\frac{\theta}{\theta_0} = 0.98$$

$$\theta = T - T_\infty = (0.98)(180) = 176$$
$$T = 176 + 150 = 326°F$$

We compute the energy lost by the slab by using Fig. 4-15. For this calculation

$$\frac{h^2\alpha\tau}{k^2} = \frac{(100)^2(3.66)(0.0167)}{(120)^2} = 0.043$$

$$\frac{hL}{k} = \frac{(100)(0.0833)}{120} = 0.0695$$

**Fig. 4-16** Dimensionless heat loss $Q/Q_0$ of an infinite cylinder of radius $r_o$ with time, from Gröber [6].

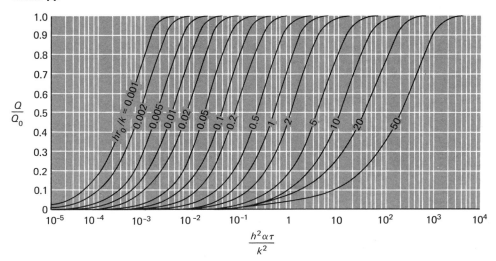

For unit area

$$Q_0 = \rho C V \theta_i = \rho C A 2 L \theta_i$$
$$\rho = 169 \text{ lb}_m/\text{ft}^3$$
$$C = 0.21 \text{ Btu/lb}_m\text{-}°F$$
$$Q_0 = (169)(0.21)(1)(\tfrac{2}{12})(300) = 1770 \text{ Btu}$$

From Fig. 4-15

$$\frac{Q}{Q_0} = 0.47$$

so that    $Q = (0.47)(1770) = 833 \text{ Btu}$    $(8.79 \times 10^5 \text{ J})$

## Example 4-5

A long aluminum cylinder 2 in. in diameter and initially at 450°F is suddenly exposed to a convection environment, where $h = 100$ Btu/hr-ft$^2$-°F. Calculate the temperature at a radius of $\frac{1}{2}$ in. and the heat lost per unit length of the cylinder 1 min after the cylinder is exposed to the environment at 150°F.

**Solution**    The Heisler charts of Figs. 4-9 and 4-12 are used in the solution (similar to Example 4-4) of this problem. We first use Fig. 4-9 to calculate the axis temperature of the cylinder and then use Fig. 4-12 to calculate the temperature at the specified radial position. From the conditions of the problem,

**Fig. 4-17** Dimensionless heat loss $Q/Q_0$ of a sphere of radius $r_o$ with time, from Gröber [6].

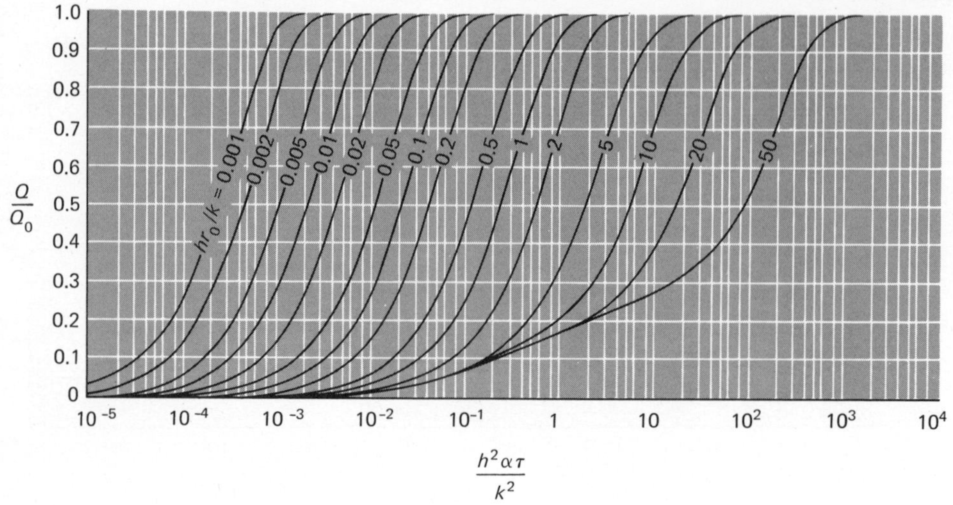

$$\theta_i = T_i - T_\infty = 450 - 150 = 300$$
$$\alpha = 3.66 \text{ ft}^2/\text{hr}$$
$$r_o = 1 \text{ in.} = 0.0833 \text{ ft}$$
$$\tau = 1 \text{ min} = 0.0167 \text{ hr}$$
$$k = 120 \text{ Btu/hr-ft-°F}$$
$$h = 100 \text{ Btu/hr-ft}^2\text{-°F}$$

$$\frac{\alpha\tau}{r_o{}^2} = \frac{(3.66)(0.0167)}{(0.0833)^2} = 8.79$$
$$\frac{k}{hr_o} = \frac{120}{(100)(0.0833)} = 14.4$$
$$\frac{r}{r_o} = \frac{0.5}{1.0} = 0.5$$

From Fig. 4-9

$$\frac{\theta_0}{\theta_i} = 0.3$$
$$\theta_0 = T_0 - T_\infty = (0.3)(300) = 90$$

From Fig. 4-12, at $r/r_o = 0.5$,

$$\frac{\theta}{\theta_0} = 0.98$$
$$\theta = T - T_\infty = (0.98)(90) = 88$$
$$T = 88 + 150 = 238°\text{F}$$

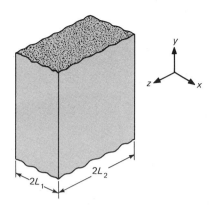

Fig. 4-18   Infinite rectangular bar.

We compute the energy lost by the cylinder by using Fig. 4-16. For this calculation

$$\frac{h^2\alpha\tau}{k^2} = \frac{(100)^2(3.66)(0.0167)}{(120)^2} = 0.043$$

$$\frac{hr_o}{k} = \frac{(100)(0.0833)}{120} = 0.0695$$

For unit length

$$Q_0 = \rho C V \theta_i = \rho C \pi r_o^2 L \theta_i$$
$$Q_0 = (169)(0.21)\pi(0.0833)^2(1)(300) = 233 \text{ Btu}$$

From Fig. 4-16

$$\frac{Q}{Q_0} = 0.71$$

so that    $Q = (0.71)(233) = 165 \text{ Btu}$    $(1.74 \times 10^5 \text{ J})$

## 4-5□MULTIPLE - DIMENSION SYSTEMS

The Heisler charts discussed above may be used to obtain the temperature distribution in the infinite plate of thickness $2L$, in the long cylinder, or in the sphere. When a wall whose height and depth dimensions are not large compared with the thickness or a cylinder whose length is not large compared with its diameter is encountered, additional space coordinates are necessary to specify the temperature, the above charts no longer apply, and we are forced to seek another method of solution. Fortunately, it is possible to combine the solutions for the one-dimensional systems in a very straightforward way to obtain solutions for the multidimensional problems.

It is clear that the infinite rectangular bar in Fig. 4-18 can be formed from two infinite plates of thickness $2L_1$ and $2L_2$, respectively. The differential equation governing this situation would be

$$\frac{\partial^2 T}{\partial x^2} + \frac{\partial^2 T}{\partial z^2} = \frac{1}{\alpha}\frac{\partial T}{\partial \tau} \tag{4-23}$$

and to use the separation of variables method to effect a solution, we should assume a product solution of the form

$$T(x, z, \tau) = X(x)Z(z)\Theta(\tau)$$

It can be shown that the dimensionless temperature distribution may be expressed as a product of the solutions for two plate problems of thickness $2L_1$ and $2L_2$, respectively.

$$\left(\frac{T - T_\infty}{T_i - T_\infty}\right)_{\text{bar}} = \left(\frac{T - T_\infty}{T_i - T_\infty}\right)_{2L_1\,\text{plate}} \left(\frac{T - T_\infty}{T_i - T_\infty}\right)_{2L_2\,\text{plate}} \tag{4-24}$$

where $T_i$ is the initial temperature of the bar and $T_\infty$ is the environment temperature.

For two infinite plates the respective differential equations would be

$$\frac{\partial^2 T_1}{\partial x^2} = \frac{1}{\alpha}\frac{\partial T_1}{\partial \tau} \qquad \frac{\partial^2 T_2}{\partial z^2} = \frac{1}{\alpha}\frac{\partial T_2}{\partial \tau} \tag{4-25}$$

and the product solutions assumed would be

$$T_1 = T_1(x,\tau) \qquad T_2 = T_2(z,\tau) \tag{4-26}$$

We shall now show that the product solution to Eq. (4-23) can be formed from a simple product of the functions $(T_1, T_2)$, that is,

$$T(x,z,\tau) = T_1(x,\tau)T_2(z,\tau) \tag{4-27}$$

The appropriate derivatives for substitution in Eq. (4-23) are obtained from Eq. (4-27) as

$$\frac{\partial^2 T}{\partial x^2} = T_2\frac{\partial^2 T_1}{\partial x^2} \qquad\qquad \frac{\partial^2 T}{\partial z^2} = T_1\frac{\partial^2 T_2}{\partial z^2}$$

$$\frac{\partial T}{\partial \tau} = T_1\frac{\partial T_2}{\partial \tau} + T_2\frac{\partial T_1}{\partial \tau}$$

Using Eqs. (4-25), we have

$$\frac{\partial T}{\partial \tau} = \alpha T_1\frac{\partial^2 T_2}{\partial z^2} + \alpha T_2\frac{\partial^2 T_1}{\partial x^2}$$

Substituting these relations in Eq. (4-23) gives

$$T_2\frac{\partial^2 T_1}{\partial x^2} + T_1\frac{\partial^2 T_2}{\partial z^2} = \frac{1}{\alpha}\left[\alpha T_1\frac{\partial^2 T_2}{\partial z^2} + \alpha T_2\frac{\partial^2 T_1}{\partial x^2}\right]$$

or the assumed product solution of Eq. (4-27) does indeed satisfy the original differential equation (4-23). This means that the dimensionless temperature distribution for the infinite rectangular bar may be expressed as a product of the solutions for two plate problems of thickness $2L_1$ and $2L_2$, respectively, as indicated by Eq. (4-24).

In a manner similar to that described above, the solution for a three-dimensional block may be expressed as a product of three infinite-plate solutions for plates having the thickness of the three sides of the block. Similarly, a solution for a cylinder of finite length could be expressed as a product of solutions of the infinite cylinder and an infinite plate having a thickness equal to the length of the cylinder. Combinations could also be made with the infinite-cylinder and -plate solutions to obtain temperature distributions in semi-infinite bars and cylinders. Some of the combinations are summarized in Fig. 4-19, where

$C(\Theta)$ = solution for infinite cylinder
$P(X)$ = solution for infinite plate
$S(X)$ = solution for semi-infinite solid

The following examples illustrate the use of the various charts for calculating temperatures in multidimensional systems.

## Example 4-6

A short aluminum cylinder 2 in. in diameter and 4 in. long is initially at a uniform temperature of 450°F. It is suddenly subjected to a convection environment at 150°F, where $h = 100$ Btu/hr-ft²-°F. Calculate the temperature at a radial position of $\frac{1}{2}$ in. and a distance of $\frac{1}{4}$ in. from one end of the cylinder 1 min after exposure to the environment.

**Solution**    To solve this problem, we combine the solutions from the Heisler charts for an infinite cylinder and an infinite slab in accordance with the combination shown in Fig. 4-19.

For the corresponding infinite-slab problem

$$L = 2 \text{ in.} = 0.167 \text{ ft}$$

The $x$ position is measured from the center of the slab so that

$$x = 2 - 0.25 = 1.75 \text{ in.}$$

and
$$\frac{x}{L} = \frac{1.75}{2.0} = 0.875$$

The properties of aluminum are

$\alpha = 3.66 \text{ ft}^2/\text{hr}$
$k = 120 \text{ Btu/hr-ft-°F}$

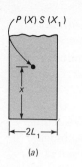

$P(X)S(X_1)$

$x$

$2L_1$

$(a)$

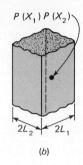

$P(X_1)P(X_2)$

$2L_2$  $2L_1$

$(b)$

**Fig. 4-19 Product solutions for temperatures in multidimensional systems. (a) Semi-infinite plate; (b) infinite rectangular bar; (c) semi-infinite rectangular bar; (d) rectangular parallelepiped; (e) semi-infinite cylinder; (f) short cylinder.**

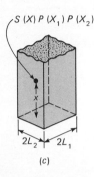

$S(X)P(X_1)P(X_2)$

$x$

$2L_2$  $2L_1$

$(c)$

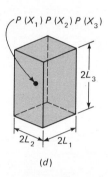

$P(X_1)P(X_2)P(X_3)$

$2L_3$

$2L_2$  $2L_1$

$(d)$

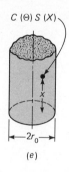

$C(\Theta)S(X)$

$x$

$2r_0$

$(e)$

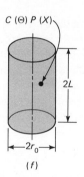

$C(\Theta)P(X)$

$2L$

$2r_0$

$(f)$

so that the parameters for use with Figs. 4-8 and 4-11 are

$$\frac{k}{hL} = \frac{120}{(100)(0.167)} = 7.2$$

$$\frac{\alpha\tau}{L^2} = \frac{(3.66)(0.0167)}{(0.167)^2} = 2.19$$

From Fig. 4-8

$$\frac{\theta_0}{\theta_i} = 0.8$$

and from Fig. 4-11

$$\frac{\theta}{\theta_0} = 0.95$$

so that

$$\left(\frac{\theta}{\theta_i}\right)_{\text{plate}} = (0.8)(0.95) = 0.76 = P(X)$$

For the cylinder

$$r_o = 1 \text{ in.} = 0.0833 \text{ ft}$$
$$\frac{r}{r_o} = \frac{0.5}{1.0} = 0.5$$
$$\frac{k}{hr_o} = \frac{120}{(100)(0.0833)} = 14.4$$
$$\frac{\alpha\tau}{r_o{}^2} = \frac{(3.66)(0.0167)}{(0.0833)^2} = 8.79$$

From Fig. 4-9

$$\frac{\theta_0}{\theta_i} = 0.3$$

and from Fig. 4-12

$$\frac{\theta}{\theta_0} = 0.98$$

so that

$$\left(\frac{\theta}{\theta_i}\right)_{\text{cylinder}} = (0.3)(0.98) = 0.294 = C(\Theta)$$

Combining the solutions for the plate and cylinder,

$$\left(\frac{\theta}{\theta_i}\right)_{\substack{\text{short} \\ \text{cylinder}}} = C(\Theta)P(X) = (0.294)(0.76) = 0.223$$

Thus

$$T = T_\infty + 0.223(T_i - T_\infty)$$
$$T = 150 + (0.223)(450 - 150) = 217°F$$

## Example 4-7

A semi-infinite cylinder 2 in. in diameter is initially at a uniform temperature of 450°F. It is suddenly subjected to a convection boundary condition at 150°F, where $h = 100$ Btu/hr-ft²-°F. Calculate the temperatures at the axis and surface of the cylinder 6 in. from the end 1 min after exposure to the environment.

**Solution**   To solve this problem we combine the solutions for the infinite cylinder and the semi-infinite slab in accordance with the combination shown in Fig. 4-19.

For the corresponding semi-infinite slab problem

$$x = 6 \text{ in.} = 0.5 \text{ ft}$$

The properties of aluminum are

$\alpha = 3.66$ ft²/hr
$k = 120$ Btu/hr-ft-°F

so that the parameters for use with Fig. 4-7 are

$$\frac{h\sqrt{\alpha\tau}}{k} = \frac{100\sqrt{(3.66)(0.0167)}}{120} = 0.206$$

$$\frac{x}{2\sqrt{\alpha\tau}} = \frac{0.5}{2\sqrt{(3.66)(0.0167)}} = 1.01$$

From Fig. 4-7

$$\left(\frac{\theta}{\theta_i}\right)_{\substack{\text{semi-infinite}\\\text{plate}}} = 1 - 0.016 = 0.984 = S(X)$$

For the infinite-cylinder solution we seek both the axis and surface temperature ratios. The significant parameters for use with Fig. 4-9 are

$$r_o = 1 \text{ in.} = 0.0833 \text{ ft}$$

$$\frac{k}{hr_o} = 14.4$$

$$\frac{\alpha\tau}{r_o^2} = 8.79$$

From Fig. 4-9

$$\frac{\theta_0}{\theta_i} = 0.3$$

This is the axis temperature ratio. To find the surface temperature ratio, we enter Fig. 4-12, using

$$\frac{r}{r_o} = 1.0$$

so that

$$\frac{\theta}{\theta_0} = 0.97$$

Thus

$$C(\Theta) = 0.3 \qquad\qquad \text{at } r = 0$$
$$C(\Theta) = (0.3)(0.97) = 0.29 \qquad \text{at } r = r_o$$

Combining the solutions for the semi-infinite plate and infinite cylinder,

$$\left(\frac{\theta}{\theta_i}\right)_{\substack{\text{semi-infinite}\\\text{plate}}} = C(\Theta)S(X)$$

$$= (0.3)(0.984) = 0.295 \qquad \text{at } r = 0$$
$$= (0.29)(0.984) = 0.285 \qquad \text{at } r = r_o$$

The corresponding temperatures are

$$T = 150 + (0.295)(450 - 150) = 239°\text{F} \qquad r = 0, x = 6 \text{ in.}$$
$$T = 150 + (0.285)(450 - 150) = 235°\text{F} \qquad r = 1 \text{ in.}, x = 6 \text{ in.}$$

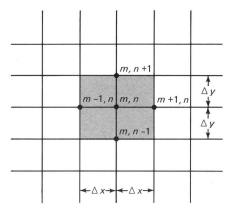

**Fig. 4-20** Nomenclature for numerical solution of two-dimensional unsteady-state conduction problem.

## 4-6□TRANSIENT NUMERICAL METHOD

The charts described above are very useful for calculating temperatures in certain regular-shaped solids under transient heat-flow conditions. Unfortunately, many geometric shapes of practical interest do not fall into these categories, and in addition, one is frequently faced with problems in which the boundary conditions vary with time. These transient boundary conditions as well as the geometric shape of the body can be of such form that a mathematical solution is not possible. In these cases, the problems are best handled by a numerical technique, with hand calculators or computers, depending on the accuracy required and economic situation involved. Regardless of whether hand calculation or computers are used, the setup for the problem is the same. It is this setup which we now wish to describe. For ease in discussion we limit the analysis to two-dimensional systems for the present. An extension to three dimensions can then be made very easily.

Consider a two-dimensional body divided into increments as shown in Fig. 4-20. The subscript $m$ denotes the $x$ position, and the subscript $n$ denotes the $y$ position. Within the solid body the differential equation which governs the heat flow is

$$k\left(\frac{\partial^2 T}{\partial x^2} + \frac{\partial^2 T}{\partial y^2}\right) = \rho C \frac{\partial T}{\partial \tau} \tag{4-28}$$

assuming constant properties. We recall from Chap. 3 that the second partial derivatives may be approximated by

$$\frac{\partial^2 T}{\partial x^2} \approx \frac{1}{(\Delta x)^2}\left(T_{m+1,n} + T_{m-1,n} - 2T_{m,n}\right) \tag{4-29}$$

$$\frac{\partial^2 T}{\partial y^2} \approx \frac{1}{(\Delta y)^2}\left(T_{m,n+1} + T_{m,n-1} - 2T_{m,n}\right) \tag{4-30}$$

The time derivative in Eq. (4-28) is approximated by

$$\frac{\partial T}{\partial \tau} \approx \frac{T_{m,n}^{p+1} - T_{m,n}^p}{\Delta \tau} \tag{4-31}$$

In this relation the superscripts designate the time increment. Combining the relations above, the difference equation equivalent to Eq. (4-28) is

$$\frac{T_{m+1,n}^p + T_{m-1,n}^p - 2T_{m,n}^p}{(\Delta x)^2} + \frac{T_{m,n+1}^p + T_{m,n-1}^p - 2T_{m,n}^p}{(\Delta y)^2} = \frac{1}{\alpha} \frac{T_{m,n}^{p+1} - T_{m,n}^p}{\Delta \tau} \tag{4-32}$$

Thus, if the temperatures of the various nodes are known at any particular time, the temperatures after a time increment $\Delta \tau$ may be calculated by writing an equation like Eq. (4-32) for each node and obtaining the values of $T_{m,n}^{p+1}$. The procedure may be repeated to obtain the distribution after any desired number of time increments. If the increments of space coordinates are chosen such that

$$\Delta x = \Delta y$$

the resulting equation for $T_{m,n}^{p+1}$ becomes

$$T_{m,n}^{p+1} = \frac{\alpha \, \Delta \tau}{(\Delta x)^2} \left( T_{m+1,n}^p + T_{m-1,n}^p + T_{m,n+1}^p + T_{m,n-1}^p \right) + \left[ 1 - \frac{4\alpha \, \Delta \tau}{(\Delta x)^2} \right] T_{m,n}^p \tag{4-33}$$

If the time and distance increments are conveniently chosen such that

$$\frac{(\Delta x)^2}{\alpha \, \Delta \tau} = 4 \tag{4-34}$$

it is seen that the temperature of node $(m, n)$ after a time increment is simply the arithmetic average of the four surrounding nodal temperatures at the beginning of the time increment.

When a one-dimensional system is involved, the equation becomes

$$T_m^{p+1} = \frac{\alpha \, \Delta \tau}{(\Delta x)^2} \left( T_{m+1}^p + T_{m-1}^p \right) + \left[ 1 - \frac{2\alpha \, \Delta \tau}{(\Delta x)^2} \right] T_m^p \tag{4-35}$$

and if the time and distance increments are chosen such that

$$\frac{(\Delta x)^2}{\alpha \, \Delta \tau} = 2 \tag{4-36}$$

the temperature of node $m$ after the time increment is given as the arithmetic average of the two adjacent nodal temperatures at the beginning of the time increment.

Some general remarks concerning the use of numerical methods for solution of transient conduction problems are in order at this point. We have already noted that the selection of the value of the parameter

$$M = \frac{(\Delta x)^2}{\alpha \, \Delta \tau}$$

governs the ease with which we may proceed to effect the numerical solution; the choice of a value of 4 for a two-dimensional system or a value of 2 for a one-dimensional system makes the calculation particularly easy.

Once the distance increments and the value of $M$ are established, the time increment is fixed; and we may not alter it without changing the value of either $\Delta x$ or $M$, or both. Clearly, the larger the values of $\Delta x$ and $\Delta \tau$, the more rapidly our solution will proceed. On the other hand, the smaller the value of these increments in the independent variables, the more accuracy will be obtained. At first glance one might assume that small distance increments could be used for greater accuracy in combination with large time increments to speed the solution. This is not the case, however, since the finite-difference equations limit the values of $\Delta \tau$ which may be used once $\Delta x$ is chosen. Note that if $M < 2$ in Eq. (4-35), the coefficient of $T_m{}^p$ becomes negative, and we generate a condition which will violate the second law of thermodynamics. Suppose, for example, that the adjoining nodes are equal in temperature but less than $T_m{}^p$. After the time increment $\Delta \tau$, $T_m{}^p$ may not be lower than these adjoining temperatures; otherwise heat would have to flow uphill on the temperature scale, and this is impossible. A value of $M < 2$ would produce just such an effect; so we must restrict the values of $M$ to

$$\frac{(\Delta x)^2}{\alpha \, \Delta \tau} = M \geq 2 \qquad \text{one-dimensional systems}$$

$$\frac{(\Delta x)^2}{\alpha \, \Delta \tau} = M \geq 4 \qquad \text{two-dimensional systems}$$

This restriction automatically limits our choice of $\Delta \tau$, once $\Delta x$ is established.

It so happens that the above restrictions, which are imposed in a physical sense, may also be derived on mathematical grounds. It may be shown that the finite-difference solutions will not converge unless these conditions are fulfilled. The problems of stability and convergence of numerical solutions are discussed in Refs. 7, 13, and 15 in detail.

The difference equations given above are useful for determining the internal temperature in a solid as a function of space and time. At the boundary of the solid a convection resistance to heat flow is usually involved, so that the above relations no longer apply. In general, each convection boundary condition must be handled separately, depending on the particular geometric shape under consideration. The case of the flat wall

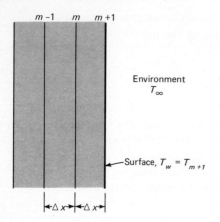

will be considered as an example.

For the one-dimensional system shown in Fig. 4-21 we may make an energy balance at the convection boundary such that

$$-kA \left. \frac{\partial T}{\partial x} \right)_{\text{wall}} = hA \left( T_w - T_\infty \right) \tag{4-37}$$

The finite-difference approximation would be given by

$$-k \frac{\Delta y}{\Delta x} \left( T_{m+1} - T_m \right) = h \, \Delta y \left( T_{m+1} - T_\infty \right)$$

or

$$T_{m+1} = \frac{T_m + \left( h \, \Delta x / k \right) T_\infty}{1 + \left( h \, \Delta x / k \right)}$$

To apply this condition, we should calculate the surface temperature $T_{m+1}$ at each time increment and then use this temperature in the nodal equations for the interior points of the solid. This is only an approximation because we have neglected the heat capacity of the element of the wall at the boundary. This approximation will work fairly well when a large number of increments in $x$ are used because the portion of the heat capacity which is neglected is then small in comparison with the total. We may take the heat capacity into account in a general way by considering the two-dimensional wall of Fig. 3-7 exposed to a convection boundary condition. We make a transient energy balance on the node $(m, n)$ by setting the sum of the energy conducted and convected into the node equal to the increase in the internal energy of the node. Thus

$$k \, \Delta y \, \frac{T^p_{m-1,n} - T^p_{m,n}}{\Delta x} + k \, \frac{\Delta x}{2} \, \frac{T^p_{m,n+1} - T^p_{m,n}}{\Delta y}$$
$$+ \, k \, \frac{\Delta x}{2} \, \frac{T^p_{m,n-1} - T^p_{m,n}}{\Delta y} + h \, \Delta y (T_\infty - T^p_{m,n})$$
$$= \rho C \, \frac{\Delta x}{2} \, \Delta y \, \frac{T^{p+1}_{m,n} - T^p_{m,n}}{\Delta \tau}$$

If $\Delta x = \Delta y$, the relation for $T^{p+1}_{m,n}$ becomes

$$T^{p+1}_{m,n} = \frac{\alpha \, \Delta \tau}{(\Delta x)^2} \left\{ 2 \, \frac{h \, \Delta x}{k} \, T_\infty + 2 T^p_{m-1,n} + T^p_{m,n+1} \right.$$
$$\left. + \, T^p_{m,n-1} + \left[ \frac{(\Delta x)^2}{\alpha \, \Delta \tau} - 2 \, \frac{h \, \Delta x}{k} - 4 \right] T^p_{m,n} \right\} \quad (4\text{-}38a)$$

The corresponding one-dimensional relation is

$$T_m{}^{p+1} = \frac{\alpha \, \Delta \tau}{(\Delta x)^2} \left\{ 2 \, \frac{h \, \Delta x}{k} \, T_\infty + 2 T^p_{m-1} + \left[ \frac{(\Delta x)^2}{\alpha \, \Delta \tau} - 2 \, \frac{h \, \Delta x}{k} - 2 \right] T_m{}^p \right\} \quad (4\text{-}38b)$$

Notice now that the selection of the parameter $(\Delta x)^2 / \alpha \, \Delta \tau$ is not so simple as it is for the interior nodal points because the heat-transfer coefficient influences the choice. It is still usually best to choose the value of this parameter so that the coefficient of $T_m{}^p$ or $T^p_{m,n}$ will be zero. These values would then be

$$\frac{(\Delta x)^2}{\alpha \, \Delta \tau} = 2 \left( \frac{h \, \Delta x}{k} + 1 \right) \qquad \text{for the one-dimensional case}$$

and
$$\frac{(\Delta x)^2}{\alpha \, \Delta \tau} = 2 \left( \frac{h \, \Delta x}{k} + 2 \right) \qquad \text{for the two-dimensional case}$$

To ensure convergence of the numerical solution, all selections of the parameter $(\Delta x)^2 / \alpha \, \Delta \tau$ must be restricted according to

$$\frac{(\Delta x)^2}{\alpha \, \Delta \tau} \geq 2 \left( \frac{h \, \Delta x}{k} + 1 \right) \qquad \text{for the one-dimensional case}$$

and
$$\frac{(\Delta x)^2}{\alpha \, \Delta \tau} \geq 2 \left( \frac{h \, \Delta x}{k} + 2 \right) \qquad \text{for the two-dimensional case}$$

### Forward and Backward Differences

The equations above have been developed on the basis of a *forward difference* technique in that the temperature of a node at a future time increment is expressed in terms of the surrounding nodal temperatures at the beginning of the time increment. The expressions are called *explicit* formulations because it is possible to write the nodal temperatures $T^{p+1}_{m,n}$ explicitly in terms of the previous nodal temperatures $T^{p}_{m,n}$. In this formulation, the calculation proceeds directly from one time increment to the next until the temperature distribution is calculated at the desired final state.

The difference equation may also be formulated by moving backward in time in the approximation of the time derivative, i.e., instead of Eq. (4-31), we would write

$$\frac{\partial T}{\partial \tau} \approx \frac{T^{p}_{m,n} - T^{p+1}_{m,n}}{\Delta \tau}$$

The space derivatives would then be expressed in terms of the temperatures at the $p + 1$ time increment and the equation equivalent to Eq. (4-32) would be

$$\frac{T^{p+1}_{m+1,n} + T^{p+1}_{m-1,n} - 2T^{p+1}_{m,n}}{(\Delta x)^2} + \frac{T^{p+1}_{m,n+1} + T^{p+1}_{m,n-1} - 2T^{p+1}_{m,n}}{(\Delta y)^2} = \frac{1}{\alpha} \frac{T^{p}_{m,n} - T^{p+1}_{m,n}}{\Delta \tau}$$

$$(4\text{-}39)$$

The equivalence to Eq. (4-33) is

$$T^{p}_{m,n} = \frac{\alpha \, \Delta \tau}{(\Delta x)^2} (T^{p+1}_{m+1,n} + T^{p+1}_{m-1,n} + T^{p+1}_{m,n+1} + T^{p+1}_{m,n-1}) + \left[ 1 - \frac{4\alpha \, \Delta \tau}{(\Delta x)^2} \right] T^{p+1}_{m,n}$$

$$(4\text{-}40)$$

We may now note that this backward difference formulation does not permit the explicit calculation of the $T^{p+1}$ in terms of $T^{p}$. Rather, a whole set of equations must be written for the entire nodal system and solved simultaneously to determine the temperatures $T^{p+1}$. Thus we say that the backward difference method produces an *implicit formulation* for the future temperatures in the transient analysis. The solution to the set of equations can be performed with the methods discussed in Chap. 3.

The advantage of an explicit forward-difference procedure is the direct calculation of future nodal temperatures; however, the stability of this calculation is governed by the selection of the values of $\Delta x$ and $\Delta \tau$. A selection of a small value of $\Delta x$ automatically forces the selection of some maximum value of $\Delta \tau$. On the other hand, no such restriction is imposed on the solution of the equations which are obtained from the implicit formulation. This means that larger time increments can be selected to speed the calculation. The obvious disadvantage of the implicit method is the larger number of calculations for each time step.

For a discussion of many applications of numerical analysis to transient heat-conduction problems the reader is referred to Refs. 4, 8, 13, 14, and 15.

It should be obvious to the reader by now that finite-difference techniques may be applied to almost any situation with just a little patience and care. Very complicated problems then become quite easy to solve when large-digital computer facilities are available. Computer programs for several heat-transfer problems of interest are given by Schenck [8]. Finite-element methods for use in conduction heat-transfer problems are discussed in Refs. 9 to 13.

## 4-7□GRAPHICAL ANALYSIS—THE SCHMIDT PLOT

A very useful graphical technique may be employed to establish transient temperature distributions when the problem is one-dimensional. The method is based on the choice of the parameter

$$\frac{(\Delta x)^2}{\alpha \, \Delta \tau} = 2 \tag{4-36}$$

so that the temperature at any node after the time increment $\Delta \tau$ is the arithmetic average of the temperatures of the adjacent nodes at the beginning of the time increment. Such an arithmetic average is very easy to construct graphically, as shown in Fig. 4-22. The value of $T_m{}^{p+1}$ is obtained by drawing a line between $T_{m-1}^p$ and $T_{m+1}^p$. Thus, to find the temperature distribution in a solid after some specified time, the solid is divided into increments of $\Delta x$. Then, using Eq. (4-36), the value of $\Delta \tau$ is obtained. This value of $\Delta \tau$, when divided into the total time, gives the number of time increments necessary to establish the desired temperature distribution. The graphical construction is repeated until the final temperature distribution is obtained. A nonintegral number of time increments is usually required, and it will probably be necessary to interpolate between the last two increments to obtain the final temperature distribution.

An example of the method is shown in Fig. 4-23, where an initial temperature distribution is given and the construction is carried out for four time increments. The boundary temperatures are maintained at constant values throughout the cooling process indicated in this example. Notice that the construction approaches the steady-state straight-line temperature distribution with increasing time.

When a convection boundary condition is involved, the construction at the boundary must be modified. Rewriting Eq. (4-30), we have

$$\left.\frac{\partial T}{\partial x}\right)_{\text{wall}} = \frac{T_w - T_\infty}{k/h} \tag{4-41}$$

and the temperature gradient at the surface is approximated by the construction shown in Fig. 4-24. A line is drawn between the temperature $T_{m+1}$ and the environment temperature $T_\infty$. The intersection of this line with the surface determines the surface temperature at that particular time. This type of construction is used at each time increment to establish the surface

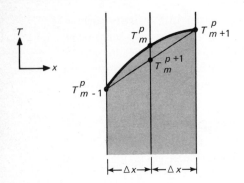

**Fig. 4-22  Graphical construction for one-dimensional unsteady-state conduction problems.**

temperature. Once this temperature is established, the construction to determine the internal temperatures in the solid proceeds as described above. An example of the construction for the convection boundary-condition problem with four time increments is shown in Fig. 4-25. In this example the temperature of the right face and the environment temperature $T_\infty$ are maintained constant. If the environment temperature changes with time, according to some known variation, this could easily be incorporated in the construction by moving the $T_\infty$ point up or down as required. In a similar fashion a variable heat-transfer coefficient could be considered by changing the value of $k/h$, according to some specified variation, and moving the environment point in or out a corresponding distance.

Refinements in the Schmidt graphical method are discussed by Jakob [5], particularly the techniques for improving accuracy at the boundary for either convection or other boundary conditions. The accuracy of the method is improved when smaller $\Delta x$ increments are taken, but this requires a larger number of time increments to obtain a temperature distribution after a given length of time. The particular problem under consideration will dictate the accuracy required and the effort to be devoted to the solution.

## REVIEW QUESTIONS
1.  What is meant by a lumped capacity. What are the physical assumptions necessary for a lumped-capacity unsteady-state analysis to apply?
2.  What is meant by a semi-infinite solid?
3.  What are the initial conditions that are imposed on the transient solutions presented in graphical form in this chapter?
4.  What boundary conditions are applied to problems in this chapter?
5.  Define the error function.
6.  Define the Biot and Fourier numbers.
7.  Describe how one-dimensional transient solutions may be used for solution of two- and three-dimensional problems.

**Fig. 4-23  Schmidt-plot construction for four time increments.**

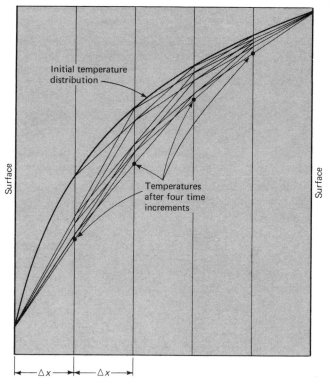

**Fig. 4-24  Graphical  technique  of  representing convection boundary condition with Schmidt plot.**

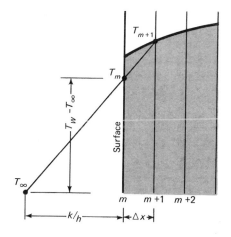

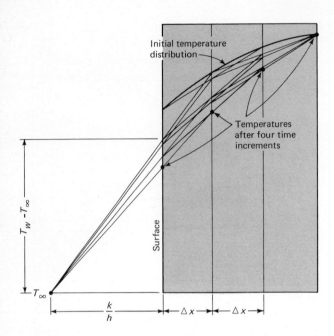

Fig. 4-25 Schmidt plot for four time increments, including convection boundary condition.

8. What are the advantages and disadvantages of the forward and backward difference formulations in the unsteady-state numerical method. Under what conditions would you choose one method over the other?

## PROBLEMS

4-1. A solid body at some initial temperature $T_0$ is suddenly placed in a room where the air temperature is $T_\infty$ and the walls of the room are very large. The heat-transfer coefficient for the convection heat loss is $h$, and the surface of the solid may be assumed black. Assuming that the temperature in the solid is uniform at any instant, write the differential equation for the variation in temperature with time, considering both radiation and convection.

4-2. A 1- by 1-ft slab of copper 2 in. thick is at a uniform temperature of 500°F and suddenly has its surface temperature lowered to 100°F. Using the thermal resistance and capacitance concepts and the lumped-capacity analysis, find the time at which the center temperature becomes 200°F. $\rho = 558$ lb$_m$/ft$^3$, $C_p = 0.091$ Btu/lb$_m$-°F, and $k = 215$ Btu/hr-ft-°F.

4-3. A piece of aluminum weighing 12 lb and initially at a temperature of 550°F is suddenly immersed in a fluid at 60°F. The convection heat-transfer coefficient is 10 Btu/hr-ft²-°F. Taking the aluminum as a sphere having the same weight as that given, estimate the time required to cool the aluminum to 200°F, using the lumped-capacity method of analysis.

4-4.   A copper sphere initially at a uniform temperature $T_0$ is immersed in a fluid. Electric heaters are placed in the fluid and so controlled that the temperature of the fluid follows a periodic variation given by

$$T_\infty - T_m = A \sin \omega\tau$$

where $T_m$ = time average mean fluid temperature
  $A$ = amplitude of temperature wave
  $\omega$ = frequency

Derive an expression for the temperature of the sphere as a function of time and the heat-transfer coefficient from the fluid to the sphere. Assume that the temperatures of the sphere and fluid are uniform at any instant of time so that the lumped-capacity method of analysis may be used.

4-5.   Two identical 3-in. cubes of copper at 800 and 100°F are brought into contact. Assuming that the blocks exchange heat only with each other and that there is no resistance to heat flow as a result of the contact of the blocks, plot the temperature of each block as a function of time, using the lumped-capacity method of analysis. That is, assume the resistance to heat transfer is the conduction resistance of the two blocks. Assume that all surfaces are insulated except those in contact.

4-6.   Repeat Prob. 4-5 for a 3-in. copper cube at 800°F in contact with a 3-in. steel cube at 100°F. Sketch the thermal circuit.

4-7.   When a sine-wave temperature distribution is impressed on the surface of a semi-infinite solid, the temperature distribution is given by

$$T_{x,\tau} - T_m = A e^{-x\sqrt{\pi n/\alpha}} \sin\left(2\pi n\tau - x\sqrt{\frac{\pi n}{\alpha}}\right)$$

where $T_{x,\tau}$ = temperature at a depth $x$ and time $\tau$ after start of temperature wave at surface
  $T_m$ = mean surface temperature
  $n$ = frequency of wave, cycles per unit time
  $A$ = amplitude of temperature wave at surface

If a sine-wave temperature distribution is impressed on the surface of a large slab of concrete such that the temperature varies from 100 to 200°F and a complete cycle is accomplished in 15 min, find the heat flow through a plane 2 in. from the surface 2 hr after the start of the initial wave.

4-8.   Using the temperature distribution of Prob. 4-7, show that the time lag between maximum points in the temperature wave at the surface and at a depth $x$ is given by

$$\Delta\tau = \frac{x}{2}\sqrt{\frac{1}{\alpha\pi n}}$$

4-9.  A 5-lb roast initially at 70°F is placed in an oven at 350°F. Assuming that the heat-transfer coefficient is 2.5 Btu/hr-ft²-°F and that the thermal properties of the roast may be approximated as those of water, estimate the time required for the center of the roast to attain a temperature of 200°F.

4-10.  A thick concrete wall having a uniform temperature of 130°F is suddenly subjected to an airstream at 50°F. The heat-transfer coefficient is 1.5 Btu/hr-ft²-°F. Calculate the temperature in the concrete slab at a depth of 3 in. after 30 min.

4-11.  A steel sphere 4 in. in diameter is suddenly immersed in a tank of oil at 50°F. The initial temperature of the sphere is 500°F. $h = 50$ Btu/hr-ft²-°F. How long will it take the center of the sphere to cool to 300°F?

4-12.  A very large slab of copper is initially at a temperature of 600°F. The surface temperature is suddenly lowered to 100°F. What is the temperature at a depth of 3 in., 4 min after the surface temperature is changed?

4-13.  On a hot summer day a concrete driveway may reach a temperature of 130°F. Suppose that a stream of water is directed on the driveway so that the surface temperature is suddenly lowered to 60°F. How long will it take to cool the concrete to 80°F at a depth of 2 in. from the surface?

4-14.  A long steel bar 2 by 4 in. is initially maintained at a uniform temperature of 500°F. It is suddenly subjected to a change such that the environment temperature is lowered to 100°F. Assuming a heat-transfer coefficient of 4 Btu/hr-ft²-°F, use the relaxation method to estimate the time required for the center temperature to reach 200°F. Check this result with a calculation, using the Heisler charts.

4-15.  A steel cylinder 4 in. in diameter and 4 in. long is initially at 500°F. It is suddenly immersed in an oil bath which is maintained at 100°F. $h = 50$ Btu/hr-ft²-°F. Find (a) the temperature at the center of the solid after 2 min, and (b) the temperature at the center of one of the circular faces after 2 min.

4-16.  A steel bar 1 in. square and 3 in. long is initially at a temperature of 500°F. It is immersed in a tank of oil maintained at 100°F. The heat-transfer coefficient is 100 Btu/hr-ft²-°F. Calculate the temperature in the center of the bar after 2 min.

4-17.  A cube of aluminum 4 in. on each side is initially at a temperature of 700°F and is immersed in a fluid at 200°F. The heat-transfer coefficient is 200 Btu/hr-ft²-°F. Calculate the temperature at the center of one face after 1 min.

4-18.  A short concrete cylinder 6 in. in diameter and 12 in. long is initially at 80°F. It is allowed to cool in an atmospheric environment in which the temperature is 30°F. Calculate the time required for the center temperature to reach 40°F if the heat-transfer coefficient is 3 Btu/hr-ft²-°F.

4-19.  Work Prob. 4-13 using the relaxation method.

4-20.   Work Prob. 4-10 using the relaxation method.

4-21.   A steel rod $\frac{1}{2}$ in. in diameter and 8 in. long has one end attached to a heat reservoir at 500°F. The bar is initially maintained at this temperature throughout. It is then subjected to an airstream at 100°F such that the convection heat-transfer coefficient is 6 Btu/hr-ft²-°F. Estimate the time required for the temperature midway along the length of the rod to attain a value of 380°F.

4-22.   A semi-infinite slab of copper is exposed to a constant heat flux at the surface of 10⁵ Btu/hr-ft². Assume that the slab is in a vacuum so that there is no convection at the surface. What is the surface temperature after 5 min if the initial temperature of the slab is 100°F? What is the temperature at a distance of 6 in. from the surface after 5 min?

4-23.   Work Prob. 4-13 using the Schmidt plot.

4-24.   Work Prob. 4-10 using the Schmidt plot.

4-25.   A concrete slab 6 in. thick, having a thermal conductivity of 0.5 Btu/hr-ft-°F, has one face insulated and the other face exposed to an environment. The slab is initially uniform in temperature at 600°F, and the environment temperature is suddenly lowered to 200°F. The heat-transfer coefficient is proportional to the fourth root of the temperature difference between the surface and environment and has a value of 2 Btu/hr-ft²-°F at time zero. The environment temperature increases linearly with time and has a value of 400°F after 20 min. Using the Schmidt plot, obtain the temperature distribution in the slab after 5, 10, 15, and 20 min.

4-26.   An infinite plate of thickness $2L$ is suddenly exposed to a constant-temperature radiation heat source or sink of temperature $T_s$. The plate has a uniform initial temperature of $T_i$. The radiation heat loss from each side of the plate is given by

$$q = \sigma \epsilon A (T^4 - T_s{}^4)$$

where $\sigma$ and $\epsilon$ are constants and $A$ is the surface area. Assuming that the plate behaves as a lumped capacity, that is, $k \rightarrow \infty$, derive an expression for the temperature of the plate as a function of time.

4-27.   A stainless-steel rod (18% Cr, 8% Ni) $\frac{1}{4}$ in. in diameter is initially at a uniform temperature of 100°F and is suddenly immersed in a liquid at 300°F with $h = 25$ Btu/hr-ft²-°F. Using the lumped-capacity method of analysis, calculate the time necessary for the rod temperature to reach 250°F.

4-28.   A large slab of copper is initially at a uniform temperature of 200°F. Its surface temperature is suddenly lowered to 100°F. Calculate the heat-transfer rate through a plane 3 in. from the surface, 5 sec after the surface temperature is lowered.

4-29.   A 2-in.-diameter copper sphere is initially at a uniform temperature of 500°F. It is suddenly exposed to an environment at 100°F, having a heat-transfer coefficient $h = 5$ Btu/hr-ft²-°F. Using the lumped-capacity method of analysis, calculate the time necessary for the sphere temperature to reach 200°F.

4-30.    The stainless-steel plate shown is initially at a uniform temperature of 300°F and is suddenly exposed to a convection environment at 100°F with $h = 3$ Btu/hr-ft²-°F. Using numerical techniques, calculate the time necessary for the temperature at a depth of 0.25 in. to reach 150°F. Employ a digital computer for this solution if possible.

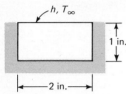

4-31.    An infinite plate having a thickness of 1.0 in. is initially at a temperature of 300°F, and the surface temperature is suddenly lowered to 100°F. The thermal diffusivity of the material is .07 ft²/hr. Calculate the centerplane temperature after a time of 1 min by summing the first four terms of Eq. (4-3). Check the answer using the Heisler charts.

4-32.    What error would result from using the first four terms of Eq. (4-3) to compute the temperature at $\tau = 0$ and $x = L$?

4-33.    A large slab of aluminum at a uniform temperature of 100°F is suddenly exposed to a constant surface heat flux of 5000 Btu/hr-ft². What is the temperature at a depth of 1 in. after a time of 2 min?

4-34.    For the slab in Prob. 4-33, how long would it take for the temperature to reach 300°F at the depth of 1 in.?

4-35.    The two-dimensional body of Fig. 3-6 has the initial surface and internal temperature as shown and calculated in Table 3-2. At time zero the 500°F face is suddenly lowered to 100°F. Taking $\Delta x = \Delta y = 0.5$ ft and $\alpha = 0.5$ ft²/hr calculate the temperatures at nodes 1, 2, 3, 4 after 30 min. Perform the calculation using both a forward and backward difference method. For the backward difference method use only two time increments.

4-36.    Develop a backward difference formulation for a boundary node subjected to a convection environment.

4-37.    A fused quartz sphere has a thermal diffusivity of 0.037 ft²/hr, a diameter of 1 in., and a thermal conductivity of 0.88 Btu/hr-ft-°F. The sphere is initially at a uniform temperature of 80°F and is suddenly subjected to a convection environment at 400°F. The convection heat-transfer coefficient is 20 Btu/hr-ft²-°F. Calculate the temperatures at the center and at a radius of $\frac{1}{4}$ in. after a time of 4 min.

## REFERENCES

1.    Schneider, P. J.: "Conduction Heat Transfer," Addison-Wesley Publishing Company, Inc., Reading, Mass., 1955.

2. Heisler, M. P.: Temperature Charts for Induction and Constant Temperature Heating, *Trans. ASME*, vol. 69, pp. 227–236, 1947.
3. Jahnke, E., and F. Emde: "Tables of Functions," Dover Publications, Inc., New York, 1945.
4. Dusinberre, G. M.: "Heat Transfer Calculations by Finite Differences," International Textbook Company, Scranton, Pa., 1961.
5. Jakob, M.: "Heat Transfer," vol. 1, John Wiley & Sons, Inc., New York, 1949.
6. Gröber, H., S. Erk, and U. Grigull: "Fundamentals of Heat Transfer," McGraw-Hill Book Company, New York, 1961.
7. Schneider, P. J.: "Temperature Response Charts," John Wiley & Sons, Inc., New York, 1963.
8. Schenck, H.: "Fortran Methods in Heat Flow," The Ronald Press Company, New York, 1963.
9. Richardson, P. D., and Y. M. Shum: Use of Finite-Element Methods in Solution of Transient Heat Conduction Problems, *ASME Paper* 69-WA/HT-36.
10. Emery, A. F., and W. W. Carson: "Evaluation of Use of the Finite Element Method in Computation of Temperature, *ASME Paper* 69-WA/HT-38.
11. Wilson, E. L., and R. E. Nickell: Application of the Finite Element Method to Heat Conduction Analysis, *Nucl. Eng. Design*, vol. 4, pp. 276–286, 1966.
12. Zienkiewicz, O. C.: "The Finite Element Method in Structural and Continuum Mechanics," McGraw-Hill Book Company, New York, 1967.
13. Myers, G. E.: "Conduction Heat Transfer," McGraw-Hill Book Company, New York, 1972.
14. Arpaci, V. S.: "Conduction Heat Transfer," Addison-Wesley Publishing Company, Inc., Reading, Mass., 1966.
15. Ozisik, M. N.: "Boundary Value Problems of Heat Conduction," International Textbook Company, Scranton, Pa., 1968.

# CHAPTER

# 5

# PRINCIPLES
# OF CONVECTION

## 5-1□INTRODUCTION

The preceding chapters have considered the mechanism and calculation of conduction heat transfer. Convection was considered only insofar as it related to the boundary conditions imposed on a conduction problem. We now wish to examine the methods of calculating convection heat transfer and, in particular, the ways of predicting the value of the convection heat-transfer coefficient $h$. The subject of convection heat transfer requires an energy balance along with an analysis of the fluid dynamics of the problems concerned. Our discussion in this chapter will first consider some of the simple relations of fluid dynamics and boundary-layer analysis which are important for a basic understanding of convection heat transfer. Next, we shall impose an energy balance on the flow system and determine the influence of the flow on the temperature gradients in the fluid. Finally, having obtained a knowledge of the temperature distribution, the heat-transfer rate from a heated surface to a fluid which is forced over it may be determined.

Our development in this chapter is primarily analytical in character, and is concerned only with forced-convection flow systems. Subsequent chapters will present empirical relations for calculating forced-convection heat transfer and will also treat the subjects of natural convection and boiling and condensation heat transfer.

## 5-2□VISCOUS FLOW

Consider the flow over a flat plate as shown in Figs. 5-1 and 5-2. Beginning at the leading edge of the plate, a region develops where the influence of viscous forces is felt. These viscous forces are described in terms of a shear stress $\tau$ between the fluid layers. If this stress is assumed to be proportional to the normal velocity gradient, we have the defining equation for the viscosity,

$$\tau = \mu \frac{du}{dy} \tag{5-1}$$

**Fig. 5-1    Sketch showing different boundary-layer flow regimes on a flat plate.**

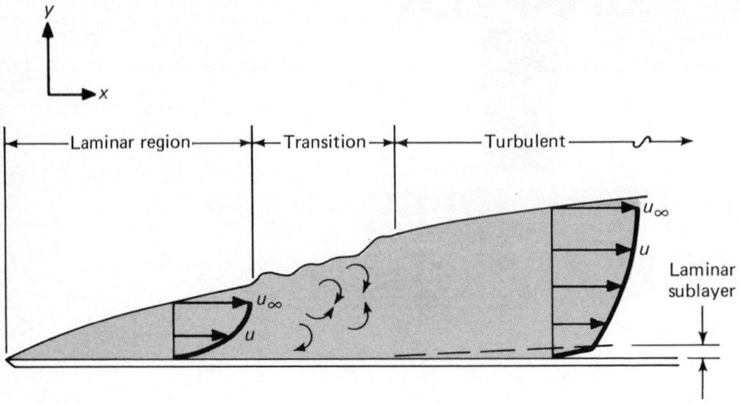

The constant of proportionality $\mu$ is called the dynamic viscosity. A typical set of units is $lb_f$-sec/ft²; however, there are many sets of units used for the viscosity, and care must be taken to select the proper group which will be consistent with the formulation at hand.

The region of flow which develops from the leading edge of the plate in which the effects of viscosity are observed is called the boundary layer. Some arbitrary point is used to designate the $y$ position where the boundary layer ends; this point is usually chosen as the $y$ coordinate, where the velocity becomes 99 percent of the free-stream value.

Initially, the boundary-layer development is laminar, but at some critical distance from the leading edge, depending on the flow field and fluid properties, small disturbances in the flow begin to become amplified, and a transition process takes place until the flow becomes turbulent. The turbulent-flow region may be pictured as a random churning action with chunks of fluid moving to and fro in all directions. The transition from laminar to turbulent flow occurs when

$$\frac{u_\infty x}{\nu} = \frac{\rho u_\infty x}{\mu} > 5 \times 10^5$$

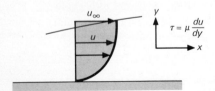

**Fig. 5-2    Laminar velocity profile on a flat plate.**

where $u_\infty$ = free-stream velocity

$x$ = distance from leading edge

$\nu = \mu/\rho$ = kinematic viscosity

This particular grouping of terms is called the Reynolds number, and is dimensionless if a consistent set of units is used for all the properties.

$$\mathrm{Re}_x = \frac{u_\infty x}{\nu} \qquad (5\text{-}2)$$

Although the critical Reynolds number for transition on a flat plate is usually taken as $5 \times 10^5$ for most analysis purposes, the critical value in a practical situation is strongly dependent on the surface roughness conditions and the "turbulence level" of the free stream. The normal range for the beginning of transition is between $5 \times 10^5$ and $10^6$. With very large disturbances present in the flow, transition may begin with Reynolds numbers as low as $10^5$, and for flows which are very free from fluctuations, it may not start until $\mathrm{Re} = 2 \times 10^6$ or more. In reality, the transition process is one which covers a range of Reynolds numbers, with transition being complete and with developed turbulent flow usually observed at Reynolds numbers twice the value at which transition began.

The relative shapes for the velocity profiles in laminar and turbulent flow are indicated in Fig. 5-1. The laminar profile is approximately parabolic in shape, while the turbulent profile has a portion near the wall which is very nearly linear. This linear portion is said to be due to a laminar sublayer which hugs the surface very closely. Outside this sublayer the velocity profile is relatively flat in comparison with the laminar profile.

The physical mechanism of viscosity is one of momentum exchange. Consider the laminar-flow situation. Molecules may move from one lamina to another, carrying with them a momentum corresponding to the velocity of the flow. There is a net momentum transport from regions of high velocity to regions of low velocity, thus creating a force in the direction of the flow. This force is the viscous shear stress which is calculated with Eq. (5-1).

The rate at which the momentum transfer takes place is dependent on the rate at which the molecules move across the fluid layers. In a gas, the molecules would move about with some average speed proportional to the square root of the absolute temperature since, in the kinetic theory of gases, we identify temperature with the mean kinetic energy of a molecule. The faster the molecules move, the more momentum they will transport. Hence we should expect the viscosity of a gas to be approximately proportional to the square root of temperature, and this expectation is corroborated fairly well by experiment. The viscosities of some typical fluids are given in App. A.

In the turbulent-flow region distinct fluid layers are no longer observed, and we are forced to seek a somewhat different concept for viscous action. A qualitative picture of the turbulent-flow process may be obtained by

**Fig. 5-3** (a) Velocity profile for laminar flow in a tube. (b) Velocity profile for turbulent tube flow.

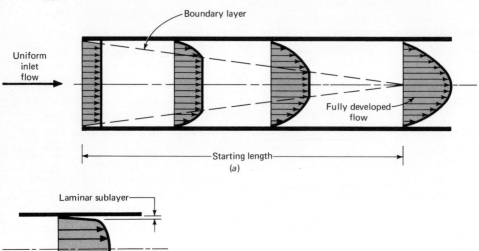

imagining macroscopic chunks of fluid transporting energy and momentum instead of microscopic transport on the basis of individual molecules. Naturally, we should expect the larger mass of the macroscopic elements of fluid to transport more energy and momentum than the individual molecules, and we should also expect a larger viscous shear force in turbulent flow than in laminar flow (and a larger thermal conductivity as well). This expectation is verified by experiment, and it is this larger viscous action in turbulent flow which causes the flat velocity profile indicated in Fig. 5-1.

Consider the flow in a tube as shown in Fig. 5-3. A boundary layer develops at the entrance, as shown. Eventually the boundary layer fills the entire tube, and the flow is said to be fully developed. If the flow is laminar, a parabolic velocity profile is experienced as shown in Fig. 5-3a. When the flow is turbulent, a somewhat blunter profile is observed as in Fig. 5-3b. In a tube, the Reynolds number is again used as a criterion for laminar and turbulent flow. For

$$\mathrm{Re}_d = \frac{u_m d}{\nu} > 2300 \tag{5-3}$$

the flow is usually observed to be turbulent.

Again, a range of Reynolds numbers for transition may be observed, depending on the pipe roughness and smoothness of the flow. The generally accepted range for transition is

$$2000 < \mathrm{Re}_d < 4000$$

although laminar flow has been maintained up to Reynolds numbers of 25,000 in carefully controlled laboratory conditions.

The continuity relation for one-dimensional flow in a tube is

$$\dot{m} = \rho u_m A \tag{5-4}$$

where $\dot{m}$ = mass rate of flow
$\quad u_m$ = mean velocity
$\quad A$ = cross-sectional area

We define the mass velocity as

$$\text{Mass velocity} = G = \frac{\dot{m}}{A} = \rho u_m \tag{5-5}$$

so that the Reynolds number may be written

$$\mathrm{Re}_d = \frac{Gd}{\mu} \tag{5-6}$$

Equation (5-6) is sometimes more convenient to use than Eq. (5-3).

## 5-3□INVISCID FLOW

Although no real fluid is inviscid, there are instances where the fluid may be treated as such, and it is worthwhile to present some of the equations which apply in these circumstances. For example, in the flat-plate problem discussed above, the flow at a sufficiently large distance from the plate will behave as a nonviscous flow system. The reason for this behavior is that the velocity gradients normal to the flow direction are very small, and hence the viscous shear forces are small.

If a balance of forces is made on an element of incompressible fluid and these forces set equal to the change in momentum of the fluid element, the Bernoulli equation for flow along a streamline results.

$$\frac{p}{\rho} + \frac{1}{2}\frac{V^2}{g_c} = \text{const} \tag{5-7a}$$

or in differential form,

$$\frac{dp}{\rho} + \frac{V\,dV}{g_c} = 0 \tag{5-7b}$$

where $\rho$ = fluid density
$\quad p$ = pressure at the particular point in flow
$\quad V$ = velocity of flow at that point

The Bernoulli equation is sometimes considered an energy equation because the $V^2/2g_c$ term represents kinetic energy and the pressure represents potential energy; however, it must be remembered that these terms are derived on the basis of a dynamic analysis, so that the equation is fundamentally a dynamic equation. In fact, the concept of kinetic energy is based on a dynamic analysis.

When the fluid is compressible, an energy equation must be written which will take into account changes in internal thermal energy of the system and the corresponding changes in temperature. For a one-dimensional flow system this equation is the steady-flow energy equation for a control volume,

$$i_1 + \frac{1}{2g_c} V_1{}^2 + Q = i_2 + \frac{1}{2g_c} V_2{}^2 + Wk \qquad (5\text{-}8)$$

where $i$ is the enthalpy defined by

$$i = e + pv \qquad (5\text{-}9)$$

and where $e$ = internal energy
$Q$ = heat added to control volume
$Wk$ = net external work done in the process
$v$ = specific volume of fluid

(The symbol $i$ is used to denote the enthalpy instead of the customary $h$ to avoid confusion with the heat-transfer coefficient.) The subscripts 1 and 2 refer to entrance and exit conditions to the control volume. To calculate pressure drop in compressible flow, it is necessary to specify the equation of state of the fluid, viz., for an ideal gas,

$$p = \rho RT$$
$$\Delta e = c_v \, \Delta T$$
$$\Delta i = c_p \, \Delta T$$

In addition, the process must be specified. For example, reversible adiabatic flow through a nozzle yields the following familiar expressions relating the properties at some point in the flow to the Mach number and the stagnation properties, i.e., the properties where the velocity is zero:

$$\frac{T_0}{T} = 1 + \frac{\gamma - 1}{2} M^2$$
$$\frac{p_0}{p} = \left(1 + \frac{\gamma - 1}{2} M^2\right)^{\gamma/(\gamma - 1)}$$
$$\frac{\rho_0}{\rho} = \left(1 + \frac{\gamma - 1}{2} M^2\right)^{1/(\gamma - 1)}$$

where $T_0, p_0, \rho_0$ = stagnation properties
$\gamma$ = ratio of specific heats $c_p/c_v$
$M$ = Mach number

$$M = \frac{V}{a}$$

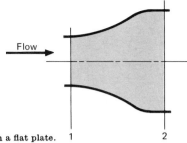

Flow

Fig. 5-1   Sketch showing different boundary-layer flow regimes on a flat plate.   1        2

where $a$ is the local velocity of sound, which may be calculated from

$$a = \sqrt{\gamma g_c R T}$$

for a perfect gas.†

### Example 5-1

Eighty gpm of water at 70°F flows through the diffuser arrangement shown in the figure. The diameter at section 1 is $1\frac{1}{4}$ in., and the diameter at section 2 is 3 in. Determine the increase in static pressure between sections 1 and 2. Assume frictionless flow.

**Solution**   We first determine the mass-flow rate

$$\dot{m} = \frac{(80)(8.33)}{60} = 11.1 \ \text{lb}_m/\text{sec}$$

since there is 8.33 $\text{lb}_m$ of water per gallon. The cross-sectional area of the flow at section 1 is

$$A_1 = \frac{\pi d_1^2}{4} = \frac{\pi(1.25)^2}{(4)(144)} = 0.0085 \ \text{ft}^2$$

and at section 2

$$A_2 = \frac{\pi d_2^2}{4} = \frac{\pi(3)^2}{(4)(144)} = 0.0492 \ \text{ft}^2$$

The flow velocities are

$$u = \frac{\dot{m}}{\rho A}$$

$$u_1 = \frac{11.1}{(62.4)(0.0085)} = 20.9 \ \text{ft/sec}$$

$$u_2 = \frac{11.1}{(62.4)(0.0492)} = 3.62 \ \text{ft/sec}$$

† The isentropic flow formulas are derived in Ref. 5, p. 53.

We may obtain the pressure difference from the Bernoulli equation.

$$\frac{p_2 - p_1}{\rho} = \frac{1}{2g_c}(u_1{}^2 - u_2{}^2)$$

$$p_2 - p_1 = \frac{62.4}{(2)(32.2)}[(20.9)^2 - (3.62)^2]$$

$$= 411 \text{ lb}_f/\text{ft}^2 = 2.85 \text{ psi} \qquad (1.97 \times 10^4 \text{ N/m}^2)$$

## Example 5-2

Air at 500°F and 100 psia is expanded isentropically from a tank until the velocity is 800 ft/sec. Determine the static temperature of the air at the high-velocity condition. Assume $c_p = 0.24$ Btu/lb$_m$-°F and $\gamma = 1.4$ for air.
**Solution**   We may write the steady-flow energy equation [Eq. (5-8)] as

$$i_1 = i_2 + \frac{u_2{}^2}{2g_c}$$

since the initial velocity is very small and the process is adiabatic. In terms of temperature,

$$c_p(T_1 - T_2) = \frac{u_2{}^2}{2g_c}$$

$$(0.24)(500 - T_2) = \frac{(800)^2}{(2)(32.2)(778)}$$

$$T_2 = 447°F$$

If the exit pressure is desired, we may use the isentropic relation between pressure and temperature.

$$\frac{p_2}{p_1} = \left(\frac{T_2}{T_1}\right)^{\gamma/(\gamma-1)}$$

$$p_2 = 100 \times (\tfrac{907}{960})^{3.5} = 82 \text{ psia} \qquad (5.65 \times 10^5 \text{ N/m}^2)$$

## 5-4□LAMINAR BOUNDARY LAYER ON A FLAT PLATE

Consider the elemental control volume shown in Fig. 5-4. We shall derive the equation of motion for the boundary layer by making a force-and-momentum balance on this element. To simplify the analysis we assume:

1. The fluid is incompressible and the flow is steady.
2. There are no pressure variations in the direction perpendicular to the plate.
3. The viscosity is constant.
4. Viscous-shear forces in the $y$ direction are negligible.
5) PHYSICAL PROPERTIES OF FLUID REMAIN CONSTANT

We apply Newton's second law of motion,

$$\sum F_x = \frac{d(mV)_x}{d\tau}$$

**Fig. 5-4   Elemental control volume for force balance on laminar boundary layer.**

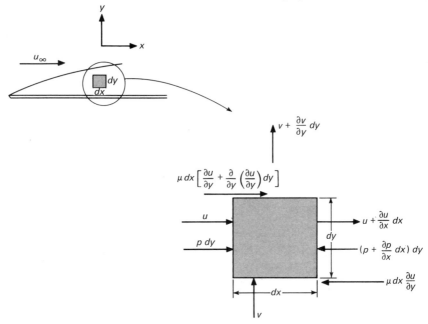

The above form of Newton's second law of motion applies to a system of constant mass. In fluid dynamics it is not usually convenient to work with elements of mass; rather, we deal with elemental control volumes such as that shown in Fig. 5-4, where mass may flow in or out of the different sides of the volume, which is fixed in space. For this system the force balance is then written

$$\Sigma F_x = \text{increase in momentum flux in } x \text{ direction}$$

The momentum flux in the $x$ direction is the product of the mass flow through a particular side of the control volume and the $x$ component of velocity at that point.

The mass entering the left face of the element per unit time is

$$\rho u \, dy$$

assuming unit depth in the $z$ direction. Thus the momentum entering the left face per unit time is

$$\rho u \, dy \, u = \rho u^2 \, dy \quad (1)$$

The mass flow leaving the right face is

$$\rho \left( u + \frac{\partial u}{\partial x} \, dx \right) dy \quad (2)$$

and the momentum leaving the right face is

$$\rho \left( u + \frac{\partial u}{\partial x} dx \right)^2 dy \quad \text{--(2)}$$

The mass flow entering the bottom face is

$$\rho v \, dx$$

and the mass flow leaving the top face is

$$\rho \left( v + \frac{\partial v}{\partial y} dy \right) dx$$

A mass balance on the element yields

$$\rho u \, dy + \rho v \, dx = \rho \left( u + \frac{\partial u}{\partial x} dx \right) dy + \rho \left( v + \frac{\partial v}{\partial y} dy \right) dx$$

or

$$\frac{\partial u}{\partial x} + \frac{\partial v}{\partial y} = 0 \qquad (5\text{-}10)$$

This is the mass continuity equation for the boundary layer.

Returning to the momentum-and-force analysis, the momentum *in the x direction* which enters the bottom face is

$$\rho v u \, dx \quad \text{--(3)}$$

and the momentum *in the x direction* which leaves the top face is

$$\rho \left( v + \frac{\partial v}{\partial y} dy \right) \left( u + \frac{\partial u}{\partial y} dy \right) dx \quad \text{--(4)}$$

We are interested only in the momentum in the $x$ direction because the forces considered in the analysis are those in the $x$ direction. These forces are those due to viscous shear and the pressure forces on the element. The pressure force on the left face is $p \, dy$, and that on the right is $-[p + (\partial p / \partial x) \, dx] \, dy$, so that the net pressure force in the direction of motion is

$$- \frac{\partial p}{\partial x} dx \, dy \quad \text{--(5)}$$

The viscous-shear force on the bottom face is

$$- \mu \frac{\partial u}{\partial y} dx$$

and the shear force on the top is

$$\mu \, dx \left[ \frac{\partial u}{\partial y} + \frac{\partial}{\partial y} \left( \frac{\partial u}{\partial y} \right) dy \right]$$

The net viscous-shear force in the direction of motion is the sum of the above.

$$\text{Net viscous-shear force} = \mu \frac{\partial^2 u}{\partial y^2} dx \, dy \quad \text{--(6)}$$

Fig. 5-5  Elemental control volume for integral momentum analysis of laminar boundary layer.

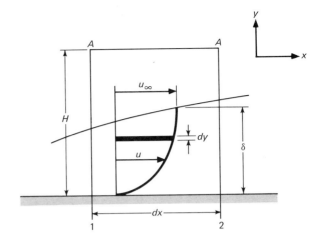

Fig. 5-5  Elemental control volume for integral momentum analysis of laminar boundary layer.

Equating the sum of the viscous shear and pressure forces to the net momentum transfer in the x direction, we have

$$\mu \frac{\partial^2 u}{\partial y^2}\, dx\, dy - \frac{\partial p}{\partial x}\, dx\, dy = \rho \left( u + \frac{\partial u}{\partial x}\, dx \right)^2 dy - \rho u^2\, dy$$

$$+ \rho \left( v + \frac{\partial v}{\partial y}\, dy \right)\left( u + \frac{\partial u}{\partial y}\, dy \right) dx - \rho v u\, dx \quad (5\text{-}10A)$$

*NET VISCOUS SHEAR FORCE*  *NET MOMENTVM*

*NET PRESSURE FORCE*

Clearing terms, making use of the continuity relation (5-10), and neglecting second-order differentials gives

$$\rho \left( u \frac{\partial u}{\partial x} + v \frac{\partial u}{\partial y} \right) = \mu \frac{\partial^2 u}{\partial y^2} - \frac{\partial p}{\partial x} \qquad (5\text{-}11)$$

This is the momentum equation of the laminar boundary layer with constant properties. The equation may be solved exactly for many boundary conditions, and the reader is referred to the treatise by Schlichting [1] for details on the various methods employed in the solutions. In App. B we have included the classical method for obtaining an exact solution to Eq. (5-11) for laminar flow over a flat plate. For the development in this chapter we shall be satisfied with an approximate analysis which furnishes an easier solution without a loss in physical understanding of the processes involved. The approximate method is due to von Kármán [2].

Consider the boundary-layer flow system shown in Fig. 5-5. The free-stream velocity outside the boundary layer is $u_\infty$, and the boundary-layer thickness is $\delta$. We wish to make a momentum-and-force balance on the control volume bounded by the planes 1, 2, A-A, and the solid wall. The velocity components normal to the wall are neglected, and only those in the $x$ direction are considered. We assume that the control volume is sufficiently high so that it always encloses the boundary layer; that is, $H > \delta$.

The mass flow through plane 1 is

$$\int_0^H \rho u \, dy \qquad\qquad (a)$$

and the momentum flow through plane 1 is

$$\int_0^H \rho u^2 \, dy \qquad\qquad (b)$$

The momentum flow through plane 2 is

$$\int_0^H \rho u^2 \, dy + \frac{d}{dx}\left(\int_0^H \rho u^2 \, dy \, dx\right) \qquad\qquad (c)$$

and the mass flow through plane 2 is

$$\int_0^H \rho u \, dy + \frac{d}{dx}\left(\int_0^H \rho u \, dy\right) dx \qquad\qquad (d)$$

Considering the conservation of mass and the fact that no mass can enter the control volume through the solid wall, the additional mass flow in expression $(d)$ over that in $(a)$ must enter through plane $A$-$A$. This mass flow carries with it a momentum in the $x$ direction equal to

$$u_\infty \frac{d}{dx}\left(\int_0^H \rho u \, dy\right) dx \qquad (e)$$

The net momentum flow out of the control volume is therefore

$$\frac{d}{dx}\left(\int_0^H \rho u^2 \, dy\right) dx - u_\infty \frac{d}{dx}\left(\int_0^H \rho u \, dy\right) dx \qquad (f)$$

This expression may be put in a somewhat more useful form by recalling the product formula from the differential calculus.

$$d(\eta\phi) = \eta \, d\phi + \phi \, d\eta$$

or
$$\eta \, d\phi = d(\eta\phi) - \phi \, d\eta$$

In the momentum expression given above, the integral

$$\int_0^H \rho u \, dy$$

is the $\phi$ function and $u_\infty$ is the $\eta$ function. Thus

$$u_\infty \frac{d}{dx}\left(\int_0^H \rho u \, dy\right) dx = \frac{d}{dx}\left(u_\infty \int_0^H \rho u \, dy\right) dx - \frac{du_\infty}{dx}\left(\int_0^H \rho u \, dy\right) dx$$

$$= \frac{d}{dx}\left(\int_0^H \rho u u_\infty \, dy\right) dx - \frac{du_\infty}{dx}\left(\int_0^H \rho u \, dy\right) dx \qquad (5\text{-}12)$$

The $u_\infty$ may be placed inside the integral since it is not a function of $y$ and thus may be treated as a constant insofar as an integral with respect to $y$ is concerned.

Returning to the analysis, the force on plane 1 is the pressure force $pH$ and that on plane 2 is $[p + (dp/dx)\, dx]H$. The shear force at the wall is

$$-\tau_w\, dx = -\mu\, dx \left.\frac{\partial u}{\partial y}\right)_{y=0}$$

There is no shear force at plane $A$-$A$ since the velocity gradient is zero outside the boundary layer. Setting the forces on the element equal to the net increase in momentum and collecting terms gives

$$-\tau_w - \frac{dp}{dx} H = -\rho \frac{d}{dx} \int_0^H (u_\infty - u)u\, dy + \frac{du_\infty}{dx} \int_0^H \rho u\, dy \qquad (5\text{-}13)$$

This is the integral momentum equation of the boundary layer. If the pressure is constant throughout the flow,

$$\frac{dp}{dx} = 0 = -\rho u_\infty \frac{du_\infty}{dx} \qquad (5\text{-}14)$$

since the pressure and free-stream velocity are related by the Bernoulli equation. For the constant-pressure condition the integral boundary-layer equation becomes

$$\rho \frac{d}{dx} \int_0^\delta (u_\infty - u)u\, dy = \tau_w = \mu \left.\frac{\partial u}{\partial y}\right)_{y=0} \qquad (5\text{-}15)$$

The upper limit on the integral has been changed to $\delta$ because the integrand is zero for $y > \delta$ since $u = u_\infty$ for $y > \delta$.

If the velocity profile were known, the appropriate function could be inserted in Eq. (5-15) to obtain an expression for the boundary-layer thickness. For our approximate analysis we first write down some conditions which the velocity function must satisfy.

$$u = 0 \qquad \text{at } y = 0 \qquad\qquad (a)$$
$$u = u_\infty \qquad \text{at } y = \delta \qquad\qquad (b)$$
$$\frac{\partial u}{\partial y} = 0 \qquad \text{at } y = \delta \qquad\qquad (c)$$

For a constant-pressure condition Eq. (5-11) yields

$$\frac{\partial^2 u}{\partial y^2} = 0 \qquad \text{at } y = 0 \qquad\qquad (d)$$

since the velocities $u$ and $v$ are zero at $y = 0$. We assume that the velocity profiles at various $x$ positions are similar; i.e., they have the same functional dependence on the $y$ coordinate. There are four conditions to satisfy. The simplest function which we can choose to satisfy these conditions is a polynomial with four arbitrary constants. Thus

$$u = C_1 + C_2 y + C_3 y^2 + C_4 y^3 \tag{5-16}$$

Applying the four conditions (a) to (d),

$$\frac{u}{u_\infty} = \frac{3}{2}\frac{y}{\delta} - \frac{1}{2}\left(\frac{y}{\delta}\right)^3 \tag{5-17}$$

Inserting the expression for the velocity into Eq. (5-15),

$$\frac{d}{dx}\left\{\rho u_\infty^2 \int_0^\delta \left[\frac{3}{2}\frac{y}{\delta} - \frac{1}{2}\left(\frac{y}{\delta}\right)^3\right]\left[1 - \frac{3}{2}\frac{y}{\delta} + \frac{1}{2}\left(\frac{y}{\delta}\right)^3\right]dy\right\} = \mu\left(\frac{\partial u}{\partial y}\right)_{y=0} = \frac{3}{2}\frac{\mu u_\infty}{\delta}$$

Carrying out the integration,

$$\frac{d}{dx}\left(\frac{39}{280}\rho u_\infty^2\,\delta\right) = \frac{3}{2}\frac{\mu u_\infty}{\delta}$$

Since $\rho$ and $u_\infty$ are constants, the variables may be separated to give

$$\delta\,d\delta = \frac{140}{13}\frac{\mu}{\rho u_\infty}\,dx = \frac{140}{13}\frac{\nu}{u_\infty}\,dx$$

and

$$\frac{\delta^2}{2} = \frac{140}{13}\frac{\nu x}{u_\infty} + \text{const}$$

At $x = 0$, $\delta = 0$, so that

$$\delta = 4.64\sqrt{\frac{\nu x}{u_\infty}} \tag{5-18}$$

This may be written in terms of the Reynolds number as

$$\frac{\delta}{x} = \frac{4.64}{\text{Re}_x^{\frac{1}{2}}} \tag{5-19}$$

where

$$\text{Re}_x = \frac{u_\infty x}{\nu}$$

The exact solution of the boundary-layer equations as given in App. B yields

$$\frac{\delta}{x} = \frac{5.0}{\text{Re}_x^{\frac{1}{2}}} \tag{5-19a}$$

## Example 5-3

Air at 70°F and 14.7 psia flows over a flat plate at a speed of 5 ft/sec. Calculate the boundary-layer thickness at distances of 6 in. and 1 ft from the leading edge of the plate. Calculate the mass flow which enters the boundary layer between $x = 6$ in. and $x = 1$ ft. The viscosity of air at 70°F is 0.044 $\text{lb}_m$/hr-ft. Assume unit depth in the $z$ direction.

**Solution**    The density of air is calculated from

$$\rho = \frac{p}{RT} = \frac{(14.7)(144)}{(53.35)(530)} = 0.0748 \text{ lb}_m/\text{ft}^3$$

The Reynolds number is calculated as

At $x = 6$ in.: $\mathrm{Re}_x = \frac{\rho u_\infty x}{\mu} = \frac{(0.0748)(5)(3600)(0.5)}{0.044} = 15,300$

At $x = 12$ in.:                     $\mathrm{Re}_x = 30,600$

The boundary-layer thickness is calculated from Eq. (5-19).

$$\delta = \frac{4.64x}{\mathrm{Re}_x^{\frac{1}{2}}}$$

At $x = 6$ in.:            $\delta = \frac{(4.64)(0.5)}{(15,300)^{\frac{1}{2}}} = 0.0188 \text{ ft}$

At $x = 12$ in.:            $\delta = \frac{(4.64)(1)}{(30,600)^{\frac{1}{2}}} = 0.0259 \text{ ft}$

To calculate the mass flow which enters the boundary layer from the free stream between $x = 6$ in. and $x = 1$ ft, we simply take the difference between the mass flow in the boundary layer at these two $x$ positions. At any $x$ position the mass flow in the boundary layer is given by the integral

$$\int_0^\delta \rho u \, dy$$

where the velocity is given by Eq. (5-17),

$$u = u_\infty \left[ \frac{3}{2}\frac{y}{\delta} - \frac{1}{2}\left(\frac{y}{\delta}\right)^3 \right]$$

Evaluating the integral with this velocity distribution, we have

$$\int_0^\delta \rho u_\infty \left[ \frac{3}{2}\frac{y}{\delta} - \frac{1}{2}\left(\frac{y}{\delta}\right)^3 \right] dy = \frac{5}{8}\rho u_\infty \delta$$

Thus the mass flow entering the boundary layer is

$$\tfrac{5}{8}\rho u_\infty [\delta]_{x=0.5}^{x=1.0} = \tfrac{5}{8}(0.0748)(5)(0.0259 - 0.0188)$$
$$= 1.66 \times 10^{-3} \text{ lb}_m/\text{sec} \qquad (0.753 \times 10^{-3} \text{ kg/s})$$

## 5-5□ENERGY EQUATION OF THE BOUNDARY LAYER

The foregoing analysis considered the fluid dynamics of a laminar-boundary-layer flow system. We shall now develop the energy equation for this system and then proceed to an integral method of solution.

Consider the elemental control volume shown in Fig. 5-6. To simplify

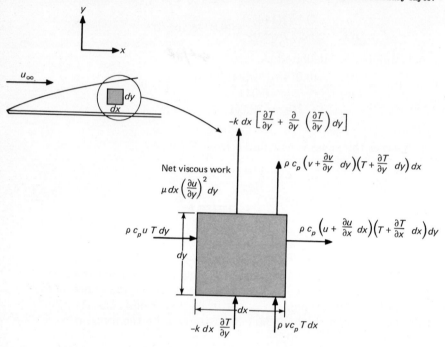

Fig. 5-6  Elemental control volume for energy analysis of laminar boundary layer.

the analysis we assume

1.  Incompressible steady flow
2.  Constant viscosity, thermal conductivity, and specific heat
3.  Negligible heat conduction in the direction of flow ($x$ direction)

Then, for the element shown, the energy balance may be written

Energy convected in left face + energy convected in bottom face
  + heat conducted in bottom face + net viscous work done on element
  = energy convected out right face + energy convected out top face
                                                        + heat conducted out top face

The convective and conduction energy quantities are indicated in Fig. 5-6, and the energy term for the viscous work may be derived as follows: The viscous work may be computed as a product of the net viscous-shear force and the distance this force moves in unit time. The viscous-shear force is the product of the shear stress and the area $dx$,

$$\mu \frac{\partial u}{\partial y} \, dx$$

and the distance through which it moves per unit time in respect to the elemental control volume $dx\,dy$ is

$$\frac{\partial u}{\partial y}\,dy$$

so that the net viscous energy delivered to the element is

$$\mu\left(\frac{\partial u}{\partial y}\right)^2 dx\,dy$$

Writing the energy balance corresponding to the quantities shown in Fig. 5-6, assuming unit depth in the $z$ direction, and neglecting second-order differentials yields

$$\rho c_p\left[u\frac{\partial T}{\partial x} + v\frac{\partial T}{\partial y} + T\left(\frac{\partial u}{\partial x} + \frac{\partial v}{\partial y}\right)\right]dx\,dy = k\frac{\partial^2 T}{\partial y^2}\,dx\,dy + \mu\left(\frac{\partial u}{\partial y}\right)^2 dx\,dy \quad (5\text{-}9a)$$

Using the continuity relation,

$$\frac{\partial u}{\partial x} + \frac{\partial v}{\partial y} = 0 \tag{5-10}$$

and dividing by $\rho c_p$ gives

$$u\frac{\partial T}{\partial x} + v\frac{\partial T}{\partial y} = \alpha\frac{\partial^2 T}{\partial y^2} + \frac{\mu}{\rho c_p}\left(\frac{\partial u}{\partial y}\right)^2 \quad \text{FOR HIGH VELOCITY} \tag{5-20}$$

ENERGY TO ELEMENT          ENERGY AWAY FROM

This is the energy equation of the laminar boundary layer. The left side represents the net transport of energy into the control volume, and the right side represents the sum of the net heat conducted out of the control volume and the net viscous work done on the element. The viscous-work term is of importance only at high velocities since its magnitude will be small compared with the other terms when low-velocity flow is studied. This may be shown with an order-of-magnitude analysis of the two terms on the right side of Eq. (5-20). For this order-of-magnitude analysis we might consider the velocity as having the order of the free-stream velocity $u_\infty$ and the $y$ dimension of the order of $\delta$. Thus

$$u \sim u_\infty$$
$$y \sim \delta$$

so that

$$\alpha\frac{\partial^2 T}{\partial y^2} \sim \alpha\frac{T}{\delta^2}$$

$$\frac{\mu}{\rho c_p}\left(\frac{\partial u}{\partial y}\right)^2 \sim \frac{\mu}{\rho c_p}\frac{u_\infty^2}{\delta^2}$$

If the ratio of these quantities is small, i.e.,

$$\frac{\mu}{\rho c_p \alpha}\frac{u_\infty^2}{T} \ll 1$$

the viscous dissipation is small in comparison with the conduction term.

As an example, consider the flow of air at

$$u_\infty = 200 \text{ ft/sec}$$
$$T = 70°F$$
$$p = 14.7 \text{ psia}$$

The grouping of terms becomes

$$\frac{\mu}{\rho c_p \alpha} \frac{u_\infty^2}{T} = 0.0087 \ll 1$$

indicating that the viscous dissipation is small. Thus, for low-velocity incompressible flow, we have

$$u \frac{\partial T}{\partial x} + v \frac{\partial T}{\partial y} = \alpha \frac{\partial^2 T}{\partial y^2} \tag{5-21}$$

In reality, our derivation of the energy equation has been a simplified one, and several terms have been left out of the analysis because they are small in comparison with others. In this way we arrive at the boundary-layer approximation immediately, without resorting to a cumbersome elimination process to obtain the final simplified relation. The general derivation of the boundary-layer energy equation is very involved and quite beyond the scope of our discussion. The interested reader should consult the books by Schlichting [1] and Liepmann and Roshko [5] for more information.

There is a striking similarity between Eq. (5-21) and the momentum equation for constant pressure,

$$u \frac{\partial u}{\partial x} + v \frac{\partial u}{\partial y} = \nu \frac{\partial^2 u}{\partial y^2} \tag{5-22}$$

The solution to the two equations will have exactly the same form when $\alpha = \nu$. Thus we should expect that the relative magnitudes of the thermal diffusivity and kinematic viscosity would have an important influence on convection heat transfer since these magnitudes relate the velocity distribution to the temperature distribution. This is exactly the case, and we shall see the role which these parameters play in the subsequent discussion.

## 5-6□THE THERMAL BOUNDARY LAYER

Just as the hydrodynamic boundary layer was defined as that region of the flow where viscous forces are felt, a thermal boundary layer may be defined as that region where temperature gradients are present in the flow. These temperature gradients would result from a heat-exchange process between the fluid and the wall.

Consider the system shown in Fig. 5-7. The temperature of the wall is $T_w$, the temperature of the fluid outside the thermal boundary layer is $T_\infty$,

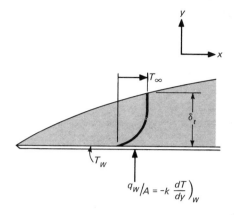

**Fig. 5-7** Temperature profile in the thermal boundary layer.

and the thickness of the thermal boundary layer is designated as $\delta_t$. At the wall, the velocity is zero, and the heat transfer into the fluid takes place by conduction. Thus, the local heat flux per unit area, $q''$, is

$$q'' = -k \frac{\partial T}{\partial y}\bigg)_{\text{wall}} \tag{5-23}$$

From Newton's law of cooling [Eq. (1-8)]

$$\frac{q''}{A} = h(T_w - T_\infty) \tag{5-24}$$

where $h$ is the convection heat-transfer coefficient. Combining these equations,

$$h = \frac{-k(\partial T/\partial y)_{\text{wall}}}{T_w - T_\infty} \tag{5-25}$$

so that we only need to find the temperature gradient at the wall in order to evaluate the heat-transfer coefficient. This means that we must obtain an expression for the temperature distribution. To do this an approach similar to that used in the momentum analysis of the boundary layer is followed.

The conditions which the temperature distribution must satisfy are

$$T = T_w \qquad \text{at } y = 0 \tag{a}$$

$$\frac{\partial T}{\partial y} = 0 \qquad \text{at } y = \delta_t \tag{b}$$

$$T = T_\infty \qquad \text{at } y = \delta_t \tag{c}$$

and by writing Eq. (5-21) at $y = 0$ with no viscous heating,

$$\frac{\partial^2 T}{\partial y^2} = 0 \qquad \text{at } y = 0 \tag{d}$$

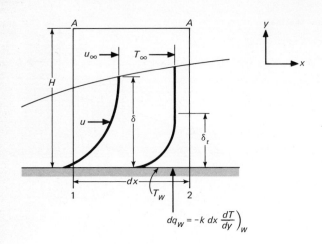

since the velocities must be zero at the wall.

Conditions $(a)$ to $(d)$ may be fitted to a cubic polynomial as in the case of the velocity profile, so that

$$\frac{\theta}{\theta_\infty} = \frac{T - T_w}{T_\infty - T_w} = \frac{3}{2}\frac{y}{\delta_t} - \frac{1}{2}\left(\frac{y}{\delta_t}\right)^3 \tag{5-26}$$

where $\theta = T - T_w$. There now remains the problem of finding an expression for $\delta_t$, the thermal-boundary-layer thickness. This may be obtained by an integral analysis of the energy equation for the boundary layer.

Consider the control volume bounded by the planes 1, 2, $A$-$A$, and the wall as shown in Fig. 5-8. It is assumed that the thermal boundary layer is thinner than the hydrodynamic boundary layer, as shown. The wall temperature is $T_w$, the free-stream temperature is $T_\infty$, and the heat given up to the fluid over the length $dx$ is $dq_w$. We wish to make the energy balance

Energy convected in + viscous work within element
$$+ \text{ heat transfer at wall} = \text{energy convected out} \tag{5-27}$$

The energy convected in through plane 1 is

$$\rho c_p \int_0^H uT\, dy \quad (A)$$

and the energy convected out through plane 2 is

$$\rho c_p \left(\int_0^H uT\, dy\right) + \frac{d}{dx}\left(\rho c_p \int_0^H uT\, dy\right) dx \quad (B)$$

The mass flow through plane $A$-$A$ is

$$\frac{d}{dx}\left(\int_0^H \rho u\, dy\right) dx$$

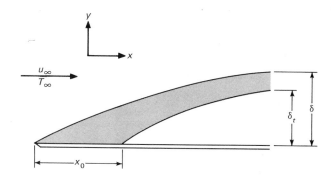

**Fig. 5-9 Hydrodynamic and thermal boundary layers on a flat plate. Heating starts at $x = x_0$.**

and this carries with it an energy equal to

$$c_p T_\infty \frac{d}{dx} \left( \int_0^H \rho u \, dy \right) dx$$

The net viscous work done within the element is

$$\mu \left[ \int_0^H \left( \frac{du}{dy} \right)^2 dy \right] dx$$

and the heat transfer at the wall is

$$dq_w = -k \, dx \, \frac{\partial T}{\partial y} \bigg)_w$$

Combining these energy quantities according to Eq. (5-27) and collecting terms gives

$$\frac{d}{dx} \left[ \int_0^H (T_\infty - T) u \, dy \right] + \frac{\mu}{\rho c_p} \left[ \int_0^H \left( \frac{du}{dy} \right)^2 dy \right] = \alpha \frac{\partial T}{\partial y} \bigg)_w \qquad (5\text{-}28)$$

This is the integral energy equation of the boundary layer for constant properties and constant free-stream temperature $T_\infty$.

To calculate the heat transfer at the wall, we need to derive an expression for the thermal-boundary-layer thickness which may be used in conjunction with Eqs. (5-25) and (5-26) to determine the heat-transfer coefficient. For now, we neglect the viscous-dissipation term; this term is very small unless the velocity of the flow field becomes very large, and the calculation of high-velocity heat transfer will be considered later.

The plate under consideration need not be heated over its entire length. The situation which we shall analyze is shown in Fig. 5-9, where the hydrodynamic boundary layer develops from the leading edge of the plate, while heating does not begin until $x = x_0$.

Inserting the temperature distribution Eq. (5-26) and the velocity distribution Eq. (5-17) into Eq. (5-28) and neglecting the viscous-dissipation term gives

$$\frac{d}{dx}\left[\int_0^H (T_\infty - T)u\,dy\right] = \frac{d}{dx}\left[\int_0^H (\theta_\infty - \theta)u\,dy\right]$$

$$= \theta_\infty u_\infty \frac{d}{dx}\left\{\int_0^H \left[1 - \frac{3}{2}\frac{y}{\delta_t} + \frac{1}{2}\left(\frac{y}{\delta_t}\right)^3\right]\left[\frac{3}{2}\frac{y}{\delta} - \frac{1}{2}\left(\frac{y}{\delta}\right)^3\right]dy\right\}$$

$$= \alpha\left.\frac{\partial T}{\partial y}\right)_{y=0} = \frac{3\alpha\theta_\infty}{2\delta_t}$$

Let us assume that the thermal boundary layer is thinner than the hydro-dynamic boundary layer. Then we only need to carry out the integration to $y = \delta_t$ since the integrand is zero for $y > \delta_t$. Performing the necessary algebraic manipulation, carrying out the integration, and making the substitution $\zeta = \delta_t/\delta$ yields

$$\theta_\infty u_\infty \frac{d}{dx}\left[\delta\left(\tfrac{3}{20}\zeta^2 - \tfrac{3}{280}\zeta^4\right)\right] = \frac{3}{2}\frac{\alpha\theta_\infty}{\delta\zeta} \tag{5-29}$$

Since $\delta_t < \delta$, $\zeta < 1$, and the term involving $\zeta^4$ is small compared with the $\zeta^2$ term, we neglect the $\zeta^4$ term and write

$$\tfrac{3}{20}\theta_\infty u_\infty \frac{d}{dx}(\delta\zeta^2) = \frac{3}{2}\frac{\alpha\theta_\infty}{\zeta\delta} \tag{5-30}$$

Performing the differentiation,

$$\tfrac{1}{10}u_\infty\left(2\delta\zeta\frac{d\zeta}{dx} + \zeta^2\frac{d\delta}{dx}\right) = \frac{\alpha}{\delta\zeta}$$

or

$$\tfrac{1}{10}u_\infty\left(2\delta^2\zeta^2\frac{d\zeta}{dx} + \zeta^3\delta\frac{d\delta}{dx}\right) = \alpha$$

But

$$\delta\,d\delta = \frac{140}{13}\frac{\nu}{u_\infty}dx$$

and

$$\delta^2 = \frac{280}{13}\frac{\nu x}{u_\infty}$$

so that we have

$$\zeta^3 + 4x\zeta^2\frac{d\zeta}{dx} = \frac{13}{14}\frac{\alpha}{\nu} \tag{5-31}$$

Noting that

$$\zeta^2\frac{d\zeta}{dx} = \frac{1}{3}\frac{d}{dx}\zeta^3$$

Eq. (5-31) is a linear differential equation of the first order in $\zeta^3$, and the solution is

$$\zeta^3 = Cx^{-\frac{3}{4}} + \frac{13}{14}\frac{\alpha}{\nu} \qquad (5\text{-}31\,B)$$

When the boundary condition

$$\delta_t = 0 \qquad \text{at } x = x_0$$
$$\zeta = 0 \qquad \text{at } x = x_0$$

is applied, the final solution becomes

$$\zeta = \frac{\delta_t}{\delta} = \frac{1}{1.026} \Pr^{-\frac{1}{3}} \left[ 1 - \left( \frac{x_0}{x} \right)^{\frac{3}{4}} \right]^{\frac{1}{3}} \tag{5-32}$$

where

$$\Pr = \frac{\nu}{\alpha} \tag{5-33}$$

has been introduced. The ratio $\nu/\alpha$ is called the Prandtl number after Ludwig Prandtl, the German scientist who introduced the concepts of boundary-layer theory.

When the plate is heated over the entire length, $x_0 = 0$, and

$$\frac{\delta_t}{\delta} = \zeta = \frac{1}{1.026} \Pr^{-\frac{1}{3}} \tag{5-34}$$

In the foregoing analysis the assumption was made that $\zeta < 1$. This assumption is satisfactory for fluids having Prandtl numbers greater than about 0.7. Fortunately, most gases and liquids fall within this category. Liquid metals are a notable exception, however, since they have Prandtl numbers of the order of 0.01.

The Prandtl number $\nu/\alpha$ has been found to be the parameter which relates the relative thicknesses of the hydrodynamic and thermal boundary layers. The kinematic viscosity of a fluid conveys information about the rate at which momentum may diffuse through the fluid because of molecular motion. The thermal diffusivity tells us the same thing in regard to the diffusion of heat in the fluid. Thus the ratio of these two quantities should express the relative magnitudes of diffusion of momentum and heat in the fluid. But these diffusion rates are precisely the quantities that determine how thick the boundary layers will be for a given external flow field; large diffusivities mean that the viscous or temperature influence is felt farther out in the flow field. The Prandtl number is thus the connecting link between the velocity field and the temperature field.

The Prandtl number is dimensionless when a consistent set of units is used.

$$\Pr = \frac{\nu}{\alpha} = \frac{\mu/\rho}{k/\rho c_p} = \frac{c_p \mu}{k} \tag{5-35}$$

A typical set of units for the parameters would be $\mu$ in $\text{lb}_m/\text{hr-ft}$, $c_p$ in $\text{Btu}/\text{lb}_m\text{-}°\text{F}$, $k$ in $\text{Btu/hr-ft-}°\text{F}$.

Returning now to the analysis,

$$h = \frac{-k(\partial T/\partial y)_w}{T_w - T_\infty} = \frac{3}{2}\frac{k}{\delta_t} = \frac{3}{2}\frac{k}{\zeta\delta} \qquad (5\text{-}36)$$

Substituting for the hydrodynamic-boundary-layer thickness from Eq. (5-19) and using Eq. (5-32) gives

$$h_x = 0.332k\ \mathrm{Pr}^{\frac{1}{3}}\left(\frac{u_\infty}{\nu x}\right)^{\frac{1}{2}}\left[1 - \left(\frac{x_0}{x}\right)^{\frac{3}{4}}\right]^{-\frac{1}{3}} \qquad (5\text{-}37)$$

The equation may be nondimensionalized by multiplying both sides by $x/k$, producing the dimensionless group on the left side,

$$\mathrm{Nu}_x = \frac{h_x x}{k} \qquad (5\text{-}38)$$

called the Nusselt number after Wilhelm Nusselt, who made significant contributions to the theory of convection heat transfer. Finally,

$$\mathrm{Nu}_x = 0.332\ \mathrm{Pr}^{\frac{1}{3}}\ \mathrm{Re}_x^{\frac{1}{2}}\left[1 - \left(\frac{x_0}{x}\right)^{\frac{3}{4}}\right]^{-\frac{1}{3}} \qquad (5\text{-}39)$$

*FOR ANY CASE*

or, for the plate heated over its entire length, $x_0 = 0$ and

$$\mathrm{Nu}_x = 0.332\ \mathrm{Pr}^{\frac{1}{3}}\ \mathrm{Re}_x^{\frac{1}{2}} \qquad (5\text{-}40)$$

Equations (5-37), (5-39), and (5-40) express the local values of the heat-transfer coefficient in terms of the distance from the leading edge of the plate and the fluid properties. For the case where $x_0 = 0$ the average heat-transfer coefficient and Nusselt number may be obtained by integrating over the length of the plate.

$$\bar{h} = \frac{\int_0^L h_x\, dx}{\int_0^L dx} = 2h_{x=L} \qquad (5\text{-}41)$$

$$\overline{\mathrm{Nu}}_L = \frac{\bar{h}L}{k} = 2\ \mathrm{Nu}_{x=L} \qquad (5\text{-}42)$$

The reader should carry out the integrations to verify these results.

The foregoing analysis was based on the assumption that the fluid properties were constant throughout the flow. When there is an appreciable variation between wall and free-stream conditions, it is recommended that the properties be evaluated at the so-called "film" temperature $T_f$, defined as the arithmetic mean between the wall and free-stream temperature.

$$T_f = \frac{T_w + T_\infty}{2} \qquad (5\text{-}43)$$

An exact solution to the energy equation is given in App. B. The results of the exact analysis are the same as those of the approximate analysis given above.

## Example 5-4

For the flow system in Example 5-3 assume that the plate is heated over its entire length to a temperature of 120°F. Calculate the heat transferred in (a) the first 6 in. of the plate, and (b) the first foot of the plate.

**Solution**   The total heat transfer over a certain length of the plate is desired; so we wish to calculate average heat-transfer coefficients. For this purpose we use Eqs. (5-40) and (5-41), evaluating the properties at the film temperature. From App. A the properties are

$$\nu = 0.637 \text{ ft}^2/\text{hr}$$
$$k = 0.0154 \text{ Btu/hr-ft-°F}$$
$$\text{Pr} = 0.71$$

At $x = 6$ in.:

$$\text{Re}_x = \frac{u_\infty x}{\nu} = \frac{(5)(3600)(0.5)}{0.637} = 14,130$$

$$\text{Nu}_x = \frac{h_x x}{k} = 0.332 \text{ Re}_x^{\frac{1}{2}} \text{ Pr}^{\frac{1}{3}}$$

$$= (0.332)(14,130)^{\frac{1}{2}}(0.71)^{\frac{1}{3}} = 35.2$$

$$h_x = \text{Nu}_x \frac{k}{x} = \frac{(35.2)(0.0154)}{0.5} = 1.085 \text{ Btu/hr-ft}^2\text{-°F}$$

The average value of the heat-transfer coefficient is twice this value.

$$\bar{h} = (2)(1.085) = 2.17 \text{ Btu/hr-ft}^2\text{-°F}$$

The heat flow is

$$q = \bar{h} A (T_w - T_\infty)$$

If we assume unit depth in the $z$ direction,

$$q = (2.17)(0.5)(120 - 70) = 54.3 \text{ Btu/hr} \qquad (15.9 \text{ W})$$

At $x = 12$ in.:

$$\text{Re}_x = \frac{(5)(3600)(1.0)}{0.637} = 28,200$$

$$\text{Nu}_x = (0.332)(28,200)^{\frac{1}{2}}(0.71)^{\frac{1}{3}} = 49.8$$

$$h_x = \frac{(49.8)(0.0154)}{1.0} = 0.768$$

$$\bar{h} = (2)(0.768) = 1.536 \text{ Btu/hr-ft}^2\text{-°F}$$

$$q = (1.536)(1.0)(120 - 70) = 76.8 \text{ Btu/hr} \qquad (22.5 \text{ W})$$

## 5-7☐THE RELATION BETWEEN FLUID FRICTION AND HEAT TRANSFER

We have already seen that the temperature and flow fields are related. Now we seek an expression whereby the frictional resistance may be directly related to heat transfer.

The shear stress at the wall may be expressed in terms of a friction coefficient $C_f$:

$$\tau_w = C_f \frac{\rho u_\infty^2}{2} \tag{5-44}$$

Equation (5-44) is the defining equation for the friction coefficient. The shear stress may also be calculated from the relation

$$\tau_w = \mu \left.\frac{\partial u}{\partial y}\right)_w$$

Using the velocity distribution given by Eq. (5-17),

$$\tau_w = \frac{3}{2} \frac{\mu u_\infty}{\delta}$$

and making use of the relation derived for the boundary-layer thickness,

$$\tau_w = \frac{3}{2} \frac{\mu u_\infty}{4.64} \left(\frac{u_\infty}{\nu x}\right)^{\frac{1}{2}} \tag{5-45}$$

Combining Eqs. (5-44) and (5-45),

$$\frac{C_{fx}}{2} = \frac{3}{2} \frac{\mu u_\infty}{4.64} \left(\frac{u_\infty}{\nu x}\right)^{\frac{1}{2}} \frac{1}{\rho u_\infty^2} = 0.323 \, \mathrm{Re}_x^{-\frac{1}{2}} \tag{5-46}$$

Equation (5-40) may be rewritten in the following form:

$$\frac{\mathrm{Nu}_x}{\mathrm{Re}_x \, \mathrm{Pr}} = \frac{h_x}{\rho c_p u_\infty} = 0.332 \, \mathrm{Pr}^{-\frac{2}{3}} \, \mathrm{Re}_x^{-\frac{1}{2}}$$

The group on the left is called the Stanton number,

$$\mathrm{St}_x = \frac{h_x}{\rho c_p u_\infty}$$

so that    $\mathrm{St}_x \, \mathrm{Pr}^{\frac{2}{3}} = 0.332 \, \mathrm{Re}_x^{-\frac{1}{2}} \tag{5-47}$

Upon comparing Eqs. (5-46) and (5-47), we note that the right sides of the equations are alike except for a difference of about 3 percent in the constant, which is the result of the approximate nature of the integral boundary-layer analysis. We shall recognize this approximation and write

$$\mathrm{St}_x \, \mathrm{Pr}^{\frac{2}{3}} = \frac{C_{fx}}{2} \tag{5-48}$$

Equation (5-48) expresses the relation between fluid friction and heat transfer for laminar flow on a flat plate. The heat-transfer coefficient could thus be determined by making measurements of the frictional drag on a plate under conditions in which no heat transfer is involved. This relation between fluid friction and heat transfer is called the Reynolds analogy.

It turns out that Eq. (5-48) can also be applied to turbulent flow over a flat plate and in a modified way to turbulent flow in a tube. It does not apply to laminar tube flow. In general, a more rigorous treatment of the governing equations is necessary when embarking on new applications of the heat-transfer–fluid-friction analogy and the results do not always take the simple form of Eq. (5-48). The interested reader may consult the book by Knudsen and Katz [6] for more information on this important subject. At this point, the simple analogy developed above has served to amplify our understanding of the physical processes in convection and to reinforce the notion that heat-transfer and viscous-transport processes are related at both the microscopic and macroscopic level.

### Example 5-5

For the flow system in Example 5-4, compute the drag force exerted on the first 1 ft of the plate using the analogy between fluid friction and heat transfer.

**Solution**  We use Eq. (5-48) to compute the friction coefficient, and then calculate the drag force.

$$\mathrm{St}_x \, \mathrm{Pr}^{\frac{2}{3}} = \frac{C_{fx}}{2}$$

If the average friction coefficient is desired,

$$\overline{\mathrm{St}} \, \mathrm{Pr}^{\frac{2}{3}} = \frac{\overline{C_f}}{2}$$

$$\overline{\mathrm{St}} = \frac{\bar{h}}{\rho c_p u_\infty} = \frac{1.536}{(0.0716)(0.24)(5)(3600)} = 0.00497$$

$$(0.00497)(0.71)^{\frac{2}{3}} = \frac{\overline{C_f}}{2}$$

$$\overline{C_f} = 0.00788$$

The average shear stress at the wall is computed from Eq. (5-44).

$$\overline{\tau_w} = \overline{C_f} \, \frac{\rho u_\infty{}^2}{2}$$

$$= \frac{(0.00788)(0.0716)(5)^2}{(2)(32.2)} = 2.19 \times 10^{-4} \, \mathrm{lb}_f/\mathrm{ft}^2$$

The drag force is the product of this shear stress and the area,

$$D = (2.19 \times 10^{-4})(1) = 2.19 \times 10^{-4} \, \mathrm{lb}_f \qquad (9.74 \times 10^{-4} \, \mathrm{N})$$

Fig. 5-10   Velocity profile in turbulent boundary layer on a flat plate.

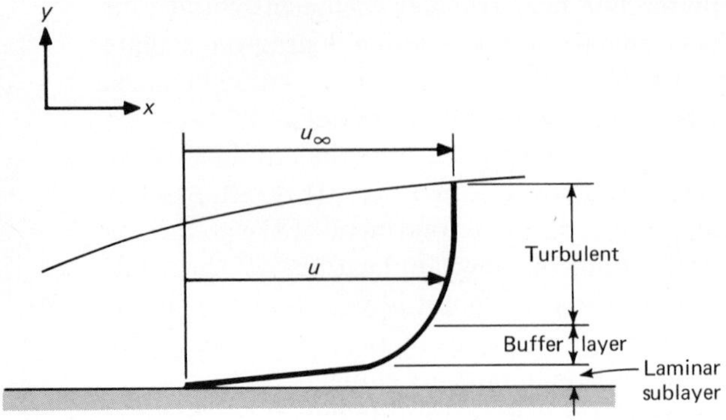

## 5-8□TURBULENT - BOUNDARY - LAYER  HEAT  TRANSFER

Consider a portion of a turbulent boundary layer as shown in Fig. 5-10. A very thin region near the plate surface has a laminar character, and the viscous action and heat transfer take place under circumstances like those in laminar flow. Farther out, at larger $y$ distances from the plate, some turbulent action is experienced, but the molecular viscous action and heat conduction are still important. This region is called the buffer layer. Still farther out, the flow is fully turbulent, and the main momentum and heat-exchange mechanism is one involving macroscopic lumps of fluid moving about in the flow. In this fully turbulent region we speak of *eddy viscosity* and *eddy thermal conductivity*. These eddy properties may be ten times as large as the molecular values.

The physical mechanism of heat transfer in turbulent flow is quite similar to that in laminar flow; the primary difference is that one must deal with the eddy properties instead of the ordinary thermal conductivity and viscosity. The main difficulty in an analytical treatment is that these eddy properties vary across the boundary layer, and the specific variation can be determined only from experimental data. This is an important point. All analyses of turbulent flow must eventually rely on experimental data because there is no completely adequate theory to predict turbulent-flow behavior.

We could attack the turbulent-heat-transfer problem in one of two ways, either by taking experimentally measured velocity profiles and performing an integral boundary-layer analysis or by taking experimentally measured friction coefficients and applying the analogy between fluid friction and heat transfer which was developed above. This second method is the easier to apply, but we need to lay a little more groundwork before proceeding further.

To use the analogy between fluid friction and heat transfer, the Prandtl number for the turbulent-flow situation must be computed. It has been noted that both the viscous action and thermal conduction are increased because of the eddy motion in turbulent flow. Furthermore, it is not unreasonable to expect that both will be increased in the same proportion so that we may use the Prandtl number calculated with the ordinary fluid properties. This assumption is a good one since heat-transfer calculations based on the fluid-friction analogy for turbulent flow match experimental data very well.

Experimental measurements show that the local friction coefficient for turbulent flow is

$$C_{fx} = 0.0576 \, \text{Re}_x^{-\frac{1}{5}} \tag{5-49}$$

for Reynolds numbers between $5 \times 10^5$ and $10^7$. Applying the fluid-friction analogy [Eq. (5-48)],

$$\text{St}_x \, \text{Pr}^{\frac{2}{3}} = 0.0288 \, \text{Re}_x^{-\frac{1}{5}} \tag{5-50}$$

The laminar-flow heat transfer was expressed in terms of the Stanton number as

$$\text{St}_x \, \text{Pr}^{\frac{2}{3}} = 0.332 \, \text{Re}_x^{-\frac{1}{2}} \tag{5-51}$$

To obtain an expression for the average heat-transfer coefficient over a length $L$ of the plate when both the laminar and turbulent boundary layers are present, we must integrate the two expressions above. For purposes of this integration it is assumed that the critical Reynolds number for transition from laminar to turbulent flow is $5 \times 10^5$. Then

$$\overline{\text{St}} \, \text{Pr}^{\frac{2}{3}} = \frac{\int_0^{x_c} 0.332 \sqrt{\dfrac{\nu}{u_\infty x}} \, dx + \int_{x_c}^{L} 0.0288 \left(\dfrac{\nu}{u_\infty x}\right)^{\frac{1}{5}} dx}{L} \tag{5-52}$$

where $\quad x_c = 5 \times 10^5 \dfrac{\nu}{u_\infty} \quad$ since $\quad \text{Re}_{\text{crit}} = 5 \times 10^5$

Carrying out the integration,

$$\overline{\text{St}} \, \text{Pr}^{\frac{2}{3}} = 0.036 \, \text{Re}_{x=L}^{-\frac{1}{5}} - 836 \, \text{Re}_{x=L}^{-1} \tag{5-53}$$

If the Nusselt number is used to express the average heat-transfer coefficient, then

$$\overline{\text{Nu}}_L = \frac{\bar{h}L}{k} = \text{Pr}^{\frac{1}{3}} \, (0.036 \, \text{Re}_L^{0.8} - 836) \tag{5-54}$$

## Example 5-6

Air at 70°F and 14.7 psia flows over a flat plate at 100 ft/sec. The plate is 2.5 ft long and is maintained at 130°F. Assuming unit depth in the $z$ direction, calculate the heat transfer from the plate.

**Fig. 5-11** Control volume for energy analysis in tube flow.

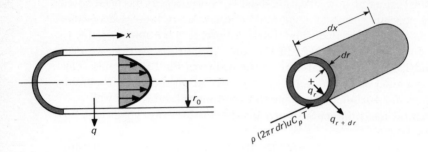

**Solution**   We evaluate the properties of air at the film temperature.

$$T_f = (70 + 130)/2 = 100°F$$
$$\rho = 0.071 \text{ lb}_m/\text{ft}^3$$
$$\mu = 0.046 \text{ lb}_m/\text{hr-ft}$$
$$\text{Pr} = 0.711$$
$$k = 0.0155 \text{ Btu/hr-ft-°F}$$
$$c_p = 0.24 \text{ Btu/lb}_m\text{-°F}$$

The Reynolds number is

$$\text{Re}_L = \frac{\rho u_\infty L}{\mu} = \frac{(0.071)(100)(3600)(2.5)}{0.046}$$
$$= 1.39 \times 10^6$$

and the boundary layer is turbulent since the Reynolds number is greater than $5 \times 10^5$. To calculate the heat transfer we use Eq. (5-54).

$$\overline{\text{Nu}}_L = \frac{\bar{h}L}{k} = \text{Pr}^{\frac{1}{3}}(0.036 \text{ Re}_L{}^{0.8} - 836)$$
$$= (0.711)^{\frac{1}{3}}[(0.036)(1.39 \times 10^6)^{0.8} - 836] = 1890$$
$$\bar{h} = \overline{\text{Nu}}_L\frac{k}{L} = \frac{(1890)(0.0155)}{2.5} = 11.7 \text{ Btu/hr-ft}^2\text{-°F}$$
$$q = \bar{h}A(T_w - T_\infty) = (11.7)(2.5)(130 - 70) = 1755 \text{ Btu/hr} \qquad \text{(514 W)}$$

## 5-9□HEAT TRANSFER IN LAMINAR TUBE FLOW

Consider the tube-flow system shown in Fig. 5-11. We wish to calculate the heat transfer under developed flow conditions when the flow remains laminar. The wall temperature is $T_w$, the radius of the tube is $r_o$, and the velocity at the center of the tube is $u_0$. It is assumed that the pressure is uniform at any cross section. The velocity distribution may be derived by considering the fluid element shown in Fig. 5-12. The pressure forces are

**Fig. 5-12  Force balance on
fluid element in tube flow.**

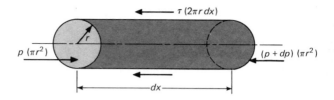

balanced by the viscous-shear forces so that

$$\pi r^2 \, dp = \tau 2\pi r \, dx = 2\pi r \mu \, dx \, \frac{du}{dr}$$

or

$$du = \frac{1}{2\mu} r \frac{dp}{dx} \, dr$$

and

$$u = \frac{1}{4\mu} \frac{dp}{dx} r^2 + \text{const} \tag{5-55}$$

With the boundary condition

$$u = 0 \qquad \text{at } r = r_o$$

$$u = \frac{1}{4\mu} \frac{dp}{dx} (r^2 - r_o{}^2)$$

The velocity at the center of the tube is given by

$$u_0 = -\frac{r_o{}^2}{4\mu} \frac{dp}{dx} \tag{5-56}$$

so that the velocity distribution may be written

$$\frac{u}{u_0} = 1 - \frac{r^2}{r_o{}^2} \tag{5-57}$$

which is the familiar parabolic distribution for laminar tube flow. Now consider the heat-transfer process for such a flow system. To simplify the analysis, we shall assume that there is a constant heat flux at the tube wall; i.e.,

$$\frac{dq_w}{dx} = \text{const}$$

The heat flow conducted into the annular element is

$$dq_r = -k2\pi r \, dx \, \frac{\partial T}{\partial r}$$

and the heat conducted out is

$$dq_{r+dr} = -k2\pi (r + dr) \, dx \left( \frac{\partial T}{\partial r} + \frac{\partial^2 T}{\partial r^2} \, dr \right)$$

The net heat convected out of the element is

$$2\pi r \, dr \, \rho c_p u \frac{\partial T}{\partial x} \, dx$$

The energy balance is

Net energy convected out = net heat conducted in

or, neglecting second-order differentials,

$$r\rho c_p u \frac{\partial T}{\partial x} \, dx \, dr = k \left( \frac{\partial T}{\partial r} + r \frac{\partial^2 T}{\partial r^2} \right) dx \, dr$$

which may be rewritten

$$\frac{1}{ur} \frac{\partial}{\partial r} \left( r \frac{\partial T}{\partial r} \right) = \frac{1}{\alpha} \frac{\partial T}{\partial x} \qquad (5\text{-}58)$$

We assume that the heat flux at the wall is constant, so that the average fluid temperature must increase linearly with $x$, or

$$\frac{\partial T}{\partial x} = \text{const}$$

This means that the temperature profiles will be similar at various $x$ distances along the tube. The boundary conditions on Eq. (5-58) are

$$\frac{\partial T}{\partial r} = 0 \qquad \text{at } r = 0$$

$$-k \frac{\partial T}{\partial r} \bigg)_{r=r_o} = q_w = \text{const}$$

To obtain the solution to Eq. (5-58), the velocity distribution given by Eq. (5-57) must be inserted. It is assumed that the temperature and velocity fields are independent; i.e., a temperature gradient does not affect the calculation of the velocity profile. This is equivalent to specifying that the properties remain constant in the flow. With the substitution of the velocity profile, Eq. (5-58) becomes

$$\frac{\partial}{\partial r} \left( r \frac{\partial T}{\partial r} \right) = \frac{1}{\alpha} \frac{\partial T}{\partial x} u_0 \left( 1 - \frac{r^2}{r_o^2} \right) r$$

Integration yields

$$r \frac{\partial T}{\partial r} = \frac{1}{\alpha} \frac{\partial T}{\partial x} u_0 \left( \frac{r^2}{2} - \frac{r^4}{4r_o^2} \right) + C_1$$

and a second integration gives

$$T = \frac{1}{\alpha} \frac{\partial T}{\partial x} u_0 \left( \frac{r^2}{4} - \frac{r^4}{16r_o^2} \right) + C_1 \ln r + C_2$$

Applying the first boundary condition, we find that

$$C_1 = 0$$

The second boundary condition has been satisfied by noting that the axial temperature gradient $\partial T/\partial x$ is constant. The temperature distribution may finally be written in terms of the temperature at the center of the tube.

$$T = T_c \quad \text{at } r = 0 \quad \text{so that } C_2 = T_c$$

$$T - T_c = \frac{1}{\alpha} \frac{\partial T}{\partial x} \frac{u_0 r_o{}^2}{4} \left[ \left(\frac{r}{r_o}\right)^2 - \frac{1}{4}\left(\frac{r}{r_o}\right)^4 \right] \qquad (5\text{-}59)$$

## The Bulk Temperature

In tube flow the convection heat-transfer coefficient is usually defined by

$$\text{Local heat flux} = q'' = h(T_w - T_b) \qquad (5\text{-}60)$$

where $T_w$ = wall temperature

$T_b$ = so-called bulk temperature, or energy-average fluid temperature across tube, which may be calculated from

$$T_b = \bar{T} = \frac{\displaystyle\int_0^{r_o} \rho 2\pi r \, dr \, u c_p T}{\displaystyle\int_0^{r_o} \rho 2\pi r \, dr \, u c_p} \qquad (5\text{-}61)$$

The reason for using the bulk temperature in the definition of heat-transfer coefficients for tube flow may be explained as follows. In a tube flow there is no easily discernible "free stream" condition as is present in the flow over a flat plate. Even the centerline temperature $T_c$ is not easily expressed in terms of the inlet flow variables and the heat transfer. For most tube or channel flow heat-transfer problems the topic of central interest is the total energy transferred to the fluid in either an elemental length of the tube or over the entire length of the channel. At any $x$ position the temperature that is indicative of the total energy of the flow is an integrated mass-energy average temperature over the entire flow area. The numerator of Eq. (5-61) represents the total energy flow through the tube and the denominator represents the product of mass flow and specific heat integrated over the flow area. The bulk temperature is thus representative of the total energy of the flow at the particular location. For this reason, the bulk temperature is sometimes referred to as the "mixing-cup" temperature, since it is the temperature the fluid would assume if placed in a mixing chamber and allowed to come to equilibrium. For the temperature distribution given in Eq. (5-59), the bulk temperature is a linear function of $x$ since the heat flux at the tube wall is constant. Calculating the bulk temperature from Eq. (5-61), we have

$$T_b = T_c + \frac{7}{96} \frac{u_0 r_o{}^2}{\alpha} \frac{\partial T}{\partial x} \qquad (5\text{-}62)$$

and for the wall temperature

$$T_w = T_c + \frac{3}{16} \frac{u_0 r_o{}^2}{\alpha} \frac{\partial T}{\partial x} \tag{5-63}$$

The heat-transfer coefficient is calculated from

$$q = hA(T_w - T_b) = kA \left(\frac{\partial T}{\partial r}\right)_{r=r_o}$$

$$h = \frac{k(\partial T/\partial r)_{r=r_o}}{T_w - T_b} \tag{5-64}$$

The temperature gradient is given by

$$\frac{\partial T}{\partial r}\Big)_{r=r_o} = \frac{u_0}{\alpha} \frac{\partial T}{\partial x} \left(\frac{r}{2} - \frac{r^3}{4r_o{}^2}\right)_{r=r_o} = \frac{u_0 r_o}{4\alpha} \frac{\partial T}{\partial x} \tag{5-65}$$

Substituting Eqs. (5-62), (5-63), and (5-65) in Eq. (5-64) gives

$$h = \frac{24}{11} \frac{k}{r_o} = \frac{48}{11} \frac{k}{d_o}$$

or expressing the result in terms of the Nusselt number,

$$\mathrm{Nu}_d = \frac{hd_o}{k} = 4.364 \tag{5-66}$$

in agreement with an exact calculation by Sellars, Tribus, and Klein [3], which considers the temperature profile as it develops. Some empirical relations for calculating heat transfer in laminar tube flow will be presented in Chap. 6.

### 5-10□TURBULENT FLOW IN A TUBE

The developed velocity profile for turbulent flow in a tube will appear as shown in Fig. 5-13. A laminar sublayer, or "film," occupies the space near the surface, while the central core of the flow is turbulent. To determine the heat transfer analytically for this situation, we require, as usual, a knowledge of the temperature distribution in the flow. To obtain this temperature distribution, the analysis must take into consideration the effect of the turbulent eddies in the transfer of heat and momentum. We shall use an approximate analysis which relates the conduction and transport of heat to the transport of momentum in the flow, i.e., viscous effects.

The heat flow across a fluid element in laminar flow may be expressed by

$$\frac{q}{A} = -k \frac{dT}{dy}$$

Dividing both sides of the equation by $\rho c_p$,

$$\frac{q}{\rho c_p A} = -\alpha \frac{dT}{dy}$$

**Fig. 5-13** Velocity profile in turbulent tube flow.

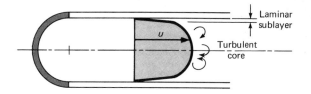

It will be recalled that $\alpha$ is the molecular diffusivity of heat. In turbulent flow one might assume that the heat transport could be represented by

$$\frac{q}{\rho c_p A} = -(\alpha + \epsilon_H) \frac{dT}{dy} \tag{5-67}$$

where $\epsilon_H$ is an eddy diffusivity of heat.

Equation (5-67) expresses the total heat conduction as a sum of the molecular conduction and the macroscopic eddy conduction. In a similar fashion, the shear stress in turbulent flow could be written

$$\frac{\tau}{\rho} = \left(\frac{\mu}{\rho} + \epsilon_M\right) \frac{du}{dy} = (\nu + \epsilon_M) \frac{du}{dy} \tag{5-68}$$

where $\epsilon_M$ is the eddy diffusivity for momentum. We now assume that the heat and momentum are transported at the same rate, that is, $\epsilon_M = \epsilon_H$ and $\nu = \alpha$, or $\Pr = 1$.

Dividing Eq. (5-67) by Eq. (5-68) gives

$$\frac{q}{c_p A \tau} du = -dT$$

An additional assumption is that the ratio of the heat transfer per unit area to the shear stress is constant across the flow field. This is consistent with the assumption that heat and momentum are transported at the same rate. Thus

$$\frac{q}{A\tau} = \text{const} = \frac{q_w}{A_w \tau_w}$$

Then, integrating Eq. (5-69) between wall conditions and mean bulk conditions gives

$$\frac{q_w}{A_w \tau_w c_p} \int_{u=0}^{u=u_m} du = \int_{T_w}^{T_b} - dT$$

$$\frac{q_w u_m}{A_w \tau_w c_p} = T_w - T_b \tag{5-69}$$

But the heat transfer at the wall may be expressed by

$$q_w = h A_w (T_w - T_b)$$

and the shear stress may be calculated from

$$\tau_w = \frac{\Delta P(\pi d_o^2)}{4\pi\, d_o L} = \frac{\Delta P}{4}\frac{d_o}{L}$$

The pressure drop may be expressed in terms of a friction factor $f$ by

$$\Delta P = f\frac{L}{d}\,\rho\,\frac{u_m^2}{2} \tag{5-70}$$

so that

$$\tau_w = \frac{f}{8}\rho u_m^2 \tag{5-71}$$

Substituting the expressions for $\tau_w$ and $q_w$ in Eq. (5-69) gives

$$\mathrm{St} = \frac{h}{\rho c_p u_m} = \frac{\mathrm{Nu}_d}{\mathrm{Re}_d\,\mathrm{Pr}} = \frac{f}{8} \tag{5-72}$$

Equation (5-72) is called the Reynolds analogy for tube flow. It relates the heat-transfer rate to the frictional loss in tube flow and is in fair agreement with experiments when used with gases whose Prandtl numbers are close to unity. (Recall that $\mathrm{Pr} = 1$ was one of the assumptions in the analysis.) An empirical formula for the turbulent-friction factor up to Reynolds numbers of about $2 \times 10^5$ for the flow in smooth tubes is

$$f = \frac{0.316}{\mathrm{Re}_d^{\frac{1}{4}}} \tag{5-73}$$

Inserting this expression in Eq. (5-72) gives

$$\frac{\mathrm{Nu}_d}{\mathrm{Re}_d\,\mathrm{Pr}} = 0.0395\,\mathrm{Re}_d^{-\frac{1}{4}}$$

or

$$\mathrm{Nu}_d = 0.0395\,\mathrm{Re}_d^{\frac{3}{4}} \tag{5-74}$$

since we assumed the Prandtl number to be unity. This derivation of the relation for turbulent heat transfer in smooth tubes is highly restrictive because of the $\mathrm{Pr} \sim 1.0$ assumption. The heat-transfer–fluid-friction analogy of Sec. 5-7 indicated a Prandtl number dependence of $\mathrm{Pr}^{\frac{2}{3}}$ for the flat plate problem and, as it turns out, this dependence works fairly well for turbulent tube flow. Equations (5-72) and (5-74) may be modified by this factor to yield

$$\mathrm{St}\,\mathrm{Pr}^{\frac{2}{3}} = \frac{f}{8} \tag{5-72a}$$

$$\mathrm{Nu}_d = 0.0395\,\mathrm{Re}_d^{\frac{3}{4}}\,\mathrm{Pr}^{\frac{1}{3}} \tag{5-74a}$$

As we shall see in Chap. 6, Eq. (5-74a) predicts heat-transfer coefficients that are somewhat higher than those observed in experiments. The purpose of the discussion at this point has been to show that one may arrive at a relation for turbulent heat transfer in a fairly simple analytical fashion. As we have indicated earlier, a rigorous development of the Reynolds analogy between heat transfer and fluid friction involves considerations beyond the scope of our discussion and the simple path of reasoning chosen here is offered for the purpose of indicating the general nature of the physical processes.

For calculation purposes, the correct relation to use for turbulent flow in a smooth tube is Eq. (6-1), which we list here for comparison.

$$\mathrm{Nu}_d = 0.023 \, \mathrm{Re}^{0.8} \, \mathrm{Pr}^{0.4} \tag{6-1}$$

All properties in Eq. (6-1) are evaluated at the bulk temperature.

## 5-11□HEAT TRANSFER IN HIGH - SPEED FLOW

Our previous analysis of boundary-layer heat transfer (Sec. 5-6) neglected the effects of viscous dissipation within the boundary layer. When the free-stream velocity is very high, as in high-speed aircraft, these dissipation effects must be considered. We begin our analysis by considering the adiabatic case, i.e., a perfectly insulated wall. In this case the wall temperature may be considerably higher than the free-stream temperature even though no heat transfer takes place. This high temperature results from two situations: (1) the increase in temperature of the fluid as it is brought to rest at the plate surface while the kinetic energy of the flow is converted to internal thermal energy, and (2) the heating effect due to viscous dissipation. Consider the first of these situations. The kinetic energy of the gas is converted to thermal energy as the gas is brought to rest, and this process is described by the steady-flow energy equation for an adiabatic process.

$$i_0 = i_\infty + \frac{1}{2g_c} u_\infty^2 \tag{5-75}$$

where $i_0$ is the stagnation enthalpy of the gas. This equation may be written in terms of temperature as

$$c_p(T_0 - T_\infty) = \frac{1}{2g_c} u_\infty^2$$

where $T_0$ is the stagnation temperature and $T_\infty$ is the static free-stream temperature. Expressed in terms of the free-stream Mach number,

$$\frac{T_0}{T_\infty} = 1 + \frac{\gamma - 1}{2} M_\infty^2 \tag{5-76}$$

where $M_\infty$ is the Mach number, defined as $M_\infty = u_\infty/a$, and $a$ is the acoustic velocity, which, for a perfect gas, may be calculated with

$$a = \sqrt{\gamma g_c R T} \tag{5-77}$$

where $R$ is the gas constant.

In the actual case of a boundary-layer flow problem, the fluid is not brought to rest reversibly since the viscous action is basically an irreversible process in a thermodynamic sense. In addition, not all the free-stream kinetic energy is converted to thermal energy—part is lost as heat, and part is dissipated in the form of viscous work. To take into account the irreversibilities in the boundary-layer flow system, a *recovery factor* is defined by

$$r = \frac{T_{aw} - T_\infty}{T_0 - T_\infty} \tag{5-78}$$

where $T_{aw}$ is the actual adiabatic wall temperature and $T_\infty$ is the static temperature of the free stream. The recovery factor may be determined experimentally, or for some flow systems, analytical calculations may be made.

The boundary-layer energy equation

$$u \frac{\partial T}{\partial x} + v \frac{\partial T}{\partial y} = \alpha \frac{\partial^2 T}{\partial y^2} + \frac{\mu}{\rho c_p} \left( \frac{\partial u}{\partial y} \right)^2 \tag{5-20}$$

has been solved for the high-speed-flow situation, taking into account the viscous-heating term. Although the complete solution is somewhat tedious, the final results are remarkably simple. For our purposes we shall present only the results, and indicate how they may be applied. The reader is referred to App. B for an exact solution to Eq. (5-20). An excellent synopsis of the high-speed heat-transfer problem is given in a report by Eckert [4]. Some typical boundary-layer temperature profiles for an adiabatic wall in high-speed flow are given in Fig. B-3.

The essential result of the high-speed heat-transfer analysis is that heat-transfer rates may generally be calculated with the same relations used for low-speed incompressible flow when the average heat-transfer coefficient is redefined with the relation

$$q = \bar{h} A (T_w - T_{aw}) \tag{5-79}$$

Notice that the difference between the adiabatic wall temperature and the actual wall temperature is used in the definition so that the expression will yield a value of zero heat flow when the wall is at the adiabatic wall temperature. For gases with Prandtl number near unity the following relations for the recovery factor have been derived:

Laminar flow: $\qquad\qquad\qquad\qquad r = \mathrm{Pr}^{\frac{1}{2}}$ $\qquad\qquad\qquad\qquad$ (5-80)
Turbulent flow: $\qquad\qquad\qquad\quad r = \mathrm{Pr}^{\frac{1}{3}}$ $\qquad\qquad\qquad\qquad$ (5-81)

These recovery factors may be used in conjunction with Eq. (5-78) to obtain the adiabatic wall temperature.

In high-velocity boundary layers substantial temperature gradients may occur, and there will be correspondingly large property variations across the boundary layer. The constant-property heat-transfer equations may still be used if the properties are introduced at a reference temperature $T^*$ as recommended by Eckert.

$$T^* = T_\infty + 0.50(T_w - T_\infty) + 0.22(T_{aw} - T_\infty) \qquad (5\text{-}82)$$

The analogy between heat transfer and fluid friction [Eq. (5-48)] may also be used when the friction coefficient is known. Summarizing the relations used for high-speed heat-transfer calculations:

Laminar boundary layer ($\mathrm{Re}_x < 5 \times 10^5$):

$$\mathrm{St}_x^* \; \mathrm{Pr}^{*\frac{2}{3}} = 0.332 \; \mathrm{Re}_x^{*-\frac{1}{2}} \qquad (5\text{-}83)$$

Turbulent boundary layer ($5 \times 10^5 < \mathrm{Re}_x < 10^7$):

$$\mathrm{St}_x^* \; \mathrm{Pr}^{*\frac{2}{3}} = 0.0288 \; \mathrm{Re}_x^{*-\frac{1}{5}} \qquad (5\text{-}84)$$

For turbulent boundary layers with Reynolds numbers between $10^7$ and $10^9$ Eckert recommends the formula

$$C_{fx} = \frac{0.370}{(\log \mathrm{Re}_x)^{2.584}} \qquad (5\text{-}85)$$

so that $\qquad \mathrm{St}_x^* \; \mathrm{Pr}^{*\frac{2}{3}} = \dfrac{0.185}{(\log \mathrm{Re}_x^*)^{2.584}} \qquad$ for $10^7 < \mathrm{Re}_x < 10^9 \qquad (5\text{-}86)$

The superscript * in the above equations indicates that the properties are evaluated at the reference temperature given by Eq. (5-82).

To obtain an average heat-transfer coefficient, the above expressions must be integrated over the length of the plate. If the Reynolds number falls in a range such that Eq. (5-86) must be used, the integration cannot be expressed in closed form, and a numerical integration must be used. Care must be taken in performing the integration for the high-speed heat-transfer problem since the reference temperature is different for the laminar and turbulent portions of the boundary layer. This results from the different value of the recovery factor used for laminar and turbulent flow as given by Eqs. (5-80) and (5-81).

## Example 5-7

A flat plate 2 ft long and 1 ft wide is placed in a wind tunnel where the flow conditions are $M = 3$, $p = \frac{1}{20}$ atm, and $T_\infty = -40°\mathrm{F}$. How much cooling must be used to maintain the plate temperature at $100°\mathrm{F}$?

**Solution**  We must consider the laminar and turbulent portions of the boundary layer separately since the recovery factors, and hence the adiabatic wall temperatures, used to establish the heat flow will be different for the different flow regimes. It turns out that the difference is rather small in this particular problem, but we shall follow a procedure which would be used if the difference were appreciable, so that the general method of solution may be indicated.

The free-stream acoustic velocity is calculated from

$$a = \sqrt{\gamma g_c R T} = 1006 \text{ ft/sec} \qquad (307 \text{ m/s})$$

so that the free-stream velocity is

$$u_\infty = (3)(1006) = 3018 \text{ ft/sec}$$

The maximum Reynolds number is estimated by making a computation based on properties evaluated at free-stream conditions.

$$\rho_\infty = \frac{(14.7)(144)}{(20)(53.35)(420)} = 0.00472 \text{ lb}_m/\text{ft}^3 \qquad (0.0756 \text{ kg/m}^3)$$

$$\mu_\infty = 0.0366 \text{ lb}_m/\text{hr-ft}$$

$$\text{Re}_{L\infty} = \frac{(0.00472)(3018)(3600)(2)}{0.0366} = 2.8 \times 10^6$$

Thus we conclude that both laminar and turbulent boundary-layer heat transfer must be considered. The laminar heat transfer is calculated with Eq. (5-83), and the turbulent heat transfer with Eq. (5-84). Before applying these equations we must determine the reference temperatures for the laminar and turbulent conditions, and then evaluate the properties at these temperatures.

*Laminar portion*

$$T_0 = T_\infty\left(1 + \frac{\gamma - 1}{2} M_\infty^2\right) = 1175°\text{R}$$

$$r = \text{Pr}^{\frac{1}{2}}$$

Assume a Prandtl number of 0.71 so that

$$r = (0.71)^{\frac{1}{2}} = 0.843$$

$$= \frac{T_{aw} - T_\infty}{T_0 - T_\infty} = \frac{T_{aw} - 420}{1175 - 420}$$

$$T_{aw} = 1056°\text{R}$$

$$T^* = T_\infty + 0.50(T_w - T_\infty) + 0.22(T_{aw} - T_\infty)$$

$$= 630°\text{R} = 170°\text{F}$$

Checking the Prandtl number,

$$\text{Pr}^* = 0.698$$

so that the calculation is valid. If there were an appreciable difference between the value of Pr* and the value used to determine the recovery factor, the calculation would have to be repeated until agreement was reached.

The other properties to be used in the laminar heat-transfer analysis are

$$\rho^* = \frac{(14.7)(144)}{(20)(53.35)(630)} = 0.00315 \text{ lb}_m/\text{ft}^3$$

$$\mu^* = 0.0502 \text{ lb}_m/\text{hr-ft}$$

$$k^* = 0.0171 \text{ Btu/hr-ft-°F}$$

$$c_p^* = 0.24 \text{ Btu/lb}_m\text{-°F}$$

*Turbulent portion.* Assume

$$Pr = 0.71$$
$$r = Pr^{\frac{1}{3}} = (0.71)^{\frac{1}{3}} = 0.892$$
$$= \frac{T_{aw} - T_\infty}{T_0 - T_\infty} = \frac{T_{aw} - 420}{1175 - 420}$$
$$T_{aw} = 1093°R$$
$$T^* = T_\infty + 0.50(T_w - T_\infty) + 0.22(T_{aw} - T_\infty)$$
$$= 420 + 0.50(560 - 420) + 0.22(1093 - 420)$$
$$= 638°R = 178°F$$
$$Pr^* = 0.695$$

The agreement between $Pr^*$ and the assumed value is sufficiently close.

The other properties to be used in the turbulent-heat-transfer analysis are

$$\rho^* = \frac{(14.7)(144)}{(20)(53.35)(638)} = 0.00311 \ lb_m/ft^3$$
$$\mu^* = 0.0507 \ lb_m/hr\text{-}ft$$
$$k^* = 0.0173 \ Btu/hr\text{-}ft\text{-}°F$$
$$c_p^* = 0.24 \ Btu/lb_m\text{-}°F$$

*Laminar heat transfer.* Assume

$$Re_{crit}^* = 5 \times 10^5 = \frac{\rho^* u_\infty x_c}{\mu^*}$$
$$x_c = \frac{(5 \times 10^5)(0.0502)}{(0.00315)(3018)(3600)} = 0.733 \ ft$$
$$\overline{Nu^*} = \frac{\bar{h} x_c}{k^*} = 0.664 \ Re_{crit}^{*\frac{1}{2}} \ Pr^{*\frac{1}{3}}$$
$$= 0.664(5 \times 10^5)^{\frac{1}{2}}(0.71)^{\frac{1}{3}} = 418$$
$$\bar{h} = \frac{(418)(0.0171)}{0.733} = 9.76 \ Btu/hr\text{-}ft^2°F \qquad (55.4 \ W/m^2\text{-}°C)$$

This is the average heat-transfer coefficient for the laminar portion of the boundary layer, and the heat transfer is calculated from

$$q = \bar{h}A(T_w - T_{aw})$$
$$= (9.76)(0.733)(1)(560 - 1056)$$
$$= -3550 \ Btu/hr \qquad (-1040 \ W)$$

so that 3550 Btu/hr of cooling is required in the laminar region of the plate.

*Turbulent heat transfer.* To determine the turbulent heat transfer we must determine an expression for the local heat-transfer coefficient from

$$St_x^* \ Pr^{*\frac{2}{3}} = 0.0288 \ Re_x^{*-\frac{1}{5}}$$

and then integrate from $x = 0.733$ ft to $x = 2$ ft to determine the total heat transfer.

$$h_x = Pr^{*-\frac{2}{3}} \rho^* u_\infty c_p^* 0.0288 \left( \frac{\rho^* u_\infty x}{\mu^*} \right)^{-\frac{1}{5}}$$
$$= 20.1 x^{-\frac{1}{5}}$$

The average heat-transfer coefficient may then be obtained from

$$\bar{h} = \frac{\int_{0.733}^{2} h_x \, dx}{\int_{0.733}^{2} dx} = 19.05 \text{ Btu/hr-ft}^2\text{-}°\text{F} \qquad (108.1 \text{ W/m}^2\text{-}°\text{C})$$

Using this value, we may calculate the heat transfer in the turbulent boundary-layer region of the flat plate.

$$
\begin{aligned}
q &= \bar{h}A(T_w - T_{aw}) \\
&= (19.05)(2 - 0.733)(1)(560 - 1093) \\
&= -13{,}100 \text{ Btu/hr} \qquad (-3838 \text{ W})
\end{aligned}
$$

The total amount of cooling required is the sum of heat transfers for the laminar and turbulent portions of the boundary layer.

$$\text{Total cooling} = 3550 + 13{,}100 = 16{,}650 \text{ Btu/hr} \qquad (-4878 \text{ W})$$

When very high flow velocities are encountered, the adiabatic wall temperature may become so high that dissociation of the gas will take place and there will be a very wide variation of the properties in the boundary layer. Eckert [4] recommends that these problems be treated on the basis of a heat-transfer coefficient defined in terms of *enthalpy* difference.

$$q = h_i A (i_w - i_{aw}) \qquad (5\text{-}87)$$

The enthalpy recovery factor is then defined as

$$r_i = \frac{i_{aw} - i_\infty}{i_0 - i_\infty} \qquad (5\text{-}88)$$

where $i_{aw}$ is the enthalpy at the adiabatic wall conditions. The same relations as before are used to calculate the recovery factor and heat transfer except that all properties are evaluated at a reference enthalpy $i^*$ given by

$$i^* = i_\infty + 0.5(i_w - i_\infty) + 0.22(i_{aw} - i_\infty) \qquad (5\text{-}89)$$

The Stanton number is redefined as

$$\text{St}_i = \frac{h_i}{\rho u_\infty} \qquad (5\text{-}90)$$

This Stanton number is then used in Eq. (5-83), (5-84), or (5-86) to calculate the heat-transfer coefficient. When calculating the enthalpies for use in the above relations, the *total* enthalpy must be used; i.e., chemical energy of dissociation as well as internal thermal energy must be included. The reference enthalpy method has proved successful for calculating high-speed heat transfer with an accuracy of better than 4 percent.

## REVIEW QUESTIONS

1. What is meant by a hydrodynamic boundary layer?
2. Define the Reynolds number. Why is it important?
3. What is the physical mechanism of viscous action?
4. Distinguish between laminar and turbulent flow in a physical sense.
5. What is the momentum equation for the laminar boundary layer on a flat plate? What assumptions are involved in the derivation of this equation?
6. How is the boundary-layer thickness defined?
7. What is the energy equation for the laminar boundary layer on a flat plate? What assumptions are involved in the derivation of this equation?
8. What is meant by a thermal boundary layer?
9. Define the Prandtl number. Why is it important?
10. Describe the physical mechanism of convection. How is the convection heat-transfer coefficient related to this mechanism?
11. Describe the relation between fluid friction and heat transfer.
12. Define the bulk temperature. How is it used?
13. How is the heat-transfer coefficient defined for high-speed heat-transfer calculations?

## PROBLEMS

5-1. A certain nozzle is designed to expand air from stagnation conditions of 200 psia and 400°F to 20 psia. The mass rate of flow is designed to be 600 $lb_m$/min. Suppose this nozzle is used in conjunction with a blowdown wind-tunnel facility so that the nozzle is suddenly allowed to discharge into a perfectly evacuated tank. What will the temperature of the air in the tank be when the pressure in the tank equals 20 psia? Assume that the tank is perfectly insulated and that air behaves as a perfect gas. Assume that the expansion in the nozzle is isentropic.

5-2. For water flowing over a flat plate at 60°F and 10 ft/sec, calculate the mass flow through the boundary layer at a distance of 2 in. from the leading edge of the plate.

5-3. Air at 100°F and 14.7 psia flows over a flat plate at a velocity of 100 ft/sec. How thick is the boundary layer at a distance of 1 in. from the leading edge of the plate?

5-4. Using a linear velocity profile

$$\frac{u}{u_\infty} = \frac{y}{\delta}$$

for flow over a flat plate, obtain an expression for the boundary-layer thickness as a function of $x$.

5-5.   Using the continuity relation

$$\frac{\partial u}{\partial x} + \frac{\partial v}{\partial y} = 0$$

along with the velocity distribution

$$\frac{u}{u_\infty} = \frac{3}{2}\frac{y}{\delta} - \frac{1}{2}\left(\frac{y}{\delta}\right)^3$$

and the expression for the boundary-layer thickness

$$\frac{\delta}{x} = \frac{4.64}{\sqrt{Re_x}}$$

derive an expression for the $y$ component of velocity $v$ as a function of $x$ and $y$. Calculate the value of $v$ at the outer edge of the boundary layer at distances of 6 and 12 in. from the leading edge for the conditions of Example 5-3.

5-6.   Repeat Prob. 5-5 for the linear velocity profile of Prob. 5-4.

5-7.   Air flows over a flat plate at a constant velocity of 100 ft/sec and ambient conditions of 3 psia and 70°F. The plate is heated to a constant temperature of 150°F, starting at a distance of 3 in. from the leading edge. What is the total heat transfer from the leading edge to a point 1 ft from the leading edge?

5-8.   Using the linear velocity profile in Prob. 5-4 and a cubic-parabola temperature distribution [Eq. (5-26)], obtain an expression for heat-transfer coefficient as a function of the Reynolds number for a laminar boundary layer on a flat plate.

5-9.   Air at 3 psia and 40°F enters a 1-in.-diameter tube at a velocity of 5 ft/sec. Using a flat-plate analysis, estimate the distance from the entrance that the flow becomes fully developed.

5-10.   A fluid flows between two large parallel plates. Develop an expression for the velocity distribution as a function of distance from the centerline between the two plates under developed flow conditions.

5-11.   Water at 60°F flows between two large parallel plates at a velocity of 5 ft/sec. The plates are separated by a distance of 0.6 in. Estimate the distance from the leading edge where the flow becomes fully developed.

5-12.   Air at standard conditions of 14.7 psia and 70°F flows over a flat plate at 100 ft/sec. The plate is 2 ft square and is maintained at 200°F. Calculate the heat transfer from the plate.

5-13.   Air at 70°F and 2 psia flows at a velocity of 500 ft/sec past a flat plate 3 ft long which is maintained at a constant temperature of 300°F. What is the average heat-transfer rate per unit area of plate?

5-14.   Air at 1 psia and 100°F flows across a 1-ft-square flat plate at 25 ft/sec. The plate is maintained at 150°F. Estimate the heat lost from the plate.

5-15.   Air at 200°F and atmospheric pressure flows over a horizontal flat plate at 200 ft/sec. The plate is 2 ft square and is maintained at a uniform temperature of 50°F. What is the total heat transfer?

5-16.   Plot the heat-transfer coefficient versus length for flow over a 3-ft-long flat plate under the following conditions:

(a)  Helium at 1 psia, 80°F, $u_\infty = 10$ ft/sec
(b)  Hydrogen at 1 psia, 80°F, $u_\infty = 10$ ft/sec
(c)  Air at 1 psia, 80°F, $u_\infty = 10$ ft/sec
(d)  Water at 80°F, $u_\infty = 10$ ft/sec
(e)  Helium at 20 psia, 80°F, $u_\infty = 10$ ft/sec

5-17.   Compute the drag force exerted on the plate by each of the systems in Prob. 5-16.

5-18.   Glycerine at 100°F flows past a 1-ft-square flat plate at a velocity of 5 ft/sec. The drag force is measured as 2.0 $lb_f$ (both sides of the plate). Calculate the heat-transfer coefficient for such a flow system.

5-19.   Using the velocity distribution for developed laminar flow in a tube, derive an expression for the friction factor as defined by Eq. (5-70).

5-20.   "Slug" flow in a tube may be described as that flow in which the velocity is constant across the entire flow area of the tube. Obtain an expression for the heat-transfer coefficient in this type of flow with a constant heat-flux condition maintained at the wall. Compare the results with those of Sec. 5-9. Explain the reason for the difference in answers on a physical basis.

5-21.   Assume that the velocity distribution in the turbulent core for tube flow may be represented by

$$\frac{u}{u_c} = \left(1 - \frac{r}{r_o}\right)^{\frac{1}{7}}$$

where $u_c$ is the velocity at the center of the tube and $r_o$ is the tube radius. The velocity in the laminar sublayer may be assumed to vary linearly with the radius. Using the friction factor given by Eq. (5-73), derive an equation for the thickness of the laminar sublayer. For this problem the average flow velocity may be calculated using only the turbulent velocity distribution.

5-22.   Using the velocity profile in Prob. 5-21, obtain an expression for the eddy diffusivity of momentum as a function of radius.

5-23.   In heat-exchanger applications it is frequently important to match heat-transfer requirements with pressure-drop limitations. Assuming a fixed total heat-transfer requirement and a fixed temperature difference between wall and bulk conditions as well as a fixed pressure drop through the tube, derive expressions for the length and diameter of the tube, assuming turbulent flow of a gas with the Prandtl number near unity.

5-24.   Air at 120°F and 10 psia flows through a 2-in.-diameter pipe at a velocity of 12 ft/sec. Estimate the heat-transfer coefficient for such a flow system. Assume fully developed flow and use Eq. (6-1).

5-25. Air at Mach 4 and 3 psia, 0°F, flows past a flat plate. The plate is to be maintained at a constant temperature of 200°F. If the plate is 18 in. long, how much cooling will be required to maintain this temperature?

5-26. Air flows over an isothermal flat plate maintained at a constant temperature of 150°F. The velocity of the air is 2000 ft/sec at static properties of 60°F and 1 psia. Calculate the average heat-transfer coefficient for a plate 3 ft long.

5-27. Calculate the heat transfer from a 1-ft-square plate over which air flows at 100°F and 2 psia. The plate temperature is 500°F, and the free-stream velocity is 20 ft/sec.

5-28. Calculate the drag (viscous friction) force on the plate in Prob. 5-27 under the conditions of no heat transfer. *Do not* use the analogy between fluid friction and heat transfer for this calculation; i.e., calculate the drag directly by evaluating the viscous-shear stress at the wall.

5-29. Water flows in a 1-in.-diameter pipe so that the Reynolds number based on diameter is 1500 (laminar flow is assumed). The average water bulk temperature is 100°F. Calculate the maximum water velocity in the tube. (Recall that $u_m = 0.5u_0$.) What would the heat-transfer coefficient be for such a system when the tube wall was subjected to a constant heat flux and the velocity and temperature profiles were completely developed? Evaluate properties at bulk temperature.

5-30. A slug flow is encountered in an annular flow system which is subjected to a constant heat flux at both the inner and outer surfaces. The temperature is the same at both inner and outer surfaces at identical $x$ locations. Derive an expression for the temperature distribution in such a flow system, assuming constant properties and laminar flow.

5-31. Using the energy equation given by Eq. (5-28), determine an expression for heat-transfer coefficient under the conditions

$$u = u_\infty = \text{const}$$
$$\frac{T - T_\infty}{T_w - T_\infty} = \frac{y}{\delta_t}$$

where $\delta_t$ is the thermal boundary-layer thickness.

5-32. Air at 1 psia and −40°F flows over a flat plate at Mach 4. The plate temperature is 100°F, and the plate length is 2 ft. Calculate the adiabatic wall temperature for the laminar portion of the boundary layer.

5-33. Derive an expression for the heat transfer in a laminar boundary layer on a flat plate under the condition $u = u_\infty = \text{const}$. Assume that the temperature distribution is given by the cubic-parabola relation in Eq. (5-26). This solution approximates the condition observed in the flow of a liquid metal over a flat plate.

5-34. Air at 3 psia and 70°F flows across a flat plate 2 ft long. The free-stream velocity is 100 ft/sec, and the plate is heated over its total length to a temperature of 130°F. For $x = 1$ ft calculate the value of $y$ for which $u$ will equal 75 ft/sec.

5-35. For the flow system in Prob. 5-34 calculate the value of the friction coefficient at a distance of 6 in. from the leading edge.

5-36. Air at 40°F and 10 psia flows over a flat plate at 20 ft/sec. A heater strip having a length of 1 in. is placed on the plate at a distance of 6 in. from the leading edge. Calculate the heat lost from the strip per unit depth of plate for a heater surface temperature of 150°F.

5-37. Show that $\partial^3 u / \partial y^3 = 0$ at $y = 0$ for an incompressible laminar boundary layer on a flat plate with zero pressure gradient.

5-38. Review the analytical developments of this chapter and list the restrictions which apply to the following equations: (5-21), (5-22), (5-40), (5-42), (5-53), and (5-66).

# REFERENCES

1. Schlichting, H.: "Boundary Layer Theory," 4th ed., McGraw-Hill Book Company, New York, 1960.
2. von Kármán, T.: Uber laminaire und turbulente Reibung, *Angew. Math. Mech.*, vol. 1, pp. 233–252, 1921; also *NACA Tech. Mem.* 1092, 1946.
3. Sellars, J. R., M. Tribus, and J. S. Klein: Heat Transfer to Laminar Flow in a Round Tube or Flat Conduit: The Graetz Problem Extended, *Trans. ASME*, vol. 78, p. 441, 1956.
4. Eckert, E. R. G.: Survey of Boundary Layer Heat Transfer at High Velocities and High Temperatures, *WADC Tech. Rept.* 59-624, April, 1960.
5. Liepmann, H. W., and A. Roshko: "Elements of Gasdynamics," John Wiley & Sons, Inc., New York, 1957.
6. Knudsen, J. D., and D. L. Katz: "Fluid Dynamics and Heat Transfer," McGraw-Hill Book Company, New York, 1958.

# CHAPTER

# 6

# EMPIRICAL AND PRACTICAL RELATIONS FOR FORCED-CONVECTION HEAT TRANSFER

## 6-1□INTRODUCTION

The discussion and analyses of Chap. 5 have shown how forced-convection heat transfer may be calculated for several cases of practical interest; the problems considered, however, were those which could be solved in an analytical fashion. In this way, the principles of the convection process and their relation to fluid dynamics were demonstrated, primary emphasis being devoted to a clear understanding of physical mechanism. Regrettably, it is not always possible to obtain analytical solutions to convection problems, and the individual is forced to resort to experimental methods to obtain design information, as well as to secure the more elusive data which increase his physical understanding of the heat-transfer processes.

Results of experimental data are usually expressed in the form of either empirical formulas or graphical charts so that they may be utilized with a maximum of generality. It is in the process of trying to generalize the results of his experiments, in the form of some empirical correlation, that one encounters difficulty. If an analytical solution is available for a similar problem, the correlation of data is much easier, since one may guess at the functional form of the results, and hence use the experimental data to obtain values of constants or exponents on certain significant parameters such as the Reynolds or Prandtl numbers. If an analytical solution for a similar problem is not available, the individual must resort to intuition based on his physical understanding of the problem, or shrewd inferences which he may be able to draw from the differential equations of the flow processes based upon dimensional or order-of-magnitude estimates. In any event, there is no substitute for physical insight and understanding.

To show how one might proceed to analyze a new problem to obtain an important functional relationship from the differential equations, consider the problem of determining the hydrodynamic-boundary-layer thickness for flow over a flat plate. This problem was solved in the preceding chapter, but we now wish to make an order-of-magnitude analysis of the differential equations to obtain the functional form of the solution. The momentum equation

$$u\frac{\partial u}{\partial x} + v\frac{\partial u}{\partial y} = \nu\frac{\partial^2 u}{\partial y^2}$$

must be solved in conjunction with the continuity equation

$$\frac{\partial u}{\partial x} + \frac{\partial v}{\partial y} = 0$$

Within the boundary layer we may say that the velocity $u$ is of the order of the free-stream velocity $u_\infty$. Similarly, the $y$ dimension is of the order of the boundary-layer thickness $\delta$. Thus

$$u \sim u_\infty$$
$$y \sim \delta$$

and we might write the continuity equation in an approximate form as

$$\frac{\partial u}{\partial x} + \frac{\partial v}{\partial y} = 0$$
$$\frac{u_\infty}{x} + \frac{v}{\delta} \approx 0$$

or

$$v \sim \frac{u_\infty \delta}{x}$$

Then, using this order of magnitude for $v$, the analysis of the momentum equation would yield

$$u\frac{\partial u}{\partial x} + v\frac{\partial u}{\partial y} = \nu\frac{\partial^2 u}{\partial y^2}$$
$$u_\infty \frac{u_\infty}{x} + \frac{u_\infty \delta}{x}\frac{u_\infty}{\delta} \approx \nu\frac{u_\infty}{\delta^2}$$

or

$$\delta^2 \sim \frac{\nu x}{u_\infty}$$
$$\delta \sim \sqrt{\frac{\nu x}{u_\infty}}$$

Dividing by $x$ to express the result in dimensionless form,

$$\frac{\delta}{x} \sim \sqrt{\frac{\nu}{u_\infty x}} = \frac{1}{\sqrt{\mathrm{Re}_x}}$$

This functional variation of the boundary-layer thickness with the Reynolds number and $x$ position is precisely that which was obtained in Sec. 5-4. Although this analysis is rather straightforward and does indeed yield correct results, the order-of-magnitude analysis may not always be so fortuitous when applied to more complex problems, particularly those involving turbulent- or separated-flow regions. Nevertheless, one may often obtain valuable information and physical insight by examining the order of magnitude of various terms in a governing differential equation for the particular problem at hand.

A conventional technique used in correlation of experimental data is that of dimensional analysis in which appropriate dimensionless groups such as the Reynolds and Prandtl numbers are derived from purely dimensional and functional considerations. There is, of course, the assumption of flow-field and temperature-profile similarity for geometrically similar heating surfaces. Generally speaking, the application of dimensional analysis to any new problem is extremely difficult when a previous analytical solution of some sort is not available. It is usually best to attempt an order-of-magnitude analysis such as the one above if the governing differential equations are known. In this way it may be possible to determine the significant dimensionless variables for correlating experimental data. In some complex flow and heat-transfer problems a clear physical model of the processes may not be available, and the engineer must first try to establish this model before he can correlate his experimental data.

Schlichting [6], Giedt [7], and Kline [28] discuss similarity considerations and their use in boundary-layer and heat-transfer problems.

The purpose of the foregoing discussion has not been to emphasize or even to imply any new method for solving problems, but rather to indicate the necessity of applying intuitive physical reasoning to a difficult problem and to point out the obvious advantage of using any and all information which may be available. When the problem of correlation of experimental data for a previously unsolved situation is encountered, one must frequently adopt devious methods to accomplish the task.

## 6-2□EMPIRICAL RELATIONS FOR PIPE AND TUBE FLOW

The analysis of Sec. 5-9 has shown how one might analytically attack the problem of heat transfer in fully developed laminar tube flow. The cases of undeveloped laminar flow, flow systems where the fluid properties vary widely with temperature, and turbulent-flow systems are considerably more complicated, but are of very important practical interest in the design of heat exchangers and associated heat-transfer equipment. These more complicated problems may sometimes be solved analytically, but the solutions, when possible, are very tedious. For design and engineering purposes empirical correlations are usually of greatest practical utility. In this section we

sent some of the more important and useful empirical relations and point
their limitations.

For fully developed turbulent flow in smooth tubes the following relation is recommended by Dittus and Boelter [1]:

$$\text{Nu}_d = 0.023 \ \text{Re}_d^{0.8} \ \text{Pr}^n \qquad (6\text{-}1)$$

The properties in this equation are evaluated at the fluid bulk temperature,
and the exponent $n$ has the following values:

$$n = \begin{cases} 0.4 & \text{for heating} \\ 0.3 & \text{for cooling} \end{cases}$$

One may ask the reason for the functional form of Eq. (6-1). Physical
reasoning, based on the experience gained with the analyses of Chap. 5,
would certainly indicate a dependence of the heat-transfer process on the
flow field, and hence on the Reynolds number. The relative rates of diffusion
of heat and momentum are related by the Prandtl number, so that the
Prandtl number is expected to be a significant parameter in the final solution. We can be rather confident of dependence of the heat transfer on
the Reynolds and Prandtl numbers. But the question arises as to the correct
functional form of the relation; i.e., would one necessarily expect a product
of two exponential functions of the Reynolds and Prandtl numbers? The
answer is that one might expect this functional form since it appears in the
flat-plate analytical solutions of Chap. 5, as well as the Reynolds analogy
for turbulent flow. In addition, this type of functional relation is convenient
to use when correlating experimental data, as described below.

Suppose a number of experiments are conducted taking measurements
of heat-transfer rates of various fluids in turbulent flow inside smooth tubes
under different temperature conditions. Different-diameter tubes may be
used to vary the range of the Reynolds number in addition to variations in
the mass-flow rate. We wish to generalize the results of these experiments
by arriving at one empirical equation which represents all the data. As
described above, we may anticipate that the heat-transfer data will be
dependent on the Reynolds and Prandtl numbers. An exponential function
for each of these parameters is perhaps the simplest type of relation to use,
so we assume

$$\text{Nu}_d = C \ \text{Re}_d^m \ \text{Pr}^n$$

where $C$, $m$, and $n$ are constants to be determined from the experimental
data.

A log-log plot of $\text{Nu}_d$ versus $\text{Re}_d$ is first made for one fluid to estimate
the dependence of the heat transfer on the Reynolds number, i.e., to find
an approximate value of the exponent $m$. This plot is made for one fluid
at a constant temperature, so that the influence of the Prandtl number will

**Fig. 6-1   Typical data correlation for forced convection in smooth tubes, turbulent flow.**

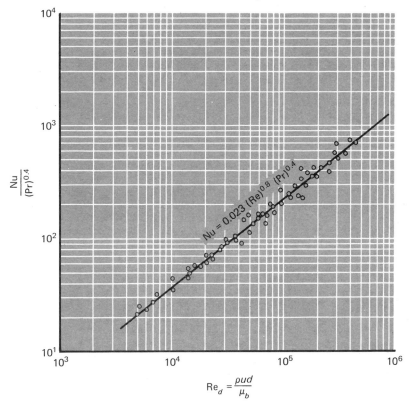

be small, since the Prandtl number will be approximately constant for the one fluid. Using this first estimate for the exponent $m$, the data for all fluids are plotted as $\log \mathrm{Nu}_d/\mathrm{Re}_d{}^m$ versus $\log \mathrm{Pr}$, and a value for the exponent $n$ is determined. Then, using this value of $n$, all the data are plotted once again as $\log \mathrm{Nu}_d/\mathrm{Pr}^n$ versus $\log \mathrm{Re}_d$, and a final value of the exponent $m$ is determined as well as a value for the constant $C$. An example of this final type of data plot is shown in Fig. 6-1. The final correlation equation usually represents the data within $\pm 25$ percent.

Equation (6-1) is valid for fully developed turbulent flow in smooth tubes for fluids with Prandtl numbers ranging from about 0.6 to 100 and with moderate temperature differences between wall and fluid conditions.

If wide temperature differences are present in the flow, there may be an appreciable change in the fluid properties between the wall of the tube and the central flow. These property variations may be evidenced by a change in the velocity profile as indicated in Fig. 6-2. The deviations from

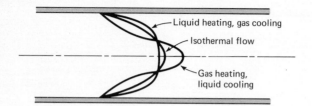

Liquid heating, gas cooling

Isothermal flow

Gas heating,
liquid cooling

Fig. 6-2  Influence of heating on velocity profile in laminar tube flow.

the velocity profile for isothermal flow as shown in this figure are a result of the fact that the viscosity of gases increases with an increase in temperature, while the viscosities of liquids decrease with an increase in temperature.

To take into account the property variations, Sieder and Tate [2] recommend the following relation:

$$\text{Nu}_d = 0.027 \text{ Re}_d{}^{0.8} \text{ Pr}^{\frac{1}{3}} \left(\frac{\mu}{\mu_w}\right)^{0.14} \tag{6-2}$$

All properties are evaluated at bulk-temperature conditions, except $\mu_w$, which is evaluated at the wall temperature.

Equations (6-1) and (6-2) apply to fully developed turbulent flow in tubes. In the entrance region the flow is not developed, and Nusselt [3] recommended the following equation:

$$\text{Nu}_d = 0.036 \text{ Re}_d{}^{0.8} \text{ Pr}^{\frac{1}{3}} \left(\frac{d}{L}\right)^{0.055} \qquad \text{for } 10 < \frac{L}{d} < 400 \tag{6-3}$$

where $L$ is the length of the tube and $d$ is the tube diameter. The properties in Eq. (6-3) are evaluated at the mean bulk temperature. Hartnett [24] has given experimental data on the thermal entrance region for water and oils. Definitive studies of turbulent heat transfer with water in smooth tubes and at uniform heat flux have been presented by Allen and Eckert [25].

Hausen [4] presents the following empirical relation for fully developed laminar flow in tubes at constant wall temperature:

$$\text{Nu}_d = 3.66 + \frac{0.0668(d/L) \text{ Re}_d \text{ Pr}}{1 + 0.04[(d/L) \text{ Re}_d \text{ Pr}]^{\frac{2}{3}}} \tag{6-4}$$

The heat-transfer coefficient calculated from this relation is the average value over the entire length of tube. Note that the Nusselt number approaches a constant value of 3.66 when the tube is sufficiently long. This situation is similar to that encountered in the constant-heat-flux problem analyzed in Chap. 5 [Eq. (5-66)], except that in this case we have a constant wall temperature instead of a linear variation with length. The temperature profile is fully developed when the Nusselt number approaches a constant value.

A somewhat simpler empirical relation was proposed by Sieder and Tate [2] for laminar heat transfer in tubes.

$$\text{Nu}_d = 1.86 \ (\text{Re}_d \ \text{Pr})^{\frac{1}{3}} \left(\frac{d}{L}\right)^{\frac{1}{3}} \left(\frac{\mu}{\mu_w}\right)^{0.14} \qquad (6\text{-}5)$$

In this formula the average heat-transfer coefficient is based on the arithmetic average of the inlet and outlet temperature differences, and all fluid properties are evaluated at the mean bulk temperature of the fluid, except $\mu_w$, which is evaluated at the wall temperature. Equation (6-5) obviously cannot be used for extremely long tubes since it would yield a zero heat-transfer coefficient. A comparison by Knudsen and Katz (Ref. 9, p. 377) of Eq. (6-5) with other relationships indicates that it is valid for

$$\text{Re}_d \ \text{Pr} \ \frac{d}{L} > 10$$

The product of the Reynolds and Prandtl numbers which occurs in the laminar-flow correlations is called the Peclet number.

$$\text{Pe} = \frac{du \ \rho c_p}{k} = \text{Re}_d \ \text{Pr} \qquad (6\text{-}6)$$

The calculation of laminar heat-transfer coefficients is frequently complicated by the presence of natural-convection effects which are superimposed on the forced-convection effects. The treatment of combined forced- and free-convection problems is discussed in Chap. 7.

The empirical correlations presented above apply to smooth tubes. Correlations are, in general, rather sparse where rough tubes are concerned, and it is recommended that the Reynolds analogy between fluid friction and heat transfer be used to effect a solution under these circumstances. Expressed in terms of the Stanton number,

$$\text{St}_b \ \text{Pr}_f^{\frac{2}{3}} = \frac{f}{8} \qquad (6\text{-}7)$$

The friction coefficient $f$ is defined by

$$\Delta p = f \frac{L}{d} \rho \frac{u_m^2}{2g_c} \qquad (6\text{-}8)$$

where $u_m$ is the mean flow velocity. Values of the friction coefficient for different roughness conditions are shown in Fig. 6-3, according to Moody [5].

Note that the relation in Eq. (6-7) is the same as Eq. (5-72), except that the Stanton number has been multiplied by $\text{Pr}^{\frac{2}{3}}$ to take into account the

**Fig. 6-3  Friction factors for pipes, from Moody [5].**

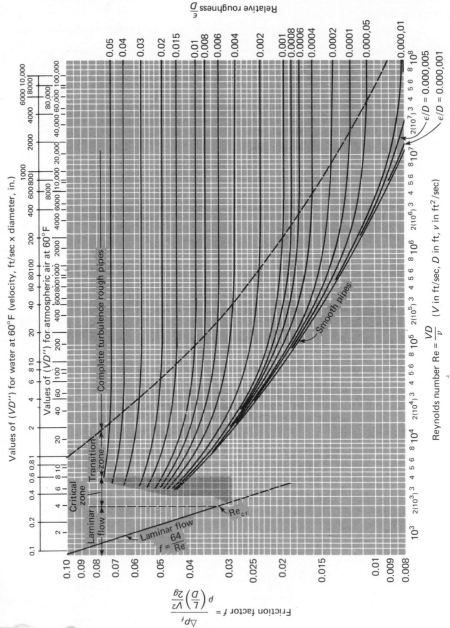

variation of the thermal properties of different fluids. This correction follows the recommendation of Colburn [15], and is based on the reasoning that fluid friction and heat transfer in tube flow are related to the Prandtl number in the same way as they are related in flat-plate flow [Eq. (5-48)]. In Eq. (6-7) the Stanton number is based on bulk temperature, while the Prandtl number and friction factor are based on properties evaluated at the film temperature. Further information on the effects of tube roughness on heat transfer is given in Refs. 27, 29, 30, and 31.

If the channel through which the fluid flows is not of circular cross section, it is recommended that the heat-transfer correlations be based on the hydraulic diameter $D_H$, defined by

$$D_H = \frac{4A}{P} \qquad (6\text{-}9)$$

where $A$ is the cross-sectional area of the flow and $P$ is the wetted perimeter. This particular grouping of terms is used because it yields the value of the physical diameter when applied to a circular cross section. The hydraulic diameter should be used in calculating the Nusselt and Reynolds numbers, and also in establishing the friction coefficient for use with the Reynolds analogy.

Although the hydraulic-diameter concept frequently yields satisfactory relations for fluid friction and heat transfer in many practical problems, there are some notable exceptions where the method does not work. Some of the problems involved in heat transfer in noncircular channels have been summarized by Irvine [20] and Knudsen and Katz [9]. The interested reader should consult these discussions for additional information.

## Example 6-1

Air at 30 psia and 400°F is heated as it flows through a 1-in.-diameter tube at a velocity of 20 ft/sec. Calculate the heat transfer per unit length of tube if a constant-heat-flux condition is maintained at the wall, and the wall temperature is 20°F above the air temperature.

**Solution**  We first calculate the Reynolds number to determine if the flow is laminar or turbulent, and then select the appropriate empirical correlation to calculate the heat transfer. The properties of the air at 400°F are

$$\rho = \frac{p}{RT} = \frac{(30)(144)}{(53.35)(860)} = 0.0941 \text{ lb}_m/\text{ft}^3$$

$$\text{Pr} = 0.69$$

$$\mu = 1.75 \times 10^{-5} \text{ lb}_m/\text{sec-ft}$$

$$k = 0.0212 \text{ Btu/hr-ft-°F}$$

$$\text{Re}_d = \frac{\rho u_m d}{\mu} = \frac{(0.0941)(20)(0.0833)}{1.75 \times 10^{-5}} = 8960$$

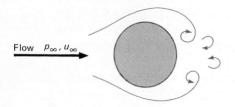

**Fig. 6-4   Cylinder in cross flow.**

Flow   $p_\infty, u_\infty$

so that the flow is turbulent. We therefore use Eq. (6-1) to calculate the heat-transfer coefficient.

$$\frac{hd}{k} = \mathrm{Nu}_d = 0.023\,\mathrm{Re}_d{}^{0.8}\,\mathrm{Pr}^{0.4}$$

$$= (0.023)(8960)^{0.8}(0.69)^{0.4} = 29.2$$

$$h = \frac{k}{d}\,\mathrm{Nu}_d = \frac{(0.0212)(29.2)}{0.0833} = 7.43\ \mathrm{Btu/hr\text{-}ft^2\text{-}°F} \qquad (42.2\ \mathrm{W/m^2\text{-}°C})$$

The heat flow per unit length is then computed from

$$\frac{q}{L} = h\pi d(T_w - T_b)$$

$$= (7.43)\pi(0.0833)(20)$$

$$= 38.9\ \mathrm{Btu/hr\text{-}ft} \qquad (37.39\ \mathrm{W/m})$$

## 6-3□FLOW ACROSS CYLINDERS AND SPHERES

While the engineer may frequently be interested in the heat-transfer characteristics of flow systems inside tubes or over flat plates, equal importance must be placed on the heat transfer which may be achieved by a cylinder in cross flow, as shown in Fig. 6-4. As would be expected, the boundary-layer development on the cylinder determines the heat-transfer characteristics. As long as the boundary layer remains laminar and well behaved, it is possible to compute the heat transfer by a method similar to the boundary-layer analysis of Chap. 5. It is necessary, however, to include the pressure gradient in the analysis since this influences the boundary-layer velocity profile to an appreciable extent. In fact, it is this pressure gradient which causes a separated-flow region to develop on the back side of the cylinder when the free-stream velocity is sufficiently large.

The phenomenon of boundary-layer separation is indicated in Fig. 6-5. The physical reasoning which explains the phenomenon in a qualitative way is as follows:   Consistent with boundary-layer theory, the pressure through the boundary layer is essentially constant at any $x$ position on the body. In the case of the cylinder, one might measure $x$ distance from the front stagnation point of the cylinder. Thus the pressure in the boundary layer should follow that of the free stream for potential flow around a cylinder, provided this behavior would not contradict some basic principle which

**Fig. 6-5** Velocity distributions indicating flow separation on a cylinder in cross
flow.

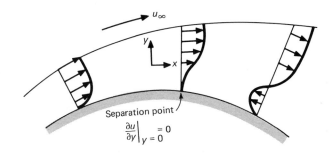

**Fig. 6-5** Velocity distributions indicating flow separation on a cylinder in cross
flow.

must apply in the boundary layer. As the flow progresses along the front side of the cylinder, the pressure would decrease and then increase along the back side of the cylinder, resulting in an increase in free-stream velocity on the front side of the cylinder and a decrease on the back side. The transverse velocity (that velocity parallel to the surface) would decrease from a value of $u_\infty$ at the outer edge of the boundary layer to zero at the surface. As the flow proceeds to the back side of the cylinder, the pressure increase causes a reduction in velocity in the free stream and throughout the boundary layer. The pressure increase and reduction in velocity are related through the Bernoulli equation written along a streamline.

$$\frac{dp}{\rho} = -d\left(\frac{u^2}{2g_c}\right)$$

Since the pressure is assumed constant throughout the boundary layer, we note that reverse flow may begin in the boundary layer near the surface; i.e., the momentum of the fluid layers near the surface is not sufficiently high to overcome the increase in pressure. When the velocity gradient at the surface becomes zero, the flow is said to have reached a separation point.

$$\text{Separation point at } \frac{\partial u}{\partial y}\bigg)_{y=0} = 0$$

This separation point is indicated in Fig. 6-5. As the flow proceeds past the separation point, reverse-flow phenomena may occur, as also shown in Fig. 6-5. Eventually, the separated-flow region on the back side of the cylinder becomes turbulent and random in motion.

The drag coefficient for bluff bodies is defined by

$$\text{Drag force} = F_D = C_D A \frac{\rho u_\infty^2}{2g_c} \tag{6-10}$$

where $C_D$ is the drag coefficient and $A$ is the *frontal area* of the body exposed to the flow, which, for a cylinder, is the product of diameter and length. The values of the drag coefficient for cylinders and spheres are given as a function of the Reynolds number in Figs. 6-6 and 6-7.

**Fig. 6-6  Drag coefficient for circular cylinders as a function of the Reynolds number, from Schlichting [6].**

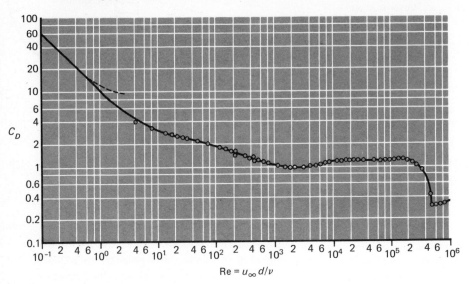

$$\text{Re} = u_\infty d/\nu$$

**Fig. 6-7  Drag coefficient for spheres as a function of the Reynolds number, from Schlichting [6].**

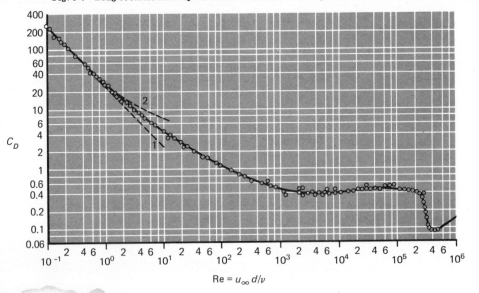

$$\text{Re} = u_\infty d/\nu$$

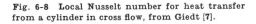

Fig. 6-8 Local Nusselt number for heat transfer from a cylinder in cross flow, from Giedt [7].

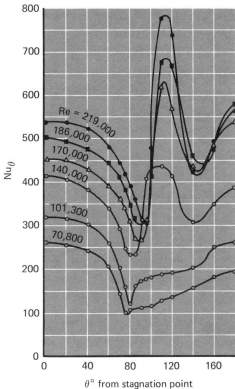

The drag force on the cylinder is a result of a combination of frictional resistance and so-called form, or pressure drag, resulting from a low-pressure region on the rear of the cylinder created by the flow-separation process. At low Reynolds numbers of the order of unity, there is no flow separation, and all the drag results from viscous friction. At Reynolds numbers of the order of 10, the friction and form drag are of the same order, while the form drag resulting from the turbulent separated-flow region predominates at Reynolds numbers greater than 1000. At Reynolds numbers of approximately $10^5$, based on diameter, the boundary-layer flow may become turbulent, resulting in a steeper velocity profile and extremely late flow separation. Consequently, the form drag is reduced, and this is represented by the break in the drag-coefficient curve at about $Re = 3 \times 10^5$. The same reasoning applies to the sphere as to the circular cylinder. Similar behavior is observed with other bluff bodies, such as elliptic cylinders and airfoils.

The flow processes discussed above obviously influence the heat transfer from a heated cylinder to a fluid stream. The detailed behavior of the heat transfer from a heated cylinder to air has been investigated by Giedt [7], and the results are summarized in Fig. 6-8. At the lower Reynolds numbers

**Table 6-1  Constants for Use with Eq. (6-11)†**

| $\mathrm{Re}_{df}$ | $C$ | $n$ |
|---|---|---|
| 0.4–4 | 0.989 | 0.330 |
| 4–40 | 0.911 | 0.385 |
| 40–4,000 | 0.683 | 0.466 |
| 4,000–40,000 | 0.193 | 0.618 |
| 40,000–400,000 | 0.0266 | 0.805 |

† Based on Refs. 8 and 9.

(70,800 and 101,300) a minimum point in the heat-transfer coefficient occurs at approximately the point of separation. There is a subsequent increase in the heat-transfer coefficient on the rear side of the cylinder, resulting from the turbulent eddy motion in the separated flow. At the higher Reynolds numbers two minimum points are observed. The first occurs at the point of transition from laminar to turbulent boundary layer, and the second minimum occurs when the turbulent boundary layer separates. There is a rapid increase in heat transfer when the boundary layer becomes turbulent, and another when the increased eddy motion at separation is encountered.

Because of the complicated nature of the flow separation processes, it is not possible to calculate analytically the average heat-transfer coefficients in crossflow; however, correlations of the experimental data of Hilpert [8] for gases and Knuden and Katz [9] for liquids indicate that the average heat-transfer coefficients may be calculated with

$$\frac{hd}{k_f} = C \left(\frac{u_\infty d}{\nu_f}\right)^n \mathrm{Pr}^{\frac{1}{3}} \tag{6-11}$$

where the constants $C$ and $n$ are tabulated in Table 6-1. The heat-transfer data for air are plotted in Fig. 6-9. Properties for use with Eq. (6-11) are evaluated at the film temperature as indicated by the subscript $f$.

Figure 6-10 shows the temperature field around heated cylinders placed in a transverse airstream. The dark lines are lines of constant temperature, made visible through the use of an interferometer. Note the separated-flow region which develops on the back side of the cylinder at the higher Reynolds numbers and the turbulent field which is present in that region.

We may note that the original correlation for gases omitted the Prandtl number term in Eq. (6-11) with little error because most diatomic gases have $\mathrm{Pr} \sim 0.7$. The introduction of the $\mathrm{Pr}^{\frac{1}{3}}$ factor follows from the previous reasoning in Chap. 5.

Fand [21] has shown that the heat-transfer coefficients from liquids to cylinders in cross flow may be better represented by the relation

$$\mathrm{Nu}_f = (0.35 + 0.56\,\mathrm{Re}_f^{0.52})\,\mathrm{Pr}_f^{0.3} \tag{6-11a}$$

**Fig. 6-9** Data for heating and cooling of air flowing normal to single cylinders, from McAdams [10].

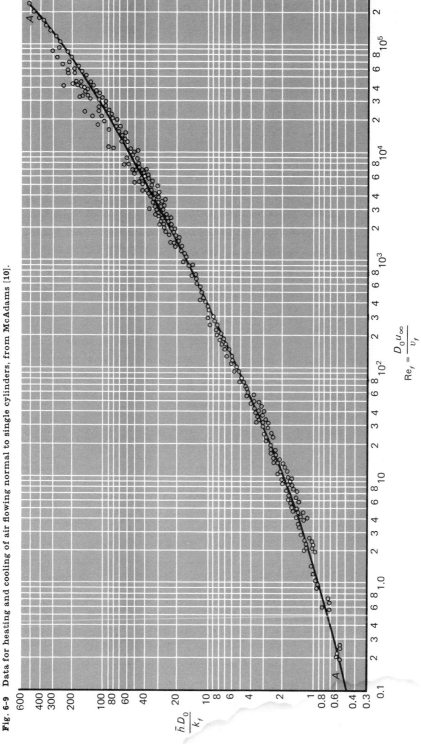

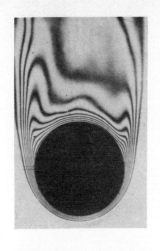

RE = 597    1.5"DIA.

RE = 1600    1.5"DIA.

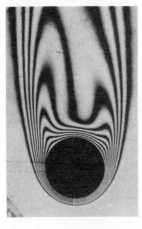

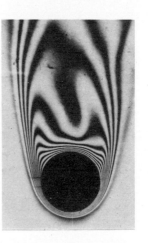

RE = 120    1.0"DIA.

RE = 218    1.0"DIA.

**Fig. 6-10** Interferometer photograph showing isotherms around heated horizontal cylinders placed in a transverse airstream. (Photographs courtesy of **E. Soehngen.**)

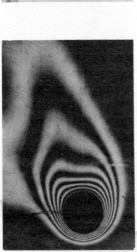

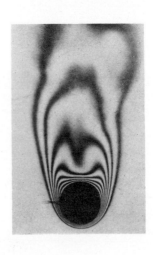

RE = 23    0.5"DIA.

RE = 85    0.5"DIA.

**Table 6-2  Constants for Heat Transfer from Noncircular Cylinders According to Jakob [22]**

| Geometry | $\mathrm{Re}_{df}$ | $C$ | $n$ |
|---|---|---|---|
| $u_\infty \rightarrow$ ◇ $d$ | $5 \times 10^3 - 10^5$ | 0.246 | 0.588 |
| $u_\infty \rightarrow$ ▢ $d$ | $5 \times 10^3 - 10^5$ | 0.102 | 0.675 |
| $u_\infty \rightarrow$ ⬡ $d$ | $5 \times 10^3 - 1.95 \times 10^4$<br>$1.95 \times 10^4 - 10^5$ | 0.160<br>0.0385 | 0.638<br>0.782 |
| $u_\infty \rightarrow$ ⬢ $d$ | $5 \times 10^3 - 10^5$ | 0.153 | 0.638 |
| $u_\infty \rightarrow$ ❘ $d$ | $4 \times 10^3 - 1.5 \times 10^4$ | 0.228 | 0.731 |

This relation is valid for $10^{-1} < \mathrm{Re}_f < 10^5$ provided excessive free-stream turbulence is not encountered.

Jakob [22] has summarized the results of experiments with heat transfer from noncircular cylinders. Equation (6-11) is employed in order to obtain an empirical correlation for gases, and the constants for use with this equation are summarized in Table 6-2.

McAdams [10] recommends the following relation for heat transfer from spheres to a flowing gas:

$$\frac{hd}{k_f} = 0.37 \left(\frac{u_\infty d}{\nu_f}\right)^{0.6} \qquad \text{for } 17 < \mathrm{Re}_d < 70{,}000 \qquad (6\text{-}12)$$

For flow of liquids past spheres, the data of Kramers [11] may be used to obtain the correlation

$$\frac{hd}{k_f} \mathrm{Pr}_f^{-0.3} = 0.97 + 0.68 \left(\frac{u_\infty d}{\nu_f}\right)^{0.5} \qquad \text{for } 1 < \mathrm{Re}_d < 2000 \quad (6\text{-}13)$$

Vliet and Leppert [19] recommend the following expression for heat transfer from spheres to oil and water over a more extended range of Reynolds numbers from 1 to 200,000:

$$\mathrm{Nu} \, \mathrm{Pr}^{-0.3} \left(\frac{\mu_w}{\mu}\right)^{0.25} = 1.2 + 0.53 \, \mathrm{Re}_d^{0.54} \qquad (6\text{-}13a)$$

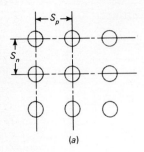

(a)

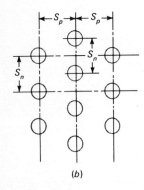

(b)

Fig. 6-11 Nomenclature for use with Table 6-3. (a) In-line tube rows; (b) staggered tube rows.

where all properties are evaluated at free-stream conditions, except $\mu_w$, which is evaluated at the surface temperature of the sphere. Equation (6-13a) represents the data of Ref. 11, as well as the more recent data of Ref. 19.

## Example 6-2

Air at 14.7 psia and 100°F flows past a 2-in.-diameter cylinder at a velocity of 150 ft/sec. The cylinder is maintained at a surface temperature of 300°F. Calculate the heat loss per unit length of the cylinder.

**Solution** We first determine the Reynolds number and then find the applicable constants from Table 6-1 for use with Eq. (6-11). The properties of the air are evaluated at film temperature

$$T_f = \frac{T_w + T_\infty}{2} = \frac{300 + 100}{2} = 200°F$$

so that the properties are

$$Pr_f = 0.693$$
$$\nu_f = 0.239 \times 10^{-3} \text{ ft}^2/\text{sec}$$
$$k_f = 0.0181 \text{ Btu/hr-ft-°F}$$
$$Re_{df} = \frac{u_\infty d}{\nu_f} = \frac{(150)(0.167)}{0.239 \times 10^{-3}} = 104{,}800$$

From Table 6-1

$$C = 0.0266$$
$$n = 0.805$$

so that

$$\frac{hd}{k_f} = 0.0266(104{,}800)^{0.805}(0.693)^{\frac{1}{3}} = 259$$

$$\bar{h} = \frac{(259)(0.0181)}{0.167} = 28.1 \text{ Btu/hr-ft}^2\text{-}°\text{F} \qquad (159.5 \text{ W/m}^2\text{-}°\text{C})$$

The heat transfer per unit length is therefore

$$\frac{q}{L} = h\pi d(T_w - T_\infty)$$
$$= (28.1)\pi(0.167)(300 - 100) = 2938 \text{ Btu/hr-ft} \qquad (2{,}824 \text{ W/m})$$

## 6-4□FLOW ACROSS TUBE BANKS

Since many heat-exchanger arrangements involve multiple rows of tubes, the heat-transfer characteristics for tube banks are of important practical interest. The heat-transfer characteristics of staggered and in-line tube banks were studied by Grimson [12], and on the basis of a correlation of the results of various investigators, he was able to represent the data in the form of Eq. (6-11). The values of the constant $C$ and the exponent $n$ are given in Table 6-3 in terms of the geometric parameters used to describe the tube-bundle arrangement. The Reynolds number is based on the maximum velocity occurring in the tube bank, i.e., the velocity through the minimum-flow area. This area will depend on the geometric tube arrangement. The nomenclature for use with Table 6-3 is shown in Fig. 6-11. The data of Table 6-3 pertain to tube banks having 10 or more rows of tubes in the direction of flow. For fewer rows the ratio of $h$ for $N$ rows deep to that for 10 rows is given in Table 6-4.

Pressure drop for flow of gases over a bank of tubes may be calculated with the relation

$$\Delta p = \frac{f'G_{max}^2 N}{\rho(2.09 \times 10^8)}\left(\frac{\mu_w}{\mu_b}\right)^{0.14} \qquad (6\text{-}14)$$

where $G_{max}$ = mass velocity at minimum flow area, $\text{lb}_m/\text{hr-ft}^2$
        $\rho$ = density evaluated at free-stream conditions, $\text{lb}_m/\text{ft}^3$
        $N$ = number of transverse rows
The empirical friction factor $f'$ is given by Jakob [18] as

$$f' = \left\{0.25 + \frac{0.118}{[(S_n - d)/d]^{1.08}}\right\} \text{Re}_{max}^{-0.16} \qquad (6\text{-}15)$$

**Table 6-3   Correlation of Grimson for Heat Transfer for Tube Banks of 10 Rows or More†**

| Arrangement | $\dfrac{S_p}{D}$ | $\dfrac{S_n}{D}$ 1.25 | | 1.5 | | 2.0 | | 3.0 | |
|---|---|---|---|---|---|---|---|---|---|
| | | $C$ | $n$ | $C$ | $n$ | $C$ | $n$ | $C$ | $n$ |
| In line | 1.25 | 0.386 | 0.592 | 0.305 | 0.608 | 0.111 | 0.704 | 0.0703 | 0.752 |
| | 1.5 | 0.407 | 0.586 | 0.278 | 0.620 | 0.112 | 0.702 | 0.0753 | 0.744 |
| | 2.0 | 0.464 | 0.570 | 0.332 | 0.602 | 0.254 | 0.632 | 0.220 | 0.648 |
| | 3.0 | 0.322 | 0.601 | 0.396 | 0.584 | 0.415 | 0.581 | 0.317 | 0.608 |
| Staggered | 0.6 | ..... | ..... | ..... | ..... | ..... | ..... | 0.236 | 0.636 |
| | 0.9 | ..... | ..... | ..... | ..... | 0.495 | 0.571 | 0.445 | 0.581 |
| | 1.0 | ..... | ..... | 0.552 | 0.558 | | | | |
| | 1.125 | ..... | ..... | ..... | ..... | 0.531 | 0.565 | 0.575 | 0.560 |
| | 1.25 | 0.575 | 0.556 | 0.561 | 0.554 | 0.576 | 0.556 | 0.579 | 0.562 |
| | 1.5 | 0.501 | 0.568 | 0.511 | 0.562 | 0.502 | 0.568 | 0.542 | 0.568 |
| | 2.0 | 0.448 | 0.572 | 0.462 | 0.568 | 0.535 | 0.556 | 0.498 | 0.570 |
| | 3.0 | 0.344 | 0.592 | 0.395 | 0.580 | 0.488 | 0.562 | 0.467 | 0.574 |

† From Ref. 12.

for staggered tube arrangements, and

$$f' = \left\{ 0.044 + \frac{0.08 S_p/d}{[(S_n - d)/d]^{0.43+1.13d/S_p}} \right\} \text{Re}_{\text{max}}^{-0.15} \tag{6-16}$$

for in-line tube arrangements.

### Example 6-3

Air at 14.7 psia and 50°F flows across a bank of tubes 15 rows high and 5 rows deep at a velocity of 20 ft/sec measured at a point in the flow before the stream enters the tube bank. The surfaces of the tubes are maintained at 150°F. The diameter of the tubes is 1 in., and the tubes are arranged in an in-line manner so that the spacing in both the direction normal and parallel to the flow is $1\frac{1}{2}$ in. Calculate the total heat transfer per unit length for the tube bank.

**Solution**   The constants for use with Eq. (6-11) may be obtained from Table 6-3. We have

$$\frac{S_p}{D} = \frac{1.5}{1} = 1.5$$

$$\frac{S_n}{D} = \frac{1.5}{1} = 1.5$$

**Table 6-4   Ratio of $h$ for $N$ Rows Deep to That for 10 Rows Deep†**

| $N$ | 1 | 2 | 3 | 4 | 5 | 6 | 7 | 8 | 9 | 10 |
|---|---|---|---|---|---|---|---|---|---|---|
| Ratio for staggered tubes | 0.68 | 0.75 | 0.83 | 0.89 | 0.92 | 0.95 | 0.97 | 0.98 | 0.99 | 1.0 |
| Ratio for in-line tubes | 0.64 | 0.80 | 0.87 | 0.90 | 0.92 | 0.94 | 0.96 | 0.98 | 0.99 | 1.0 |

† From Ref. 17.

so that $C = 0.250$ and $n = 0.620$. The properties of air are evaluated at the film temperature.

$$T_f = \frac{T_w + T_\infty}{2} = \frac{50 + 150}{2} = 100°F$$
$$\nu_f = 0.18 \times 10^{-3} \text{ ft}^2/\text{sec}$$
$$k_f = 0.0154 \text{ Btu/hr-ft-°F}; \text{Pr}_f = 0.706$$

To calculate the maximum velocity, we must determine the minimum-flow area. From Fig. 6-11 we find that the ratio of the minimum-flow area to the total frontal area is $(S_n - D)/S_n$. The maximum velocity is thus

$$u_{max} = u_\infty \frac{S_n}{S_n - D} = \frac{(20)(1.5)}{0.5} = 60 \text{ ft/sec}$$

where $u_\infty$ is the incoming velocity before entrance to the tube bank. Using the maximum velocity, the Reynolds number is computed as

$$\text{Re} = \frac{u_{max}d}{\nu_f} = \frac{(60)(0.0833)}{0.18 \times 10^{-3}} = 27{,}800$$

The heat-transfer coefficient is then calculated with Eq. (6-11).

$$\frac{hd}{k_f} = (0.278)(27{,}800)^{0.62}(0.706)^{\frac{1}{3}} = 141$$
$$h = \frac{(141)(0.0154)}{0.0833} = 26.0 \qquad (147.6 \text{ W/m}^2\text{-°C})$$

This is the heat-transfer coefficient that would be obtained if there were 10 rows of tubes in the direction of flow. Since there are only 5 rows, this value must be multiplied by the factor 0.92, as determined from Table 6-4.

The total surface area for heat transfer, considering unit length of tubes, is

$$A = N\pi d(1)$$
$$= (15)(5)\pi(0.0833) = 19.62 \text{ ft}^2 \qquad (1.823 \text{ m}^2)$$

where $N$ is the number of tubes. The total heat transfer is calculated from

$$q = hA(T_w - T_\infty)$$
$$= (0.92)(26.0)(19.62)(150 - 50) = 46{,}900 \text{ Btu/hr} \qquad (13742 \text{ W})$$

The foregoing analysis has assumed that the free-stream temperature $T_\infty$ is constant throughout the heating process in the tube bank. This turns out to be a false assumption, as may be shown by the following calculation: Using the above heat-transfer rate, we may compute the increase in the fluid temperature $\Delta T$ with the energy balance:

$$q = \dot{m}c_p \, \Delta T$$

where $\dot{m}$ is the total mass flow in the tube bank. Thus

$$46{,}900 = \rho_\infty u_\infty A c_p \, \Delta T$$

Inserting the values for the properties at 50°F,

$$46{,}900 = (0.0778)(20)(3600)(15)(1.5/12)(1)(0.24) \, \Delta T$$

and
$$\Delta T = 18.6°F$$

This value is sufficiently high so as to invalidate our previous calculation of $q$. Since the film temperature varies through the tube bank, the value of $h$ varies as well. Strictly speaking, an iterative (or numerical) analysis will be required to obtain the true heat-transfer rate because of the variation of $h$ and $(T_w - T_\infty)$ through the tube bank. A first approximation may be made by using the value of $h$ calculated above and an arithmetic-average temperature difference given by

$$(T_w - T_\infty)_{\text{avg}} \approx T_w - \left(50 + \frac{18.6}{2}\right) = 150 - 59.3 = 90.7°F$$

The resulting heat-transfer rate is

$$q = \frac{(46{,}900)(90.7)}{100} = 42{,}540 \text{ Btu/hr} \qquad (12{,}460 \text{ W})$$

This problem is better analyzed by using the log-mean temperature-difference concept discussed in Chap. 10.

## 6-5□LIQUID - METAL HEAT TRANSFER

In recent years concentrated interest has been placed on liquid-metal heat transfer because of the high heat-transfer rates which may be achieved with these media. These high heat-transfer rates result from the high thermal conductivities of liquid metals as compared with other fluids; as a consequence, they are particularly applicable to situations where large energy quantities must be removed from a relatively small space, as in a nuclear reactor. In addition, the liquid metals remain in the liquid state at higher temperatures than conventional fluids like water and various organic coolants. This also enables more compact heat-exchanger design. Liquid metals are difficult to handle because of their corrosive nature and the violent action which may result when they come into contact with water or air; even so, their advantages in certain heat-transfer applications have overshadowed their short-

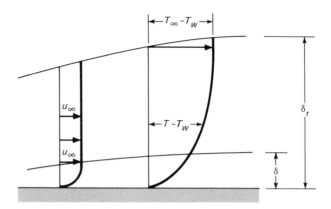

**Fig. 6-12 Boundary-layer regimes for analysis of liquid-metal heat transfer.**

comings, and suitable techniques for handling them have been developed.

Let us first consider the simple flat plate with a liquid metal flowing across it. The Prandtl number for liquid metals is very low, of the order of 0.01, so that the thermal-boundary-layer thickness should be substantially larger than the hydrodynamic-boundary-layer thickness. The situation results from the high values of thermal conductivity for liquid metals, and is depicted in Fig. 6-12. Since the ratio of $\delta/\delta_t$ is small, the velocity profile has a very blunt shape over most of the thermal boundary layer. As a first approximation, then, we might assume a slug flow model for calculation of the heat transfer; i.e., we take

$$u = u_\infty \tag{6-17}$$

throughout the thermal boundary layer for purposes of computing the energy-transport term in the integral energy equation (Sec. 5-6)

$$\frac{d}{dx}\left[\int_0^{\delta_t} (T_\infty - T)u\,dy\right] = \alpha\left(\frac{dT}{dy}\right)_w \tag{6-18}$$

The conditions on the temperature profile are the same as those in Sec. 5-6, so that we use the cubic parabola as before.

$$\frac{\theta}{\theta_\infty} = \frac{T - T_w}{T_\infty - T_w} = \frac{3}{2}\frac{y}{\delta_t} - \frac{1}{2}\left(\frac{y}{\delta_t}\right)^3 \tag{6-19}$$

Inserting Eqs. (6-17) and (6-19) in (6-18) gives

$$\theta_\infty u_\infty \frac{d}{dx}\left\{\int_0^{\delta_t}\left[1 - \frac{3}{2}\frac{y}{\delta_t} + \frac{1}{2}\left(\frac{y}{\delta_t}\right)^3\right]dy\right\} = \frac{3\alpha\theta_\infty}{2\delta_t} \tag{6-20}$$

which may be integrated to give

$$2\delta_t\,d\delta_t = \frac{8\alpha}{u_\infty}\,dx \tag{6-21}$$

The solution to this differential equation is

$$\delta_t = \sqrt{\frac{8\alpha x}{u_\infty}} \tag{6-22}$$

for a plate heated over its entire length.

The heat-transfer coefficient may be expressed by

$$h_x = \frac{-k(\partial T/\partial y)_w}{T_w - T_\infty} = \frac{3k}{2\delta_t} = \frac{3\sqrt{2}}{8} k \sqrt{\frac{u_\infty}{\alpha x}} \tag{6-23}$$

This relationship may be put in dimensionless form as

$$\mathrm{Nu}_x = \frac{h_x x}{k} = 0.530(\mathrm{Re}_x \ \mathrm{Pr})^{\frac{1}{2}} = 0.530 \ \mathrm{Pe}^{\frac{1}{2}} \tag{6-24}$$

where we have now introduced the new dimensionless grouping called the Peclet number,

$$\mathrm{Pe} = \mathrm{Re} \ \mathrm{Pr} = \frac{u_\infty x}{\alpha} \tag{6-25}$$

Using Eq. (5-19) for the hydrodynamic-boundary-layer thickness,

$$\frac{\delta}{x} = \frac{4.64}{\mathrm{Re}_x^{\frac{1}{2}}} \tag{6-26}$$

we may compute the ratio $\delta/\delta_t$.

$$\frac{\delta}{\delta_t} = \frac{4.64}{\sqrt{8}} \sqrt{\mathrm{Pr}} = 1.64 \sqrt{\mathrm{Pr}} \tag{6-27}$$

Using $\mathrm{Pr} \sim 0.01$, we obtain

$$\frac{\delta}{\delta_t} \sim 0.16$$

which is in reasonable agreement with our slug flow model.

The flow model discussed above serves to illustrate the general nature of liquid-metal heat transfer and it is important to note that the heat transfer is dependent on the Peclet number. Empirical correlations are usually expressed in terms of this parameter, three of which we present below.

Extensive data on liquid metals are given in Ref. 13, and the heat-transfer characteristics are summarized in Ref. 23. Lubarsky and Kaufman [14] recommended the following relation for calculation of heat-transfer coefficients in fully developed turbulent flow of liquid metals in smooth tubes with uniform heat flux at the wall:

$$\mathrm{Nu}_d = \frac{hd}{k} = 0.625 \ (\mathrm{Re}_d \ \mathrm{Pr})^{0.4} \tag{6-28}$$

All properties for use in Eq. (6-28) are evaluated at the bulk temperature.

Equation (6-28) is valid for $10^2 < $ Pe $< 10^4$ and for $L/D > 60$. Seban and Shimazaki [16] propose the following relation for calculation of heat transfer to liquid metals in tubes with constant wall temperature:

$$\text{Nu}_d = 5.0 + 0.025\ (\text{Re}_d\ \text{Pr})^{0.8} \tag{6-29}$$

where all properties are evaluated at the bulk temperature. Equation (6-29) is valid for Pe $> 10^2$ and $L/D > 60$.

More recent data by Skupinshi et al. [26] with sodium-potassium mixtures indicates that the following relation may be preferable to that of Eq. (6-28) for constant-heat-flux conditions:

$$\text{Nu} = 4.82 + 0.0185\ \text{Pe}^{0.827} \tag{6-28a}$$

This relation is valid for $3.6 \times 10^3 < $ Re $< 9.05 \times 10^5$ and $10^2 < $ Pe $< 10^4$.

Witte [32] has measured the heat transfer from a sphere to liquid sodium during forced convection, with the data being correlated by

$$\text{Nu} = 2 + 0.386\ (\text{Re}\ \text{Pr})^{0.5} \tag{6-30}$$

for the Reynolds number range $3.56 \times 10^4 < $ Re $< 1.525 \times 10^5$.

In general, there are many open questions concerning liquid-metal heat transfer, and the reader is referred to Refs. 13 and 23 for more information.

### Example 6-4

Ten $\text{lb}_m$/sec of liquid bismuth flows through a 2-in.-diameter stainless-steel tube. The bismuth enters the tube at 780°F, and is heated to 820°F as it passes through the tube. If a constant heat flux is maintained along the tube and the tube wall is at a temperature 40°F higher than the bismuth bulk temperature, calculate the length of tube required to effect the heat transfer.

**Solution**   Since a constant heat flux is maintained, we may use Eq. (6-28) to calculate the heat-transfer coefficient. The properties of bismuth are evaluated at the average bulk temperature, 800°F.

$\mu = 0.9 \times 10^{-3}\ \text{lb}_m/\text{ft-sec}$
$c_p = 0.0357\ \text{Btu}/\text{lb}_m\text{-°F}$
$k = 9.0\ \text{Btu/hr-ft-°F}$     $(15.6\ \text{W/m-°C})$
$\text{Pr} = 0.013$

The total heat transfer is given by

$$q = \dot{m}c_p\ \Delta T_b = (10)(0.0357)(40) = 14.3\ \text{Btu/sec}$$
$$= 51{,}400\ \text{Btu/hr}$$
$$\text{Re}_d = \frac{dG}{\mu} = \frac{(0.167)(10)}{[\pi(0.167)^2/4](0.9 \times 10^{-3})} = 84{,}800$$
$$\text{Re}_d\ \text{Pr} = \text{Pe} = (84{,}800)(0.013) = 1102$$

The heat-transfer coefficient may now be calculated from Eq. (6-28).

$$\text{Nu}_d = \frac{hd}{k} = 0.625 \, (\text{Re}_d \, \text{Pr})^{0.4}$$

$$= (0.625)(1102)^{0.4} = 10.28$$

$$h = \frac{(10.28)(9.0)}{0.167} = 555 \text{ Btu/hr-ft}^2\text{-°F} \qquad (3151 \text{ W/m}^2\text{-°C})$$

The total required surface area of the tube may now be calculated from the Newton law of cooling.

$$q = hA(T_w - T_b)$$

$$A = \frac{5.14 \times 10^4}{(555)(40)} = 2.31 \text{ ft}^2 \qquad (0.215 \text{ m}^2)$$

But $A = \pi \, dL$, so that

$$L = \frac{2.31}{\pi(0.167)} = 4.42 \text{ ft} \qquad (1.347 \text{ m})$$

## REVIEW QUESTIONS

1.   What is the Dittus-Boelter equation? When does it apply?
2.   How may heat-transfer coefficients be calculated for flow in rough pipes?
3.   What is the hydraulic diameter? When is it used?
4.   What is the form of equation used to calculate heat transfer for flow over cylinders and bluff bodies?
5.   Why does a slug flow model yield reasonable results when applied to liquid metal heat transfer?
6.   What is the Peclet number?

## PROBLEMS

6-1.   Water at the rate of 100 $\text{lb}_m$/min is heated from 100 to 110°F in a 1-in.-diameter tube whose surface is at 200°F. How long must the tube be to accomplish this heating?

6-2.   Water at the rate of 1 $\text{lb}_m$/sec is forced through a smooth 1-in.-ID tube 50 ft long. The inlet water temperature is 50°F, and the tube wall temperature is 30°F higher than the water temperature all along the length of the tube. What is the exit water temperature?

6-3.   Five gpm of water at an average temperature of 70°F flows in a 1-in.-diameter tube 20 ft long. The pressure drop is measured as 0.3 psi. A constant heat flux is imposed, and the average wall temperature is 130°F. Estimate the exit temperature of the water.

6-4.   Water at the rate of 400 $\text{lb}_m$/min is heated from 40 to 60°F by passing it through a 2-in.-ID copper tube. The tube wall temperature is maintained at 200°F. What is the length of the tube?

6-5. One $lb_m$/sec of air at 200 psia enters a duct 3 in. in diameter and 20 ft long. The duct wall is maintained at an average temperature of 400°F. The average air temperature in the duct is 500°F. Estimate the decrease in temperature of the air as it passes through the duct.

6-6. Six gpm of water at an average temperature of 52°F flows in a 1-in.-diameter tube 20 ft long. The pressure drop is measured as 0.42 psi. A constant heat flux is imposed, and the average wall temperature is 120°F. Estimate the exit temperature of the water.

6-7. Fifteen gpm of water flows through a 1-in.-ID pipe 5.0 ft long. The pressure drop is 1.0 psi through the 5.0-ft length. The pipe wall temperature is maintained at a constant temperature of 120°F by a condensing vapor, and the inlet water temperature is 70°F. Estimate the exit water temperature.

6-8. Water at the rate of 1.0 $lb_m$/sec is to be cooled from 150 to 80°F. Which would result in less pressure drop: to run the water through a $\frac{1}{2}$-in.-diameter pipe at a constant temperature of 40°F or through a constant-temperature 1-in.-diameter pipe at 70°F?

6-9. Water at the rate of 2 $lb_m$/sec is forced through a tube with 1 in. ID. The inlet water temperature is 60°F, and the outlet water temperature is 120°F. The tube wall temperature is 25°F higher than the water temperature all along the length of the tube. What is the length of the tube?

6-10. Air at 15 psia and 60°F flows through a long rectangular duct 3 by 6 in. A 6-ft section of the duct is maintained at 250°F, and the average air temperature at exit from this section is 150°F. Calculate the air-flow rate and the total heat transfer.

6-11. One hundred gpm of water at 100°F enters a 2-in.-ID pipe having a relative roughness of 0.002. If the pipe is 30 ft long and is maintained at 150°F, calculate the exit water temperature and the total heat transfer.

6-12. A heat exchanger is constructed so that hot flue gases at 800°F flow inside a 1-in.-ID copper tube with $\frac{1}{16}$ in. wall thickness. A 2-in. tube is placed around the 1-in.-diameter tube, and high-pressure water at 300°F flows in the annular space between the tubes. If the flow rate of water is 200 $lb_m$/min and the total heat transfer is $6 \times 10^4$ Btu/hr, estimate the length of the heat exchanger for a gas mass flow of 100 $lb_m$/min. Assume that the properties of the flue gas are the same as those of air at atmospheric pressure and 800°F.

6-13. Engine oil enters a $\frac{1}{2}$-in.-diameter tube 10 ft long at a temperature of 100°F. The tube wall temperature is maintained at 150°F, and the flow velocity is 1 ft/sec. Estimate the total heat transfer to the oil and the exit temperature of the oil.

6-14. Using the values of the local Nusselt number given in Fig. 6-8, obtain values for the average Nusselt number as a function of the Reynolds number. Plot the results as log Nu versus log Re, and obtain an equation which represents all the data. Compare this correlation with that given by Eq. (6-11) and Table 6-1.

6-15. Air at 10 psia and 70°F flows across a 2-in.-diameter cylinder at a velocity of 70 ft/sec. Compute the drag force exerted on the cylinder.

6-16. Water at the rate of 100 $lb_m$/min at 200°F is forced through a 1-in.-ID copper tube at a velocity of 2 ft/sec. The wall thickness is $\frac{1}{32}$ in. Air at 60°F and atmospheric pressure is forced over the outside of the tube at a velocity of 50 ft/sec in a direction normal to the axis of the tube. What is the heat loss per foot of length of the tube?

6-17. A heated cylinder at 300°F and 1 in. in diameter is placed in an atmospheric airstream at 14.7 psia and 100°F. The air velocity is 100 ft/sec. Calculate the heat loss per foot of length of the cylinder.

6-18. Assuming that a man can be approximated by a cylinder 1 ft in diameter and 6 ft high with a surface temperature of 75°F, calculate the heat he would lose while standing in a 30-mph north wind whose temperature is 30°F.

6-19. The drag coefficient for a sphere at Reynolds numbers less than 100 may be approximated by

$$C_D = b \, \mathrm{Re}^{-1}$$

where $b$ is a constant. Assuming that the Reynolds analogy between heat transfer and fluid friction applies, derive an expression for the heat loss from a sphere of diameter $d$ and temperature $T_s$, released from rest and allowed to fall in a fluid of temperature $T_\infty$. (Obtain an expression for the heat lost between the time the sphere is released and the time it reaches some velocity $v$. Assume that the Reynolds number is less than 100 during this time and that the sphere remains at a constant temperature.)

6-20. Air at 500 psia and 100°F flows across a tube bank consisting of 400 $\frac{1}{2}$-in.-OD tubes arranged in a staggered manner 20 rows high. $S_p = 1\frac{1}{2}$ in., and $S_n = 1$ in. The incoming-flow velocity is 30 ft/sec, and the tube wall temperatures are maintained constant at 400°F by a condensing vapor on the inside of the tubes. The length of the tubes is 5 ft. Estimate the exit air temperature as it leaves the tube bank.

6-21. Two $lb_m$/sec of liquid bismuth enters a 1-in.-diameter stainless-steel pipe at 750°F. The tube wall temperature is maintained constant at 850°F. Calculate the bismuth exit temperature if the tube is 2 ft long.

6-22. Five $lb_m$/sec of liquid sodium is to be heated from 250 to 300°F. A $\frac{1}{2}$-in.-diameter electrically heated tube is available (constant heat flux). If the tube wall temperature is not to exceed 400°F, calculate the minimum length required.

6-23. Assume that one-half the heat transfer from a cylinder in cross flow occurs on the front half of the cylinder. On this assumption, compare the heat transfer from a cylinder in cross flow with the heat transfer from a flat plate having a length equal to the distance from the stagnation point on the cylinder. Discuss this comparison.

6-24.   Using the slug flow model, show that the boundary-layer energy equation reduces to the same form as the transient-conduction equation for the semi-infinite solid of Sec. 4-3. Solve this equation and compare the solution with the integral analysis of Sec. 6-5.

6-25.   An oil with Pr $= 1960$, $\rho = 53.9$ lb/ft$^3$, $\nu = 0.0017$ ft$^2$/sec, and $k = 0.082$ Btu/hr-ft-°F enters a 0.1-in.-diameter tube 2 ft long. The oil entrance temperature is 68°F, the mean flow velocity is 1.0 ft/sec, and the tube wall temperature is 248°F. Calculate the heat-transfer rate.

6-26.   One lb$_m$/sec of liquid ammonia flows through a 1-in.-diameter smooth tube 8 ft long. The ammonia enters at 50°F and leaves at 100°F, and a constant heat flux is imposed on the tube wall. Calculate the average wall temperature necessary to effect the indicated heat transfer.

6-27.   Water flows over a $\frac{1}{8}$-in.-diameter sphere at 20 ft/sec. The free-stream temperature is 100°F, and the sphere is maintained at 200°F. Calculate the heat-transfer rate.

6-28.   Show that the hydraulic diameter for an annulus space is given by $D_H = d_o - d_i$.

6-29.   A 0.005-in.-diameter wire is exposed to an airstream at $-20$°F and 6.5 psia. The flow velocity is 750 ft/sec. The wire is electrically heated and has a length of $\frac{1}{2}$ in. Calculate the electric power necessary to maintain the wire surface temperature at 350°F. Express the answer in watts.

6-30.   A spherical water droplet having a diameter of 0.05 in. is allowed to fall from rest in atmospheric air at 14.7 psia and 70°F. Estimate the velocities the droplet will attain after a drop of 100, 200, and 1000 ft.

6-31.   Air at 100°F and 14.7 psia flows past a heated $\frac{1}{16}$-in.-diameter wire at a velocity of 20 ft/sec. The wire is heated to a temperature of 300°F. Calculate the heat transfer per unit length of wire.

6-32.   A tube bank uses an in-line arrangement with $S_n = S_p = 0.75$ in. and 0.25-in.-diameter tubes. Six rows of tubes are employed with a stack 50 tubes high. The surface temperature of the tubes is constant at 200°F, and atmospheric air is forced across them at an inlet velocity of 15 ft/sec before the flow enters the tube bank. Calculate the total heat transfer per unit length for the tube bank. Estimate the pressure drop for this arrangement.

6-33.   Repeat Prob. 6-32 for a staggered tube arrangement with the same values of $S_p$ and $S_n$.

6-34.   A more compact version of the tube bank in Prob. 6-32 can be achieved by reducing the $S_p$ and $S_n$ dimensions while still retaining the same number of tubes. Investigate the effect of reducing $S_p$ and $S_n$ in half, that is, $S_p = S_n = 0.375$ in. Calculate the heat transfer and pressure drop for this new arrangement.

6-35.   Atmospheric air at 70°F flows across a 2-in.-square rod at a velocity of 50 ft/sec. The velocity is normal to one of the faces of the rod.

Calculate the heat transfer per unit length for a surface temperature of 200°F.

6-36.   A short tube is $\frac{1}{4}$ in. in diameter and 6 in. long. Water enters the tube at 5 ft/sec and 100°F and a constant heat flux condition is maintained such that the tube wall temperature remains 50°F above the water bulk temperature. Calculate the heat-transfer rate and exit water temperature.

6-37.   Helium at 14.7 psia and 100°F flows across a $\frac{1}{8}$-in.-diameter cylinder which is heated to 300°F. The flow velocity is 30 ft/sec. Calculate the heat transfer per unit length of wire. How does this compare with the heat transfer for air under the same conditions?

6-38.   Calculate the heat transfer rate per unit length for flow over a 0.001-in.-diameter cylinder maintained at 150°F. Perform the calculation for (a) air at 70°F and 14.7 psia and (b) water at 70°F. $U_\infty = 20$ ft/sec.

6-39.   Compare the heat transfer results of Eqs. (6-11) and (6-11a) for water at Reynolds numbers of $10^3$, $10^4$, and $10^5$ and a film temperature of 100°F.

6-40.   Liquid Freon 12 ($CCl_2F_2$) flows inside a 0.500-in.-diameter tube at a velocity of 10 ft/sec. Calculate the heat-transfer coefficient for a bulk temperature of 50°F. How does this compare with water at the same conditions?

## REFERENCES

1.  Dittus, F. W., and L. M. K. Boelter: *Univ. Calif. (Berkeley) Pub. Eng.*, vol. 2, p. 443, 1930.
2.  Sieder, E. N., and C. E. Tate: Heat Transfer and Pressure Drop of Liquids in Tubes, *Ind. Eng. Chem.*, vol. 28, p. 1429, 1936.
3.  Nusselt, W.: Der Warmeaustausch zwischen Wand und Wasser im Rohr, *Forsch. Gebiete Ingenieurw.*, vol. 2, p. 309, 1931.
4.  Hausen, H.: Darstellung des Warmeuberganges in Rohren durch verallgemeinerte Potenzbeziehungen, *VDI Z.*, no. 4, p. 91, 1943.
5.  Moody, F. F.: Friction Factors for Pipe Flow, *Trans. ASME*, vol. 66, p. 671, 1944.
6.  Schlichting, H.: "Boundary Layer Theory," 4th ed., McGraw-Hill Book Company, New York, 1960.
7.  Giedt, W. H.: Investigation of Variation of Point Unit-heat-transfer Coefficient around a Cylinder Normal to an Air Stream, *Trans. ASME*, vol. 71, pp. 375–381, 1949.
8.  Hilpert, R.: Warmeabgabe von geheizen Drahten und Rohren, *Forsch. Gebiete Ingenieurw.*, vol. 4, p. 220, 1933.
9.  Knudsen, J. D., and D. L. Katz: "Fluid Dynamics and Heat Transfer," McGraw-Hill Book Company, New York, 1958.
10. McAdams, W. H.: "Heat Transmission," 3d ed., McGraw-Hill Book Company, New York, 1954.
11. Kramers, H.: Heat Transfer from Spheres to Flowing Media, *Physica*, vol. 12, p. 61, 1946.
12. Grimson, E. D.: Correlation and Utilization of New Data on Flow Resistance and Heat Transfer for Cross Flow of Gases over Tube Banks, *Trans. ASME*, vol. 59, pp. 583–594, 1937.
13. Lyon, R. D. (ed.): "Liquid Metals Handbook," 3d ed., Atomic Energy Com-

mission and U.S. Navy Department, Washington, D.C., 1952.

14. Lubarsky, B., and S. J. Kaufman: Review of Experimental Investigations of Liquid-metal Heat Transfer, *NACA Tech. Notes*, no. 3336, 1955.

15. Colburn, A. P.: A Method of Correlating Forced Convection Heat Transfer Data and a Comparison with Fluid Friction, *Trans. AIChE*, vol. 29, p. 174, 1933.

16. Seban, R. A., and T. T. Shimazaki: Heat Transfer to a Fluid Flowing Turbulently in a Smooth Pipe with Walls at Constant Temperature, *Trans. ASME*, vol. 73, p. 803, 1951.

17. Kays, W. M., and R. K. Lo: Basic Heat Transfer and Flow Friction Data for Gas Flow Normal to Banks of Staggered Tubes: Use of a Transient Technique, *Stanford Univ. Tech. Rept.* 15, Navy Contract N6-ONR-251 T.O. 6, 1952.

18. Jakob, M.: Heat Transfer and Flow Resistance in Cross Flow of Gases over Tube Banks, *Trans. ASME*, vol. 60, p. 384, 1938.

19. Vliet, G. C., and G. Leppert: Forced Convection Heat Transfer from an Isothermal Sphere to Water, *J. Heat Transfer*, ser. C, vol. 83, p. 163, 1961.

20. Irvine, T. R.: Noncircular Duct Convective Heat Transfer, in W. Ibele (ed.), "Modern Developments in Heat Transfer," Academic Press Inc., New York, 1963.

21. Fand, R. M.: Heat Transfer by Forced Convection from a Cylinder to Water in Crossflow, *Intern. J. Heat Mass Transfer*, vol. 8, p. 995, 1965.

22. Jakob, M.: "Heat Transfer," vol. 1, John Wiley & Sons, Inc., New York, 1949.

23. Stein, R.: Liquid Metal Heat Transfer, *Advan. Heat Transfer*, vol. 3, 1966.

24. Hartnett, J. P.: Experimental Determination of the Thermal Entrance Length for the Flow of Water and of Oil in Circular Pipes, *Trans. ASME*, vol. 77, p. 1211, 1955.

25. Allen, R. W., and E. R. G. Eckert: Friction and Heat Transfer Measurements to Turbulent Pipe Flow of Water (Pr = 7 and 8) at Uniform Wall Heat Flux, *J. Heat Transfer*, ser. C, vol. 86, p. 301, 1964.

26. Skupinshi, E., J. Tortel, and L. Vautrey: Détermination des coéfficients de convection d'un alliage sodium-potassium dans un tube circulaire, *Intern. J. Heat Mass Transfer*, vol. 8, p. 937, 1965.

27. Dipprey, D. F., and R. H. Sabersky: Heat and Momentum Transfer in Smooth and Rough Tubes at Various Prandtl Numbers, *Intern. J. Heat Mass Transfer*, vol. 6, p. 329, 1963.

28. Kline, S. J.: "Similitude and Approximation Theory," McGraw-Hill Book Company, New York, 1965.

29. Townes, H. W., and R. H. Sabersky: Experiments on the Flow Over a Rough Surface, *Intern. J. Heat Mass Transfer*, vol. 9, p. 729, 1966.

30. Gowen, R. A., and J. W. Smith: Turbulent Heat Transfer from Smooth and Rough Surfaces, *Intern. J. Heat Mass Transfer*, vol. 11, p. 1657, 1968.

31. Sheriff, N., and P. Gumley: Heat Transfer and Friction Properties of Surfaces with Discrete Roughness, *Intern. J. Heat Mass Transfer*, vol. 9, p. 1297, 1966.

32. Witte, L. C.: An Experimental Study of Forced-Convection Heat Transfer from a Sphere to Liquid Sodium, *J. Heat Transfer*, vol. 90, p. 9, 1968.

# CHAPTER 7

# NATURAL-CONVECTION SYSTEMS

## 7-1□INTRODUCTION

Our previous discussions of convection heat transfer have considered only the calculation of forced-convection systems where the fluid is forced by or through the heat-transfer surface. Natural, or free, convection is observed as a result of the motion of the fluid due to density changes arising from the heating process. A hot radiator used for heating a room is one example of a practical device which transfers heat by free convection. The movement of the fluid in free convection, whether it is a gas or a liquid, results from the buoyancy forces imposed on the fluid when its density in the proximity of the heat-transfer surface is decreased as a result of the heating process. The buoyancy forces would not be present if the fluid were not acted upon by some external force field such as gravity, although gravity is not the only type of force field which can produce the free-convection currents; a fluid enclosed in a rotating machine is acted upon by a centrifugal force field, and thus could experience free-convection currents if one or more of the surfaces in contact with the fluid were heated. The buoyancy forces which give rise to the free-convection currents are called body forces.

## 7-2□FREE - CONVECTION HEAT TRANSFER ON A VERTICAL FLAT PLATE

Consider the vertical flat plate shown in Fig. 7-1. When the plate is heated, a free-convection boundary layer is formed, as shown. The velocity profile in this boundary layer is quite unlike the velocity profile in a forced-convection boundary layer. At the wall the velocity is zero because of the no-slip condition; it increases to some maximum value and then decreases to zero at the edge of the boundary layer since the "free-stream" conditions are at rest in the free-convection system. The initial boundary-layer development is laminar; but at some distance from the leading edge, depending on the fluid properties and the temperature difference between wall and environment, turbulent eddies are formed, and transition to a turbulent

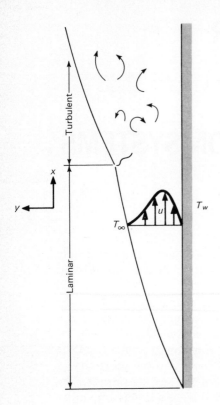

**Fig. 7-1  Boundary layer on a vertical flat plate.**

boundary layer begins. Farther up the plate the boundary layer may become fully turbulent.

To analyze the heat-transfer problem, we must first obtain the differential equation of motion for the boundary layer. For this purpose we choose the $x$ coordinate along the plate and the $y$ coordinate perpendicular to the plate as in the analyses of Chap. 5. The only new force which must be considered in the derivation is the weight of the element of fluid. As before, we equate the sum of the external forces in the $x$ direction to the change in momentum flux through the control volume $dx\, dy$. There results

$$\rho\left(u\,\frac{\partial u}{\partial x} + v\,\frac{\partial u}{\partial y}\right) = -\frac{\partial p}{\partial x} - \rho g + \mu\,\frac{\partial^2 u}{\partial y^2} \qquad (7\text{-}1)$$

where the term $-\rho g$ represents the weight force exerted on the element. The pressure gradient in the $x$ direction results from the change in elevation up the plate. Thus

$$\frac{\partial p}{\partial x} = -\rho_\infty g \tag{7-2}$$

In other words, the change in pressure over a height $dx$ is equal to the weight per unit area of the fluid element. Substituting Eq. (7-2) into Eq. (7-1),

$$\rho\left(u\frac{\partial u}{\partial x} + v\frac{\partial u}{\partial y}\right) = g(\rho_\infty - \rho) + \mu\frac{\partial^2 u}{\partial y^2} \tag{7-3}$$

The density difference $\rho_\infty - \rho$ may be expressed in terms of the volume coefficient of expansion $\beta$, defined by

$$\beta = \frac{1}{V}\left(\frac{\partial V}{\partial T}\right)_p = \frac{1}{V_\infty}\frac{V - V_\infty}{T - T_\infty} = \frac{\rho_\infty - \rho}{\rho(T - T_\infty)}$$

so that

$$\rho\left(u\frac{\partial u}{\partial x} + v\frac{\partial u}{\partial y}\right) = g\rho\beta(T - T_\infty) + \mu\frac{\partial^2 u}{\partial y^2} \tag{7-4}$$

This is the equation of motion for the free-convection boundary layer. Notice that the solution for the velocity profile demands a knowledge of the temperature distribution. The energy equation for the free-convection system is the same as that for a forced-convection system at low velocity.

$$\rho c_p\left(u\frac{\partial T}{\partial x} + v\frac{\partial T}{\partial y}\right) = k\frac{\partial^2 T}{\partial y^2} \tag{7-5}$$

The volume coefficient of expansion $\beta$ may be determined from tables of properties for the specific fluid. For ideal gases it may be calculated from (see Prob. 7-15)

$$\beta = \frac{1}{T}$$

where $T$ is the absolute temperature of the gas.

Even though the fluid motion is the result of density variations, these variations are quite small, and a satisfactory solution to the problem may be obtained by assuming incompressible flow, that is, $\rho = $ const. To effect a solution of the equation of motion, we use the integral method of analysis similar to that used in the forced-convection problem of Chap. 5. Detailed boundary-layer analyses have been presented in Refs. 13, 27, and 32.

For the free-convection system the integral momentum equation becomes

$$\frac{d}{dx}\left(\int_0^\delta \rho u^2\, dy\right) = -\tau_w + \int_0^\delta \rho g\beta(T - T_\infty)\, dy$$

$$= -\mu\left(\frac{\partial u}{\partial y}\right)_{y=0} + \int_0^\delta \rho g\beta(T - T_\infty)\, dy \tag{7-6}$$

and we observe that the functional form of both the velocity and temperature distributions must be known in order to arrive at the solution. To obtain these functions, we proceed in much the same way as in Chap. 5. The following conditions apply for the temperature distribution:

$$T = T_w \qquad \text{at } y = 0$$
$$T = T_\infty \qquad \text{at } y = \delta$$
$$\frac{\partial T}{\partial y} = 0 \qquad \text{at } y = \delta$$

so that we obtain for the temperature distribution

$$\frac{T - T_\infty}{T_w - T_\infty} = \left(1 - \frac{y}{\delta}\right)^2 \qquad (7\text{-}7)$$

Three conditions for the velocity profile are

$$u = 0 \qquad \text{at } y = 0$$
$$u = 0 \qquad \text{at } y = \delta$$
$$\frac{\partial u}{\partial y} = 0 \qquad \text{at } y = \delta$$

An additional condition may be obtained from Eq. (7-4) by noting that

$$\frac{\partial^2 u}{\partial y^2} = -g\beta \frac{T_w - T_\infty}{\nu} \qquad \text{at } y = 0$$

As in the integral analysis for forced convection problems, we assume that the velocity profiles have geometrically similar shapes at various $x$ distances along the plate. For the free convection problem, we now assume that the velocity may be represented as a polynomial function of $y$ multiplied by some arbitrary function of $x$. Thus,

$$\frac{u}{u_x} = a + by + cy^2 + dy^3$$

where $u_x$ is a fictitious velocity which is a function of $x$, and the cubic-polynomial form is chosen because there are four conditions to satisfy and this is the simplest type of function which may be used. Applying the four conditions to the velocity profile listed above,

$$\frac{u}{u_x} = \frac{\beta \delta^2 g(T_w - T_\infty)}{4 u_x \nu} \frac{y}{\delta}\left(1 - \frac{y}{\delta}\right)^2 \qquad 7\text{-}7A$$

The term involving the temperature difference, $\delta^2$, and $u_x$ may be incorporated into the function $u_x$ so that the final relation to be assumed for the velocity profile is

$$\frac{u}{u_x} = \frac{y}{\delta}\left(1 - \frac{y}{\delta}\right)^2 \qquad (7\text{-}8)$$

Fig. 7-2 Free-convection velocity profile given by Eq. (7-8).

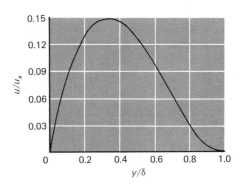

A plot of Eq. (7-8) is given in Fig. 7-2. Substituting Eqs. (7-7) and (7-8) into Eq. (7-6) and carrying out the integrations and differentiations yields

$$\frac{1}{105}\frac{d}{dx}(u_x{}^2\delta) = \tfrac{1}{3}g\beta(T_w - T_\infty)\delta - \nu\frac{u_x}{\delta} \tag{7-9}$$

The integral form of the energy equation for the free-convection system is

$$\frac{d}{dx}\left[\int_0^\delta u(T - T_\infty)\,dy\right] = -\alpha\frac{dT}{dy}\bigg)_{y=0} \tag{7-10}$$

and when the assumed velocity and temperature distributions are inserted into this equation and the operations are performed, there results

$$\frac{1}{30}(T_w - T_\infty)\frac{d}{dx}(u_x\delta) = 2\alpha\frac{T_w - T_\infty}{\delta} \tag{7-11}$$

It is clear from the reasoning which led to Eq. (7-8) that

$$u_x \sim \delta^2 \tag{7-12}$$

Inserting this type of relation in Eq. (7-9) yields the result that

$$\delta \sim x^{\frac{1}{4}} \tag{7-13}$$

We therefore assume the following exponential functional variations for $u_x$ and $\delta$:

$$u_x = C_1 x^{\frac{1}{2}} \tag{7-14}$$
$$\delta = C_2 x^{\frac{1}{4}} \tag{7-15}$$

Introducing these relations into Eqs. (7-9) and (7-11) gives

$$\frac{5}{420}C_1{}^2 C_2 x^{\frac{1}{4}} = g\beta(T_w - T_\infty)\frac{C_2}{3}x^{\frac{1}{4}} - \frac{C_1}{C_2}\nu x^{\frac{1}{4}} \tag{7-16}$$

and

$$\frac{1}{40}C_1 C_2 x^{-\frac{1}{4}} = \frac{2\alpha}{C_2}x^{-\frac{1}{4}} \tag{7-17}$$

These two equations may be solved for the constants $C_1$ and $C_2$ to give

$$C_1 = 5.17\nu \left(\frac{20}{21} + \frac{\nu}{\alpha}\right)^{-\frac{1}{2}} \left[\frac{g\beta(T_w - T_\infty)}{\nu^2}\right]^{\frac{1}{2}} \tag{7-18}$$

$$C_2 = 3.93 \left(\frac{20}{21} + \frac{\nu}{\alpha}\right)^{\frac{1}{4}} \left[\frac{g\beta(T_w - T_\infty)}{\nu^2}\right]^{-\frac{1}{4}} \left(\frac{\nu}{\alpha}\right)^{-\frac{1}{2}} \tag{7-19}$$

The resultant expression for the boundary-layer thickness is

$$\frac{\delta}{x} = 3.93 \, \mathrm{Pr}^{-\frac{1}{2}} \, (0.952 + \mathrm{Pr})^{\frac{1}{4}} \, \mathrm{Gr}_x^{-\frac{1}{4}} \tag{7-20}$$

where the Prandtl number $\mathrm{Pr} = \nu/\alpha$ has been introduced along with a new dimensionless group called the Grashof number $\mathrm{Gr}_x$

$$\mathrm{Gr}_x = \frac{g\beta(T_w - T_\infty)x^3}{\nu^2} \tag{7-21}$$

The heat-transfer coefficient may be evaluated from

$$q_w = -kA \left.\frac{dT}{dy}\right)_w = hA(T_w - T_\infty)$$

Using the temperature distribution of Eq. (7-7), one obtains

$$h = \frac{2k}{\delta}$$

or

$$\frac{hx}{k} = \mathrm{Nu}_x = 2\frac{x}{\delta} \quad \leftarrow \text{No exponent}$$

so that the dimensionless equation for the heat-transfer coefficient becomes

$$\mathrm{Nu}_x = 0.508 \, \mathrm{Pr}^{\frac{1}{2}} \, (0.952 + \mathrm{Pr})^{-\frac{1}{4}} \, \mathrm{Gr}_x^{\frac{1}{4}} \tag{7-22}$$

Equation (7-22) gives the variation of the local heat-transfer coefficient along the vertical plate. The average heat-transfer coefficient may then be obtained by performing the integration

$$\bar{h} = \frac{1}{L} \int_0^L h_x \, dx \tag{7-23}$$

For the variation given by Eq. (7-22), the average coefficient is

$$\bar{h} = \tfrac{4}{3} h_{x=L} \tag{7-24}$$

The Grashof number may be interpreted physically as a dimensionless group representing the ratio of the buoyancy forces to the viscous forces in the free-convection flow system. It has a role similar to that played by the Reynolds number in forced-convection systems and is the primary variable used as a criterion for transition from laminar to turbulent boundary-layer flow. For air in free convection on a vertical flat plate the critical Grashof number has been observed by Eckert and Soehngen [1] to be approximately $4 \times 10^8$. Values ranging between $10^8$ and $10^9$ may be observed for different fluids and environment "turbulence levels."

A very complete survey of the stability and transition of free convection boundary layers has been given by Gebhart et al. [13, 14, 15].

The foregoing analysis of free convection heat transfer on a vertical flat plate is the simplest case that may be treated mathematically, and it has served to introduce the new dimensionless variable, the Grashof number,[†] which is important in all free-convection problems. But as in some forced-convection problems, experimental measurements must be relied upon to obtain relations for heat transfer in other circumstances. These circumstances are usually those in which it is difficult to predict temperature and velocity profiles analytically. Turbulent free convection is an important example, just as is turbulent forced convection, of a problem area in which experimental data are necessary; however, the problem is more acute with free-convection flow systems than with forced-convection systems because the velocities are usually so small that they are very difficult to measure. Despite the experimental difficulties, velocity measurements have been performed using hydrogen-bubble techniques [26], hot-wire anemometry [28], and quartz-fiber anemometers. Temperature field measurements have been obtained through the use of the Zehnder-Mach interferometer. The laser anemometer [29] offers unusual promise for further free-convection measurements because it does not disturb the flow field.

An interferometer indicates lines of constant density in a fluid flow field. For a gas in free convection at low pressure these lines of constant density are equivalent to lines of constant temperature. Once the temperature field is obtained, the heat transfer from a surface in free convection may be calculated by using the temperature gradient at the surface and the thermal conductivity of the gas. Several interferometric studies of free convection have been made [1–3], and some typical photographs of the flow fields are shown in Figs. 7-3 to 7-6. Figure 7-3 shows the lines of constant temperature around a heated vertical flat plate. Notice that the lines are closest together near the plate surface, indicating a higher temperature gradient in that region. Figure 7-4 shows the lines of constant temperature around a heated horizontal cylinder in free convection, and Fig. 7-5 shows the boundary-layer interaction between a group of four horizontal cylinders. A similar phenomenon would be observed for forced-convection flow across a heated tube bank. Interferometric studies have been conducted to determine the point at which eddies are formed in the free-convection boundary layer [1], and these studies have been used in predicting the start of tran-

---

† History is not clear on the point, but it appears that the Grashof number was named for Franz Grashof a professor of applied mechanics at Karlsruhe around 1863 and one of the founding directors of *Verein deutscher Ingenieure* in 1855. He developed some early steam flow formulas but made no significant contributions to free convection [36].

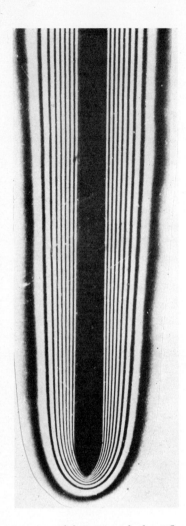

Fig. 7-3 Interferometer photograph showing lines of constant temperature around a heated vertical flat plate in free convection. (Photograph courtesy of E. Soehngen.)

sition to turbulent flow in free-convection systems.

It was mentioned above that the velocities in free convection are so small that for most systems they are difficult to measure without influencing the flow field by the insertion of a measuring device. A rough visual indication of the free-convection velocity profile is given in Fig. 7-6, where a free-convection boundary-layer wave resulting from a heat pulse near the leading edge of the plate is presented. It may be noted that the maximum points in the isotherms experience a phase lag and that a line drawn through these maximum points has the approximate shape of the free-convection velocity profile.

There are a number of references that treat the various theoretical and empirical aspects of free-convection problems. One of the most extensive discussions is given by Gebhart [13], and the interested reader may wish to consult this reference for additional information.

## 7-3□EMPIRICAL RELATIONS FOR FREE CONVECTION

Over the years it has been found that average free-convection heat-transfer coefficients can be represented in the following functional form for a variety of circumstances.

$$\overline{Nu}_f = C(\mathrm{Gr}_f \, \mathrm{Pr}_f)^m \qquad (7\text{-}25)$$

where the subscript $f$ indicates that the properties in the dimensionless groups are evaluated at the film temperature

$$T_f = \frac{T_\infty + T_w}{2}$$

The characteristic dimension to be used in the Nusselt and Grashof numbers depends on the geometry of the problem. For a vertical plate it is the height of the plate $L$; for a horizontal cylinder it is the diameter $d$, and so forth. Experimental data for free-convection problems appear in a number of references, with some conflicting results. The purpose of the sections that follow is to give these results in a summary form that may be easily used for calculation purposes. The functional form of Eq. (7-25) is used for these presentations, with the values of the constants $C$ and $m$ specified for each case.

## 7-4□FREE CONVECTION FROM VERTICAL PLANES AND CYLINDERS

### Isothermal Surfaces

For vertical surfaces, the Nusselt and Grashof numbers are formed with $L$, the height of the surface as the characteristic dimension. If the boundary-layer thickness is not large compared with the diameter of the cylinder, the heat transfer may be calculated with the same relations used for vertical plates. The general criterion is that a vertical cylinder may be treated as a flat plate [13] when

$$\frac{D}{L} \geq \frac{35}{\mathrm{Gr}_L^{\frac{1}{4}}}$$

where $D$ is the diameter of the cylinder. For *isothermal* surfaces the values of the constants are given in Table 7-1 with the appropriate references noted for further consultation. The readers attention is directed to the two sets of constants given for the turbulent case. ($\mathrm{Gr}_f \, \mathrm{Pr}_f > 10^9$.) Though there may appear to be a decided difference in these constants, a comparison by Warner and Arpaci [22] of the two relations with experimental data indicates that both sets of constants fit available data. There are some indications from the analytical work of Bayley [16], as well as heat flux measurements of Ref. 22 that the relation

$$\mathrm{Nu}_f = 0.10(\mathrm{Gr}_f \, \mathrm{Pr}_f)^{\frac{1}{3}}$$

may be the preferable one to use.

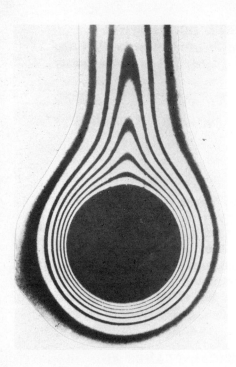

Fig. 7-4  Interferometer photograph showing lines of constant temperature around a heated horizontal cylinder in free convection. (Photograph courtesy of E. Soehngen.)

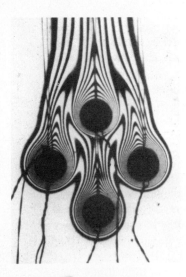

Fig. 7-5  Interferometer photograph showing the boundary-layer interaction between four heated horizontal cylinders in free convection. (Photograph courtesy of E. Soehngen.)

Fig. 7-6  Interferometer photograph showing isotherms on a heated vertical flat plate resulting from a periodic disturbance of the boundary layer. Note phase shift in maximum points of the isotherms, from Holman, Gartrell, and Soehngen [3].

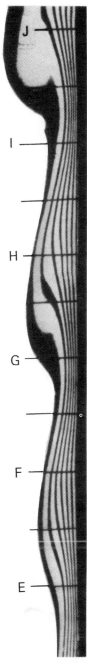

**Fig. 7-7**    Free-convection heat-transfer correlation for heat transfer from heated vertical plates, according to McAdams [4].

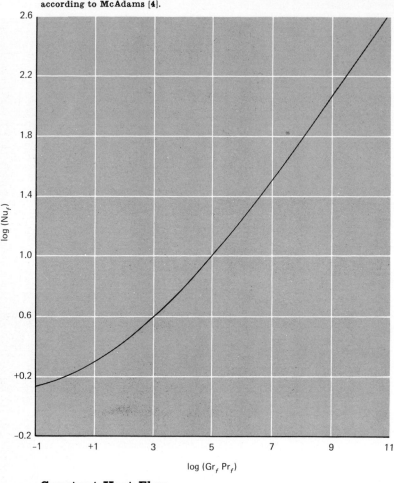

## Constant Heat Flux

Extensive experiments have been reported in Refs. 25 and 26 for free convection from vertical and inclined surfaces to water under constant heat flux conditions. In such experiments, the results are presented in terms of a modified Grashof number, $Gr^*$

$$Gr_x^* = Gr_x \, Nu_x = \frac{g\beta q_w x^4}{k\nu^2} \tag{7-26}$$

where $q_w$ is the wall heat flux in Btu/hr-ft². The *local* heat-transfer coefficients were correlated by the following relation for the laminar range

$$Nu_{x_f} = \frac{hx}{k_f} = 0.60(Gr_x^* \, Pr_f)^{\frac{1}{5}} \qquad 10^5 < Gr_x^* < 10^{11}$$

$$q_w = const \tag{7-27}$$

**Fig. 7-8**  Free-convection heat-transfer correlation for heat transfer from heated horizontal cylinders, according to McAdams [4].

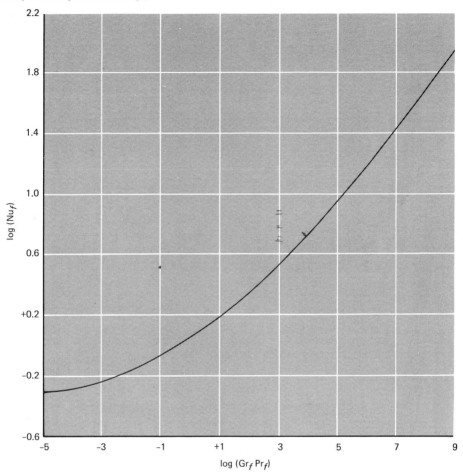

It is to be noted that the criterion for laminar flow expressed in terms of $Gr_x^*$ is not the same as that expressed in terms of $Gr_x$. Boundary-layer transition was observed to begin between $Gr_x^*\,Pr = 3 \times 10^{12}$ and $4 \times 10^{13}$ and to end between $2 \times 10^{13}$ and $10^{14}$. Fully developed turbulent flow was present by $Gr_x^*\,Pr = 10^{14}$ and the experiments were extended up to $Gr_x^*\,Pr = 10^{16}$. For the turbulent region, the local heat-transfer coefficients were correlated with

$$Nu_x = 0.568(Gr_x^*\,Pr)^{0.22} \qquad 2 \times 10^{13} < Gr_x^*\,Pr < 10^{16}$$
$$q_w = \text{const} \tag{7-28}$$

All properties in Eqs. (7-27) and (7-28) are evaluated at the local film temperature. Although these experiments were conducted for water, the resulting correlations are shown to work for air as well. The average heat-

## Table 7-1   Constants for Use with Eq. (7-25) for Isothermal Vertical Surfaces

| Geometry | $Gr_f\,Pr_f$ | $C$ | $m$ | Ref. |
|---|---|---|---|---|
| Vertical planes and cylinders | $10^{-1}\text{–}10^4$ | Use Fig. 7-7 | Use Fig. 7-7 | 4 |
|  | $10^4\text{–}10^9$ | 0.59 | $\frac{1}{4}$ | 4 |
|  | $10^9\text{–}10^{13}$ | 0.021 | $\frac{2}{5}$ | 30 |
|  | $10^9\text{–}10^{13}$ | 0.10 | $\frac{1}{3}$ | 22, 16 † |
| Horizontal cylinders | $0\text{–}10^{-5}$ | 0.4 | 0 | 4 |
|  | $10^{-5}\text{–}10^4$ | Use Fig. 7-8 | Use Fig. 7-8 | 4 |
|  | $10^4\text{–}10^9$ | 0.53 | $\frac{1}{4}$ | 4 |
|  | $10^9\text{–}10^{12}$ | 0.13 | $\frac{1}{3}$ | 4 |
| Upper surface of heated plates or lower surface of cooled plates | $10^5\text{–}2\times10^7$ | 0.54 | $\frac{1}{4}$ | 4 |
| Lower surface of heated plates or upper surface of cooled plates | $3\times10^5\text{–}3\times10^{10}$ | 0.27 | $\frac{1}{4}$ | 4 |
| Upper surface of heated plates or lower surface of cooled plates | $2\times10^7\text{–}3\times10^{10}$ | 0.14 | $\frac{1}{3}$ | 4 |

† Preferred.

transfer coefficient for the constant heat flux case may not be evaluated from Eq. (7-24) but must be obtained through a separate application of Eq. (7-23). Thus, for the laminar region, using Eq. (7-27) to evaluate $h_x$,

$$\bar{h} = \frac{1}{L}\int_0^L h_x\,dx$$

$$\bar{h} = \tfrac{5}{4}h_{x=L} \qquad q_w = \text{const}$$

## 7-5 □ FREE CONVECTION FROM HORIZONTAL CYLINDERS

McAdams [4] has assembled the data of a number of investigators to predict the average heat-transfer coefficient from horizontal cylinders. The values of the constants $C$ and $m$ are given in Table 7-1. It may be noted that the accuracy of these correlations is only about $\pm 15$ percent. The more recent work of Gebhart, Andunson, and Pera [35] presents refinements in empirical correlations for small wires, and the Prandtl number is found to influence the results in a more complicated fashion than through the simple Gr Pr product.

## 7-6 □ FREE CONVECTION FROM HORIZONTAL PLATES

The average heat-transfer coefficient from horizontal flat plates is calculated with Eq. (7-25) and the constants given in Table 7-1. The characteristic dimension $L$ for use with these relations is the length of a side for a square, the mean of the two dimensions for a rectangular surface, and $0.9d$ for a circular disk.

**Table 7-2    Simplified Equations for Free Convection from Various Surfaces to Air at Atmospheric Pressure According to McAdams [4]**

| Surface | Laminar,<br>$10^4 < \mathrm{Gr}_f \, \mathrm{Pr}_f < 10^9$ | Turbulent,<br>$\mathrm{Gr}_f \, \mathrm{Pr}_f > 10^9$ |
|---|---|---|
| Vertical planes or cylinders | $h = 0.29 \left( \dfrac{\Delta T}{L} \right)^{\frac{1}{4}}$ | $h = 0.19 (\Delta T)^{\frac{1}{3}}$ |
| Horizontal cylinders | $h = 0.27 \left( \dfrac{\Delta T}{d} \right)^{\frac{1}{4}}$ | $h = 0.18 (\Delta T)^{\frac{1}{3}}$ |
| Horizontal plates: | | |
| Heated plates facing upward or cooled plates facing downward | $h = 0.27 \left( \dfrac{\Delta T}{L} \right)^{\frac{1}{4}}$ | $h = 0.22 (\Delta T)^{\frac{1}{3}}$ |
| Heated plates facing downward or cooled plates facing upward | $h = 0.12 \left( \dfrac{\Delta T}{L} \right)^{\frac{1}{4}}$ | |

$h$ in Btu/hr-ft²-°F
$\Delta T = T_w - T_\infty$, °F
$L$ = vertical or horizontal dimension, ft
$d$ = diameter, ft

## Rectangular Solids

The heat transfer from a number of plates, cylinders, spheres, and blocks has been analyzed by King [31]. It was found that the average heat-transfer coefficient could be correlated in the form of Eq. (7-25) with $C = 0.60$ and $m = \frac{1}{4}$ for $10^4 < \mathrm{Gr} \, \mathrm{Pr} < 10^9$. In this relation, the characteristic dimension for rectangular blocks is calculated from

$$\frac{1}{L} = \frac{1}{L_h} + \frac{1}{L_v} \tag{7-29}$$

where $L_h$ and $L_v$ are the horizontal and vertical dimensions, respectively. This relation should only be used in the absence of specific data for the particular body shape.

## 7-7□SIMPLIFIED EQUATIONS FOR AIR

Simplified equations for the heat-transfer coefficient from various surfaces to air at atmospheric pressure and moderate temperatures are given in Table 7-2. These relations may be extended to higher or lower pressures by multiplying by the following factors:

$$\left( \frac{p}{14.7} \right)^{\frac{1}{2}} \quad \text{for laminar cases}$$

$$\left( \frac{p}{14.7} \right)^{\frac{2}{3}} \quad \text{for turbulent cases}$$

where $p$ is the pressure in psia. Due caution should be exercised in the use of these simplified relations because they are only approximations of the more precise equations stated earlier.

## Example 7-1

Steam at 500°F flows through a 12-in.-OD pipe which is exposed to atmospheric air at 50°F. Calculate the heat transfer per foot of length.

**Solution**   We first determine the Grashof-Prandtl number product and then select the appropriate constants from Table 7-1 for use with Eq. (7-25). The properties of air are evaluated at the film temperature.

$$T_f = \frac{T_w + T_\infty}{2} = \frac{500 + 50}{2} = 275°F$$

$\rho = 0.054 \text{ lb}_m/\text{ft}^3$     $(0.865 \text{ kg/m}^3)$
$c_p = 0.24 \text{ Btu/lb}_m\text{-°F}$     $(1005 \text{ J/kg-°C})$
$\mu = 1.57 \times 10^{-5} \text{ lb}_m/\text{ft-sec} = 0.0565 \text{ lb}_m/\text{ft-hr}$
$k = 0.0197 \text{ Btu/hr-ft-°F}$     $(0.034 \text{ W/m-°C})$
$\beta = 1.36 \times 10^{-1} \text{ °F}^{-1}$

$$\begin{aligned}
\text{Gr}_d \text{ Pr} &= \frac{c_p g \beta (T_w - T_\infty) \rho^2 d^3}{k\mu} \\
&= \frac{(0.24)(32.2)(3600)^2(1.36 \times 10^{-3})(500 - 50)(0.054)^2(1)^3}{(0.0197)(0.0565)} \\
&= 1.60 \times 10^8
\end{aligned}$$

From Table 7-1, $C = 0.53$ and $m = \frac{1}{4}$, so that

$$\begin{aligned}
\text{Nu}_d &= 0.53 \, (\text{Gr}_d \text{ Pr})^{\frac{1}{4}} \\
&= (0.53)(1.60 \times 10^8)^{\frac{1}{4}} = 59.7 \\
h &= \frac{k}{d} \text{Nu}_d = \frac{(0.0197)(59.7)}{1} = 1.175 \text{ Btu/hr-ft}^2\text{-°F} \quad (6.67 \text{ W/m}^2\text{-°C})
\end{aligned}$$

The heat transfer per unit length is then calculated from

$$\begin{aligned}
\frac{q}{L} &= h\pi d(T_w - T_\infty) \\
&= (1.175)\pi(1)(500 - 50) = 1660 \text{ Btu/hr-ft} \quad (1,596 \text{ W/m})
\end{aligned}$$

## Example 7-2

For the conditions of Example 7-1 compute the heat transfer, using the simplified relations of Table 7-2.

**Solution**   The heat-transfer coefficient is given by

$$h = 0.27 \left(\frac{\Delta T}{d}\right)^{\frac{1}{4}}$$

so that the heat transfer per unit length is

$$\begin{aligned}
\frac{q}{L} &= h\pi d(T_w - T_\infty) \\
&= 0.27\pi(\Delta T)^{\frac{1}{4}} d^{\frac{3}{4}} \\
&= (0.27)\pi(500 - 50)^{\frac{1}{4}}(1)^{\frac{3}{4}} \\
&= 1760 \text{ Btu/hr-ft} \quad (1,692 \text{ W/m})
\end{aligned}$$

Fig. 7-9   Nomenclature for free convection in enclosed vertical spaces.

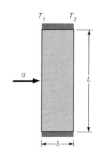

Note that the simplified relation gives a value approximately 6 percent higher than does Eq. (7-25).

## 7-8□FREE CONVECTION FROM SPHERES

Yuge [5] recommends the following empirical relation for free-convection heat transfer from spheres to air:

$$\text{Nu}_f = \frac{\bar{h}d}{k_f} = 2 + 0.392 \, \text{Gr}_{f}^{\frac{1}{4}} \qquad \text{for } 1 < \text{Gr}_f < 10^5 \qquad (7\text{-}30)$$

This equation may be modified by the introduction of the Prandtl number to give

$$\text{Nu}_f = 2 + 0.43 \, (\text{Gr}_f \, \text{Pr}_f)^{\frac{1}{4}} \qquad (7\text{-}31)$$

Properties are evaluated at the film temperature, and it is expected that this relation would be primarily applicable to calculations for free convection in gases.

## 7-9□FREE CONVECTION IN ENCLOSED SPACES

The free-convection flow phenomena inside an enclosed space are interesting examples of very complex fluid systems that may yield to analytical, empirical, and numerical solutions. Consider the system shown in Fig. 7-9 where a fluid is contained between two vertical plates separated by the distance $\delta$. As a temperature difference $\Delta T_w = T_1 - T_2$ is impressed on the fluid, a heat transfer will be experienced with the approximate flow regions shown in Fig. 7-10 according to MacGregor and Emery [18]. In this figure, the Grashof number is calculated as

$$\text{Gr}_\delta = \frac{g\beta(T_1 - T_2)\delta^3}{\nu^2} \qquad (7\text{-}31a)$$

At very low Grashof numbers, there are very minute free-convection currents and the heat transfer occurs mainly by conduction across the fluid layer. As the Grashof number is increased, different flow regimes are encountered as shown, with a progressively increasing heat transfer as expressed through the Nusselt number

$$\text{Nu}_\delta = \frac{h\delta}{k}$$

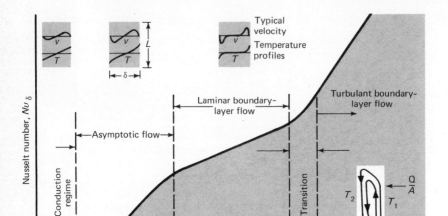

**Fig. 7-10**  Schematic diagram and flow regimes of the vertical convection layer, according to Ref. 18.

Although some open questions still remain, the experiments of Ref. 18 may be used to predict the heat transfer to a number of liquids under constant heat flux conditions. The empirical correlations obtained were

$$\mathrm{Nu}_\delta = 0.42(\mathrm{Gr}_\delta\,\mathrm{Pr})^{\frac{1}{4}}\,\mathrm{Pr}^{0.012}\left(\frac{L}{\delta}\right)^{-0.30} \qquad \text{for} \qquad q_w = \text{const} \qquad (7\text{-}32)$$

$$10^4 < \mathrm{Gr}_\delta\,\mathrm{Pr} < 10^7$$
$$1 < \mathrm{Pr} < 20{,}000$$
$$10 < L/\delta < 40$$

$$\mathrm{Nu}_\delta = 0.046\,(\mathrm{Gr}_\delta\,\mathrm{Pr})^{\frac{1}{3}} \qquad \text{for} \qquad q_w = \text{const} \qquad (7\text{-}33)$$
$$10^6 < \mathrm{Gr}_\delta\,\mathrm{Pr} < 10^9$$
$$1 < \mathrm{Pr} < 20$$
$$1 < L/\delta < 40$$

The heat flux is calculated as

$$\frac{q}{A} = q_w = h(T_1 - T_2) = \mathrm{Nu}_\delta\,\frac{k}{\delta}\,(T_1 - T_2) \qquad (7\text{-}34)$$

The results are sometimes expressed in terms of an effective or *apparent thermal conductivity* $k_e$, defined by

$$\frac{q}{A} = k_e\,\frac{T_1 - T_2}{\delta} \qquad (7\text{-}35)$$

Fig. 7-11 Benard cell pattern in
enclosed fluid layer heated from
below (33°).

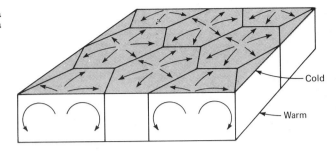

Fig. 7-11 Benard cell pattern in
enclosed fluid layer heated from
below (33°).

By comparing Eqs. (7-34) and (7-35), we see that

$$\mathrm{Nu}_\delta \equiv \frac{k_e}{k} \qquad (7\text{-}36)$$

Empirical formulas for calculation of free convection in an enclosed air space between two isothermal vertical walls have been presented by Jakob [6] in the following form:

$$\frac{k_e}{k} = \begin{cases} 0.18 \ \mathrm{Gr}_\delta^{\frac{1}{4}} \left(\frac{L}{\delta}\right)^{-\delta} & \text{for } 2000 < \mathrm{Gr}_\delta < 20{,}000 \qquad (7\text{-}37) \\[2mm] 0.065 \ \mathrm{Gr}_\delta^{\frac{1}{3}} \left(\frac{L}{\delta}\right)^{-\delta} & \text{for } 20{,}000 < \mathrm{Gr}_\delta < 11 \times 10^6 \quad (7\text{-}38) \end{cases}$$

For Grashof numbers below 2,000, the heat transfer is essentially all conduction so that $k_e/k = 1.0$ for that region. Properties for use in the foregoing equations are evaluated at the mean temperature between the two walls.

Heat transfer in horizontal enclosed spaces involves two distinct situations. If the upper plate is maintained at a higher temperature than the lower plate, the lower density fluid is above the higher density fluid and no convection currents will be experienced. In this case the heat transfer across the space will be by conduction alone and $\mathrm{Nu}_\delta = 1.0$, where $\delta$ is still the separation distance between the plates. The second, and more interesting, case is experienced when the lower plate has a higher temperature than the upper plate. For values of $\mathrm{Gr}_\delta$ below about 1,700, pure conduction is still observed and $\mathrm{Nu}_\delta = 1.0$. As convection begins, a pattern of hexagonal cells are formed as shown in Fig. 7-11. These patterns are called Benard cells [33]. Turbulence begins at about $\mathrm{Gr}_\delta = 50{,}000$ and destroys the cellular pattern. For free convection in horizontal air spaces, Jakob [7] recommends the following:

$$\frac{k_e}{k} = \begin{cases} 0.195 \ \mathrm{Gr}_\delta^{\frac{1}{4}} & \text{for } 10^4 < \mathrm{Gr}_\delta < 4 \times 10^5 \qquad (7\text{-}39) \\[1mm] 0.068 \ \mathrm{Gr}_\delta^{\frac{1}{3}} & \text{for } 4 \times 10^5 < \mathrm{Gr}_\delta \qquad\quad (7\text{-}40) \end{cases}$$

The experiments of Globe and Dropkin [8] are useful for calculation of free convection through liquids in horizontal spaces. The results of their work indicate that

$$\frac{k_e}{k} = 0.069 \ Gr_\delta^{\frac{1}{3}} \ Pr^{0.407} \qquad \text{for } 3 \times 10^5 < Gr_\delta \ Pr < 7 \times 10^9 \quad (7\text{-}41)$$

Free convection in inclined enclosures is discussed by Dropkin and Somerscales [12]. Evans and Stefany [9] have shown that transient natural-convection heating or cooling in closed vertical or horizontal cylindrical enclosures may be calculated with

$$Nu_f = 0.55 \ (Gr_f \ Pr_f)^{\frac{1}{4}} \qquad\qquad (7\text{-}42)$$

for the range $0.75 < L/d < 2.0$. The Grashof number is formed with the length of the cylinder, $L$.

Experiments with air enclosed between concentric spheres have been conducted by Bishop, Mack, and Scanlan [11]. In this work it was possible to represent the effective thermal conductivity with the relation

$$\frac{k_e}{k} = 0.106 \ Gr_\delta^{0.276} \qquad\qquad (7\text{-}43)$$

where now the gap spacing is $\delta = r_o - r_i$. The effective thermal conductivity given by Eq. (7-43) is to be used with the conventional relation for steady-state conduction in a spherical shell.

$$q = \frac{4\pi k r_i r_o \ \Delta T}{r_o - r_i} \qquad\qquad (7\text{-}44)$$

Equation (7-44) is valid for $0.25 \leq \delta/r_i \leq 1.50$ and $2.0 \times 10^4 \leq Gr_\delta \leq 3.6 \times 10^6$.

## Example 7-3

Air at atmospheric pressure is contained between two 2-ft-square vertical plates separated by $\frac{1}{2}$ in. The temperatures of the plates are 150°F and 50°F, respectively. Calculate the heat transfer across the space.

**Solution**  Evaluating the air properties at the average temperature of 100°F, we have

$$\mu = 0.0459 \ lb_m/hr\text{-}ft$$
$$k = 0.0157 \ Btu/hr\text{-}ft\text{-}°F$$
$$\rho = \frac{p}{RT} = \frac{(14.7)(144)}{(53.35)(560)} = 0.0708 \ lb_m/ft^3$$
$$\beta = \frac{1}{T} = \frac{1}{560} = 0.00178$$

The Grashof number is now calculated from Eq. (7-31) as

$$\mathrm{Gr}_\delta = \frac{(32.2)(3600)^2(0.00178)(0.0708)^2(150 - 50)(0.5/12)^3}{(0.0459)^2}$$
$$= 1.278 \times 10^4$$

We may now use Eq. (7-37) to calculate the effective thermal conductivity, with $L = 2$ ft $= 24$ in.,

$$\frac{k_e}{k} = 0.18(1.278 \times 10^4)^{\frac{1}{4}} \left(\frac{24}{0.5}\right)^{-\frac{1}{9}}$$
$$= 1.244$$

The heat transfer may now be calculated with Eq. (7-35). The area is 4 ft$^2$.

$$q = \frac{(1.244)(0.0157)(4)(150 - 50)}{(0.5/12)}$$
$$= 187.6 \text{ Btu/hr} \qquad (55 \text{ W})$$

For this problem the heat transfer is only increased by 24 percent over the value for pure conduction between the plates.

STOP AFTER HERE

## 7-10□COMBINED FREE AND FORCED CONVECTION

A number of practical situations involve convection heat transfer which is neither "forced" nor "free" in nature. The circumstances arise when a fluid is forced over a heated surface at a rather low velocity. Coupled with the forced-flow velocity is a convective velocity which is generated by the buoyancy forces resulting from a reduction in fluid density near the heated surface.

A summary of combined free- and forced-convection effects in tubes has been given by Metais and Eckert [10], and Fig. 7-12 presents the regimes for combined convection in vertical tubes. Two different combinations are indicated in this figure. Aiding flow means that the forced- and free-convection currents are in the same direction, while opposing flow means that they are in the opposite direction. The symbol UWT means uniform wall temperature, and the symbol UHF indicates data for uniform heat flux. It is fairly easy to anticipate the qualitative results of the figure. A large Reynolds number implies a large forced-flow velocity, and hence less influence of free-convection currents. The larger the value of the Grashof-Prandtl product, the more one would expect free-convection effects to prevail.

Figure 7-13 presents the regimes for combined convection in horizontal tubes. In this figure the Graetz number is defined as

$$\mathrm{Gz} = \mathrm{Re} \ \mathrm{Pr} \ \frac{d}{L} \qquad\qquad (7\text{-}45)$$

**Fig. 7-12**  Regimes of free, forced, and mixed convection for flow through vertical tubes, according to [10].

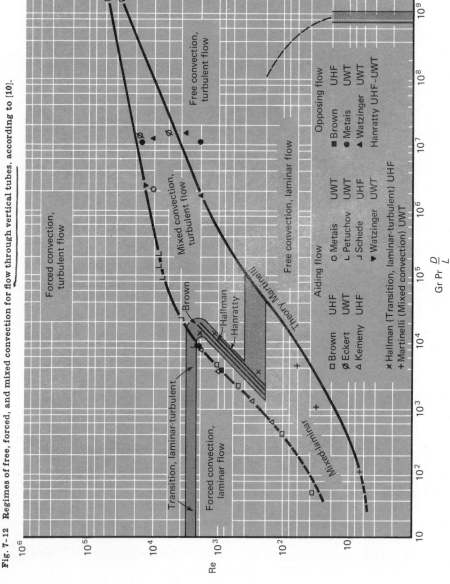

**Fig. 7-13** Regimes of free, forced, and mixed convection for flow through horizontal tubes, according to Metais and Eckert [10].

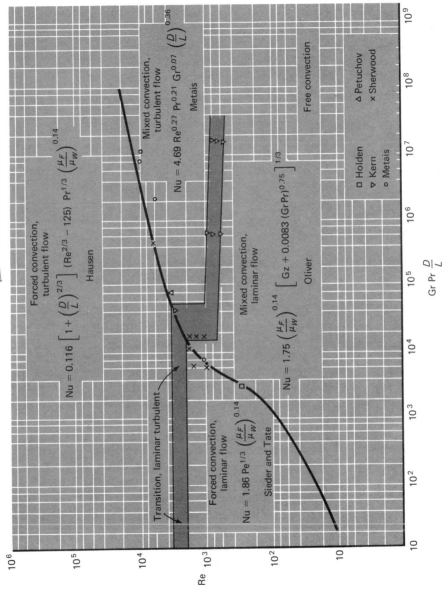

The applicable range of Figs. 7-12 and 7-13 is for

$$10^{-2} < \text{Pr}\left(\frac{d}{L}\right) < 1$$

The correlations presented in the figures are for constant wall temperature. All properties are evaluated at the film temperature.

Brown and Gauvin [17] have developed a better correlation for the mixed convection, laminar flow region of Fig. 7-13.

$$\text{Nu} = 1.75 \left(\frac{\mu_b}{\mu_w}\right)^{0.14} [\text{Gz} + 0.012 \, (\text{Gz} \, \text{Gr}^{\frac{1}{3}})^{\frac{4}{3}}]^{\frac{1}{3}} \qquad (7\text{-}45)$$

where $\mu_b$ is evaluated at the bulk temperature. This relation is preferred over that shown in Fig. 7-13.

The general notion which is applied in combined-convection analysis is that the predominance of a heat-transfer mode is governed by the fluid velocity associated with that mode. A forced-convection situation involving a fluid velocity of 100 ft/sec, for example, would be expected to overshadow most free-convection effects encountered in ordinary gravitational fields because the velocities of the free-convection currents are small in comparison with 100 ft/sec. On the other hand, a forced-flow situation at very low velocities ($\sim$1 ft/sec) might be influenced appreciably by free-convection currents. An order-of-magnitude analysis of the free-convection boundary-layer equations will indicate a general criterion for determining whether free-convection effects dominate. The criterion is that when $\text{Gr}/\text{Re}^2 > 1.0$ free convection is of primary importance. This result is in agreement with Figs. 7-12 and 7-13. Example 7-4 illustrates a situation where both free- and forced-convection effects are important.

Several additional interesting free convection phenomena are discussed in Ref. 34.

### Example 7-4

Air at 14.7 psia and 70°F is forced through a horizontal 1-in.-diameter tube at an average velocity of 1 ft/sec. The tube wall is maintained at a constant temperature of 280°F. Calculate the heat-transfer coefficient for this situation if the tube is 12 in. long.

**Solution**  For this calculation we evaluate properties at the film temperature.

$$T_f = (280 + 70)/2 = 175°\text{F} = 635°\text{R}$$
$$\rho_f = \frac{p}{RT} = \frac{(14.7)(144)}{(53.35)(635)} = 0.0625 \, \text{lb}_m/\text{ft}^3$$
$$\mu_f = 0.0501 \, \text{lb}_m/\text{hr-ft}$$
$$k_f = 0.0174 \, \text{Btu/hr-ft-°F}$$
$$\text{Pr} = 0.696$$
$$\beta_f = 0.00158°\text{F}^{-1}$$

Let us take the bulk temperature as 70°F for evaluating $\mu_b$; then $\mu_b = 0.0439 \text{ lb}_m/\text{hr-ft}$.

The significant parameters are calculated as

$$\text{Re} = \frac{\rho u d}{\mu} = \frac{(0.0625)(1)(3600)(1)}{(12)(0.0501)} = 374$$

$$\text{Gr} = \frac{\rho^2 g \beta (T_w - T_b) d^3}{\mu^2} = 1.25 \times 10^5$$

$$\text{Gr Pr}\frac{d}{L} = \frac{(1.25 \times 10^5)(0.696)(1)}{12} = 7.25 \times 10^3$$

According to Fig. 7-13, the mixed-convection laminar flow region is encountered. Thus we must use Eq. (7-45). The Graetz number is calculated as

$$\text{Gz} = \text{Re Pr}\frac{d}{L} = \frac{(374)(0.696)(1)}{12} = 21.7$$

and the numerical calculation for Eq. (7-45) becomes

$$\text{Nu} = 1.75\left(\frac{0.0439}{0.0553}\right)^{0.14}\{21.7 + (0.012)[(21.7)(1.25 \times 10^5)^{\frac{1}{3}}]^{\frac{4}{3}}\}^{\frac{1}{3}}$$

$$= 9.08$$

The average heat-transfer coefficient is then calculated as

$$\bar{h} = \frac{(9.08)(0.0174)}{\frac{1}{12}} = 1.90 \text{ Btu/hr-ft}^2\text{-}°\text{F}$$

It is of interest to compare this value with that which would be obtained for strictly laminar forced convection. The Sieder-Tate relation [Eq. (6-5)] applies, so that

$$\text{Nu} = 1.86 \,(\text{Re Pr})^{\frac{1}{3}} \left(\frac{\mu_f}{\mu_w}\right)^{0.14}\left(\frac{d}{L}\right)^{\frac{1}{3}}$$

$$= 1.86 \,\text{Gz}^{\frac{1}{3}} \left(\frac{\mu_f}{\mu_w}\right)^{0.14}$$

$$= 5.05$$

Then

$$\bar{h} = \frac{(5.05)(0.0174)}{\frac{1}{12}} = 1.055 \text{ Btu/hr-ft}^2\text{-}°\text{F}$$

Thus there would be an error of $-45$ percent if the calculation were made strictly on the basis of laminar forced convection.

## REVIEW QUESTIONS

1. Why is an analytical solution of a free-convection problem more involved than its forced-convection counterpart?

2. Define the Grashof number. What is its physical significance?

3. What is the approximate criterion for transition to turbulence in a free-convection boundary layer?

4.   What is the functional form of equation normally used for correlation of free-convection heat-transfer data?

5.   Discuss the problem of combined free and forced convection.

6.   What is the approximate criterion dividing pure conduction and free convection in an enclosed space between vertical walls?

7.   How is a modified Grashof number defined for a constant heat flux condition on a vertical plate?

## PROBLEMS

7-1.   Plot the free-convection boundary-layer thickness as a function of $x$ for a vertical plate maintained at 175°F and exposed to air at atmospheric pressure and 60°F. Consider the laminar portion only.

7-2.   Derive an expression for the maximum velocity in the free-convection boundary layer on a vertical flat plate. At what position in the boundary layer does this maximum velocity occur?

7-3.   Two vertical flat plates at 150°F are placed in a tank of water at 80°F. If the plates are 1 ft high, what is the minimum spacing which will prevent interference of the free-convection boundary layers?

7-4.   A 1-ft-square vertical plate is maintained at 150°F and is exposed to atmospheric air at 60°F. Compare the free-convection heat transfer from this plate with that which would result from forcing air over the plate at a velocity equal to the maximum velocity which occurs in the free-convection boundary layer. Discuss this comparison.

7-5.   A 4-in. length of platinum wire 0.015 in. in diameter is placed horizontally in a container of water at 100°F and is electrically heated so that the surface temperature is maintained at 200°F. Calculate the heat lost by the wire.

7-6.   Water at the rate of 100 $lb_m$/min at 200°F flows through a steel pipe 1 in. ID and $1\frac{1}{4}$ in. OD. The outside surface temperature of the pipe is 185°F, and the temperature of the surrounding air is 70°F. The room pressure is 14.7 psia, and the pipe is 50 ft long. How much heat is lost by free convection to the room?

7-7.   A horizontal pipe 3 in. in diameter is located in a room where atmospheric air is at 70°F. The surface temperature of the pipe is 460°F. Calculate the free-convection heat loss per foot of pipe.

7-8.   A horizontal $\frac{1}{2}$-in.-OD tube is heated to a surface temperature of 500°F and exposed to air at room temperature of 70°F and 14.7 psia. What is the free-convection heat transfer per unit length of tube?

7-9.   A vertical cylinder 6 ft high and 3 in. in diameter is maintained at a temperature of 200°F in an atmospheric environment of 85°F. Calculate the heat lost by free convection from this cylinder. For this calculation the cylinder may be treated as a vertical flat plate.

7-10.   A horizontal electric heater 1 in. in diameter is submerged in a light-oil bath at 100°F. The heater surface temperature is maintained at 300°F. Calculate the heat lost per foot of length of the heater.

7-11.   A 1-ft-square air-conditioning duct carries air at a temperature such that the outside temperature of the duct is maintained at 60°F and is exposed to room air at 80°F. Calculate the heat gained by the duct per foot of length.

7-12.   The outside wall of a building 20 ft high receives an average radiant heat flux from the sun of 350 Btu/hr-ft². Assuming that 30 Btu/hr-ft² is conducted through the wall, estimate the outside wall temperature. Assume the atmospheric air on the outside of the building is at 70°F.

7-13.   Assume that one-half the heat transfer by free convection from a horizontal cylinder occurs on each side of the cylinder because of symmetry considerations. Going by this assumption, compare the heat transfer on each side of the cylinder with that from a vertical flat plate having a height equal to the circumferential distance from the bottom stagnation point to the top stagnation point on the cylinder. Discuss this comparison.

7-14.   For a vertical isothermal flat plate at 200°F exposed to air at 70°F and 14.7 psia, plot the free-convection velocity profiles as a function of distance from the plate surface at $x$ positions of 6, 12, and 18 in.

7-15.   Show that

$$\beta = \frac{1}{T}$$

for an ideal gas having the equation of state

$$p = \rho RT$$

7-16.   Two 1-ft-square vertical plates are separated by a distance of 1 in. and air at 14.7 psia. The two plates are maintained at temperatures of 400 and 100°F, respectively. Calculate the heat-transfer rate across the air space.

7-17.   A horizontal air space is separated by a distance of $\frac{1}{16}$ in. Estimate the heat-transfer rate per unit area for a temperature difference of 300°F, with one plate temperature at 100°F.

7-18.   Repeat Prob. 7-17 for the case of a horizontal space filled with water.

7-19.   Develop an expression for the optimum spacing for vertical plates in air in order to achieve minimum heat transfer, assuming that the heat transfer results from pure conduction at $Gr_\delta < 2000$. Plot this optimum spacing as a function of temperature difference for air at 14.7 psia.

7-20.   Suppose the heat-transfer coefficients for forced or free convection over vertical flat plates are to be compared. Develop an approximate relation between the Reynolds and Grashof numbers such that the heat-transfer coefficients for pure forced convection and pure free convection are equal. Assume laminar flow.

7-21.   Air at 70°F and 14.7 psia is forced upward through a vertical 1-in.-diameter tube 1 ft long. Calculate the total heat-transfer rate where the tube wall is maintained at 400°F and the flow velocity is 1.5 ft/sec.

7-22.   Calculate the rate of free-convection heat loss from a 1-ft-diameter sphere maintained at 100°F and exposed to atmospheric air at 14.7 psia and 70°F.

7-23.   A 1-in.-diameter sphere at 90°F is submerged in water at 50°F. Calculate the rate of free-convection heat loss.

7-24.   Air at 14.7 psia and 100°F is forced through a horizontal $\frac{1}{4}$-in.-diameter tube at an average velocity of 10 ft/sec. The tube wall is maintained at 1000°F, and the tube is 1 ft long. Calculate the average heat-transfer coefficient. Repeat for a velocity of 100 ft/sec and a tube wall temperature of 1500°F.

7-25.   An air space in a certain building wall is 4 in. thick and 5 ft high. Estimate the free-convection heat transfer through this space for a temperature difference of 30°F.

7-26.   A double plate-glass window is constructed with a $\frac{1}{2}$-in. air space. The plate dimensions are 4 by 6 ft. Calculate the free-convection heat-transfer rate through the air space for a temperature difference of 60°F.

7-27.   A spherical balloon gondola 8 ft in diameter rises to an altitude where the ambient pressure is 0.2 psia and ambient temperature is −60°F. The outside surface of the sphere is at approximately 30°F. Estimate the free-convection heat loss from the outside of the sphere. How does this compare with the forced-convection loss from such a sphere with a low free-stream velocity of approximately 1 ft/sec?

7-28.   A 1-ft-square vertical plate is heated electrically such that a constant heat flux condition is maintained with a total heat dissipation of 20 watts. The ambient air is at 14.7 psia and 70°F. Calculate the value of the heat-transfer coefficient at heights of 6 and 12 inches. Also calculate the average heat-transfer coefficient for the plate.

7-29.   A small copper block having a square bottom 1 in. × 1 in. and a vertical height of 2 in. cools in room air at 14.7 psia and 70°F. The block is isothermal at 200°F. Calculate the heat-transfer rate.

7-30.   Two 1-ft-square vertical plates are separated by a distance of 0.5 in. and the space between them filled with water. A constant heat flux condition is imposed on the plates such that the average temperature of one is 100°F and 140°F for the other. Calculate the heat-transfer rate under these conditions. Evaluate properties at the mean temperature.

7-31.   A circular hot plate, 6-in. in diameter is maintained at 300°F in atmospheric air at 70°F. Calculate the free convection heat loss when the plate is in a horizontal position.

7-32.   A 1.0-in.-diameter sphere is maintained at 100°F and submerged in water at 60°F. Calculate the heat-transfer rate under these conditions assuming that Eq. (7-31) will apply.

7-33.   Assuming that a man may be approximated by a vertical cylinder 1 ft in diameter and 6 ft tall, estimate the free-convection heat loss for a surface temperature of 75°F in ambient air at 68°F.

## REFERENCES

1.   Eckert, E. R. G., and E. Soehngen: Interferometric Studies on the Stability and Transition to Turbulence of a Free Convection Boundary Layer, *Proc. Gen. Discussion on Heat Transfer*, ASME-IME, London, 1951.
2.   Eckert, E. R. G., and E. Soehngen: Studies on Heat Transfer in Laminar Free Convection with the Zehnder-Mach Interferometer, *UASF Tech. Rept.* 5747, December, 1948.
3.   Holman, J. P., H. E. Gartrell, and E. E. Soehngen: An Interferometric Method of Studying Boundary Layer Oscillations, *J. Heat Transfer*, ser. C, vol. 80, August, 1960.
4.   McAdams, W. H.: "Heat Transmission," 3d ed., McGraw-Hill Book Company, New York, 1954.
5.   Yuge, T.: Experiments on Heat Transfer from Spheres Including Combined Natural and Forced Convection, *J. Heat Transfer*, ser. C, vol. 82, p. 214, 1960.
6.   Jakob, M.: Free Convection through Enclosed Plane Gas Layers, *Trans. ASME*, vol. 68, p. 189, 1946.
7.   Jakob, M.: "Heat Transfer," vol. 1, John Wiley & Sons, Inc., New York, 1949.
8.   Globe, S., and D. Dropkin: *J. Heat Transfer*, February, 1959, pp. 24–28.
9.   Evans, L. B., and N. E. Stefany: An Experimental Study of Transient Heat Transfer to Liquids in Cylindrical Enclosures, AIChE paper 4, Heat Transfer Conference, Los Angeles, August, 1965.
10.   Metais, B., and E. R. G. Eckert: Forced, Mixed, and Free Convection Regimes, *J. Heat Transfer*, ser. C, vol. 86, p. 295, 1964.
11.   Bishop, E. N., L. R. Mack, and J. A. Scanlan: Heat Transfer by Natural Convection between Concentric Spheres, *Intern. J. Heat Mass Trans.*, vol. 9, p. 649, 1966.
12.   Dropkin, D., and E. Somerscales: Heat Transfer by Natural Convection in Liquids Confined by Two Parallel Plates Which Are Inclined at Various Angles with Respect to the Horizontal, *J. Heat Transfer*, vol. 87, p. 71, 1965.
13.   Gebhart, B.: "Heat Transfer," 2d ed., chap. 8, McGraw-Hill Book Company, New York, 1970.
14.   Gebhart, B.: Natural Convection Flow, Instability, and Transition, *ASME Paper* 69-HT-29, August, 1969.
15.   Mollendorf, J. C., and B. Gebhart: An Experimental Study of Vigorous Transient Natural Convection, *ASME Paper* 70-HT-2, May, 1970.
16.   Bayley, F. J.: An Analysis of Turbulent Free Convection Heat Transfer, *Proc. Inst. Mech. Engr.*, vol. 169, no. 20, p. 361, 1955.
17.   Brown, C. K., and W. H. Gauvin: "Combined Free and Forced Convection, Parts I and II, *Can. J. Chem. Eng.*, vol. 43, no. 6, pp. 306, 313, 1965.
18.   MacGregor, R. K., and A. P. Emery: Free Convection Through Vertical Plane Layers—Moderate and High Prandtl Number Fluids, *J. Heat Transfer*, vol. 91, p. 391, 1969.
19.   Newell, M. E., and F. W. Schmidt: Heat Transfer by Laminar Natural Convection Within Rectangular Enclosures, *J. Heat Transfer*, vol. 92, pp. 159–168, 1970.
20.   Husar, R. B., and E. M. Sparrow: Patterns of Free Convection Flow Adjacent to Horizontal Heated Surfaces, *Intern. J. Heat Mass Trans.*, vol. 11, p. 1206, 1968.

21. Habne, E. W. P.: Heat Transfer and Natural Convection Patterns on a Horizontal Circular Plate, *Intern. J. Heat Mass Transfer*, vol. 12, p. 651, 1969.

22. Warner, C. Y., and V. S. Arpaci: An Experimental Investigation of Turbulent Natural Convection in Air at Low Pressure along a Vertical Heated Flat Plate, *Intern. J. Heat and Mass Transfer*, vol. 11, p. 397, 1968.

23. Gunness, R. C., Jr., and B. Gebhart: Stability of Transient Natural Convection, *Phys. Fluids*, vol. 12, p. 1968, 1969.

24. Rotern, Z., and L. Claassen: Natural Convection above Unconfined Horizontal Surfaces, *J. Fluid Mech.*, vol. 39 (pt. 1), p. 173, 1969.

25. Vliet, G. C.: Natural Convection Local Heat Transfer on Constant Heat Flux Inclined Surfaces, *J. Heat Transfer*, vol. 91, p. 511, 1969.

26. Vliet, G. C., and C. K. Lin: An Experimental Study of Turbulent Natural Convection Boundary Layers, *J. Heat Transfer*, vol. 91, p. 517, 1969.

27. Ostrach, S.: "An Analysis of Laminar Free-Convection Flow and Heat Transfer about a Flat Plate Parallel to the Direction of the Generating Body Force, *NACA* TR 1111, 1953.

28. Cheesewright, R.: Turbulent Natural Convection from a Vertical Plane Surface, *J. Heat Transfer*, vol. 90, p. 1, February, 1968.

29. Goldstein, R. J., and W. F. Hagen: Turbulent Flow Measurements Utilizing the Doppler-Shift of Scattered Laser Radiation, *Phys. Fluids*, vol. 10, p. 1349, June, 1967.

30. Eckert, E. R. G., and T. W. Jackson: Analysis of Turbulent Free Convection Boundary Layer on a Flat Plate, *NACA Rept.* 1015, 1951.

31. King, W. J.: The Basic Laws and Data of Heat Transmission, *Mech. Eng.*, vol. 54, p. 347, 1932.

32. Sparrow, E. M., and J. L. Gregg: Laminar Free Convection from a Vertical Flat Plate, *Trans. ASME*, vol. 78, p. 435, 1956.

33. Benard, H.: "Les Tourbillons Cellulaires dans une Nappe Liquide Transportant de la Chaleur par Convection en Regime Permanent," *Ann. Chim. Phys.*, vol. 23, pp. 62–144, 1901.

34. *Progress in Heat and Mass Transfer*, vol. 2, Eckert Presentation Volume, Pergamon Press, New York, 1969.

35. Gebhart, B., T. Audunson, and L. Pera: *Fourth Int. Heat Transfer Conf., Paris*, August, 1970.

36. Sanders, C. J., and J. P. Holman: Franz Grashof and the Grashof Number, *Int. J. Heat Mass Transfer*, vol. 15, p. 562, 1972.

# CHAPTER

# 8

# RADIATION
# HEAT TRANSFER

## 8-1□INTRODUCTION

Preceding chapters have shown how conduction and convection heat transfer may be calculated with the aid of both mathematical analysis and empirical data. We now wish to consider the third mode of heat transfer, thermal radiation. Thermal radiation is that electromagnetic radiation emitted by a body as a result of its temperature. In this chapter, we shall first describe the nature of thermal radiation, its characteristics, and the properties which are used to describe materials insofar as the radiation is concerned. Next, the transfer of radiation through space will be considered. Finally, the overall problem of heat transfer by thermal radiation will be analyzed, including the influence of the material properties and the influence of the geometric arrangement of the bodies on the total energy which may be exchanged.

## 8-2□PHYSICAL MECHANISM

There are many types of electromagnetic radiation; thermal radiation is only one. Regardless of the type of radiation, we say that it is propagated at the speed of light, $3 \times 10^{10}$ cm/sec. This speed is equal to the product of the wavelength and frequency of the radiation,

$$c = \lambda \nu$$

where $c$ = speed of light
$\lambda$ = wavelength
$\nu$ = frequency

The unit for $\lambda$ may be centimeters, angstroms (1 Å = $10^{-8}$ cm), or the micron $\mu$ (1 $\mu$ = $10^{-6}$ m). A portion of the electromagnetic spectrum is shown in Fig. 8-1. Thermal radiation lies in the range from about 0.1 to 100 $\mu$, while the visible-light portion of the spectrum is very narrow, extending from about 0.35 to 0.75 $\mu$.

The propagation of thermal radiation takes place in the form of discrete

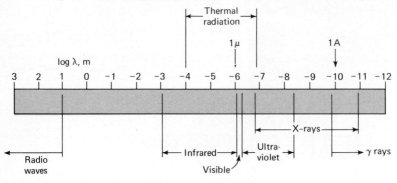

Fig. 8-1  Electromagnetic spectrum.

quanta, each quantum having an energy of

$$E = h\nu \tag{8-1}$$

where $h$ is Planck's constant and has the value

$$h = 6.625 \times 10^{-27} \text{ erg-sec}$$

A very rough physical picture of the radiation propagation may be obtained by considering each quantum as a particle having energy, mass, and momentum, just as we considered the molecules of a gas. So, in a sense, the radiation might be thought of as a "photon gas" which may flow from one place to another. Using the relativistic relation between mass and energy, expressions for the mass and momentum of the "particles" could thus be derived, viz.,

$$E = mc^2 = h\nu$$

$$m = \frac{h\nu}{c^2}$$

$$\text{Momentum} = c\frac{h\nu}{c^2} = \frac{h\nu}{c}$$

By considering the radiation as such a gas, it can be shown from thermodynamic considerations that the energy density (energy per unit volume) of the radiation is proportional to absolute temperature to the fourth power (see, for example, Ref. 1).

$$E_b = \sigma T^4 \tag{8-2}$$

Equation (8-2) is called the Stefan-Boltzmann law. $E_b$ is the energy radiated per unit time and per unit area by the ideal radiator, and $\sigma$ is the Stefan-Boltzmann constant, which has the value

$$\sigma = 0.1714 \times 10^{-8} \text{ Btu/hr-ft}^2\text{-°R}^4 \qquad (5.669 \times 10^{-8} \text{ W/m}^2\text{-°K}^4)$$

when $E_b$ is in Btu per hour per square foot and $T$ is in degrees Rankine. In the thermodynamic analysis the energy density is related to the energy radiated from a surface per unit time and per unit area. Thus the heated interior surface of an enclosure produces a certain energy density of thermal radiation in the enclosure. We are interested in radiant exchange with surfaces; hence the reason for the expression of radiation from a surface in terms of its temperature. The subscript $b$ in Eq. (8-2) denotes that this is the radiation from a blackbody. We call this blackbody radiation because materials which obey this law appear black to the eye; they appear black because they do not reflect any radiation. Thus a blackbody is also considered as one which absorbs all radiation incident upon it. $E_b$ is called the *emissive power* of a blackbody.

It is important to note at this point that the "blackness" of a surface-to-thermal radiation can be quite deceiving insofar as visual observations are concerned. A surface coated with lampblack appears black to the eye and turns out to be black for the thermal-radiation spectrum. On the other hand, snow or ice appear quite bright to the eye but are essentially "black" for long wavelength thermal radiation. Many white paints are also essentially black for long wavelength radiation. This point will be discussed further in Secs. 8-3, 8-4, 8-9, and 8-13.

## 8-3□RADIATION PROPERTIES

When radiant energy strikes a material surface, part of the radiation is reflected, part is absorbed, and part is transmitted, as shown in Fig. 8-2. We define the reflectivity $\rho$ as the fraction reflected, the absorptivity $\alpha$ as the fraction absorbed, and the transmissivity $\tau$ as the fraction transmitted. Thus

$$\rho + \alpha + \tau = 1 \qquad (8\text{-}3)$$

Most solid bodies do not transmit thermal radiation, so that for many applied problems the transmissivity may be taken as zero. Then

$$\rho + \alpha = 1$$

Two types of reflection phenomena may be observed when radiation strikes a surface. If the angle of incidence is equal to the angle of reflection, the reflection is called *specular*. On the other hand, when an incident beam is distributed uniformly in all directions after reflection, the reflection is called *diffuse*. These two types of reflection are depicted in Fig. 8-3. It is to be noted that a specular reflection presents a mirror image of the source to the observer. No real surface is either specular or diffuse. An ordinary mirror is quite specular for visible light, but would not necessarily be specular over the entire wavelength range of thermal radiation. Ordinarily, a rough surface exhibits diffuse behavior better than a highly polished surface. Similarly, a polished surface is more specular than a rough surface. The

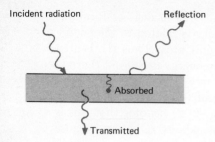

Incident radiation

Reflection

Absorbed

Transmitted

**Fig. 8-2    Sketch showing types of radiation.**

influence of surface roughness on thermal-radiation properties of materials is a matter of serious concern and remains a subject for continuing research.

The emissive power of a body $E$ is defined as the energy emitted by the body per unit area and per unit time. One may perform a thought experiment to establish a relation between the emissive power of a body and the material properties defined above. Assume that a perfectly black enclosure is available, i.e., one which absorbs all the incident radiation falling upon it, as shown schematically in Fig. 8-4. This enclosure will also emit radiation according to the $T^4$ law. Let the radiant flux arriving at some area in the enclosure be $q_i$ Btu/hr-ft². Now suppose that a body is placed inside the enclosure and allowed to come into temperature equilibrium with it. At equilibrium the energy absorbed by the body must be equal to the energy emitted; otherwise there would be an energy flow in or out of the body which

**Fig. 8-3    (a) Specular ($\phi_1 = \phi_2$) and (b) diffuse reflection.**

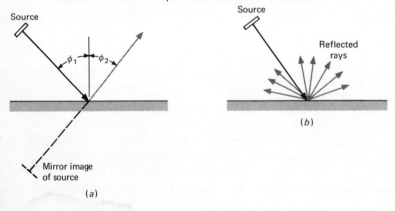

Source

$\phi_1$  $\phi_2$

Mirror image
of source

(a)

Source

Reflected
rays

(b)

**Fig. 8-4   Sketch showing model used for deriving Kirchhoff's law.**

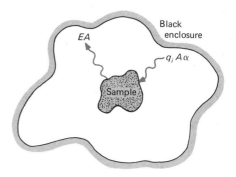

would raise or lower its temperature. At equilibrium we may write

$$EA = q_i A \alpha \tag{8-4}$$

If we now replace the body in the enclosure with a blackbody of the same size and shape and allow it to come to equilibrium with the enclosure *at the same temperature*,

$$E_b A = q_i A (1) \tag{8-5}$$

since the absorptivity of a blackbody is unity. If Eq. (8-4) is divided by Eq. (8-5),

$$\frac{E}{E_b} = \alpha$$

and we find that the ratio of the emissive power of a body to the emissive power of a blackbody *at the same temperature* is equal to the absorptivity of the body. This ratio is defined as the emissivity $\epsilon$ of the body,

$$\epsilon = \frac{E}{E_b} \tag{8-6}$$

so that

$$\epsilon = \alpha \tag{8-7}$$

Equation (8-7) is called Kirchhoff's identity. At this point we note that the emissivities and absorptivities which have been discussed are the *total* properties of the particular material; i.e., they represent the integrated behavior of the material over all wavelengths. Real substances emit less radiation than ideal black surfaces as measured by the emissivity of the material. In reality, the emissivity of a material varies with temperature and the wavelength of the radiation. A *gray body* is defined such that the monochromatic emissivity $\epsilon_\lambda$ of the body is independent of wavelength. The monochromatic emissivity is defined as the ratio of the monochromatic-emissive power of the body to the monochromatic-emissive power of a blackbody at the same wavelength and temperature. Thus

$$\epsilon_\lambda = \frac{E_\lambda}{E_{b\lambda}}$$

The total emissivity of the body may be related to the monochromatic emissivity by noting that

$$E = \int_0^\infty \epsilon_\lambda E_{b\lambda} \, d\lambda \quad \text{and} \quad E_b = \int_0^\infty E_{b\lambda} \, d\lambda = \sigma T^4$$

so that

$$\epsilon = \frac{E}{E_b} = \frac{\int_0^\infty \epsilon_\lambda E_{b\lambda} \, d\lambda}{\sigma T^4} \tag{8-8}$$

where $E_{b\lambda}$ is the emissive power of a blackbody per unit wavelength. If the gray-body condition is imposed, that is, $\epsilon_\lambda = $ const, Eq. (8-8) reduces to

$$\epsilon = \epsilon_\lambda \tag{8-9}$$

The emissivities of various substances vary widely with wavelength, temperature, and surface condition. Some typical values of the total emissivity of various surfaces are given in App. A. A very complete survey of radiation properties is given in Ref. 14.

The functional relation for $E_{b\lambda}$ was derived by Planck by introducing the quantum concept. The resulting expression is given in Eq. (8-10).

$$E_{b\lambda} = \frac{C_1 \lambda^{-5}}{e^{C_2/\lambda T} - 1} \tag{8-10}$$

where $\lambda = $ wavelength, $\mu$

$T = $ temperature, °R

$C_1 = 1.187 \times 10^8$ Btu-$\mu^4$/hr-ft$^2$    $(3.743 \times 10^8$ W-$\mu^4$/m$^2)$

$C_2 = 2.5896 \times 10^4$ $\mu$-°R    $(1.4387 \times 10^4$ $\mu$-K$)$

A plot of $E_{b\lambda}$ as a function of temperature and wavelength is given in Fig. 8-5a. Notice that the peak of the curve is shifted to the shorter wavelengths for the higher temperatures. These maximum points in the radiation curves are related by Wien's displacement law.

$$\lambda_{\max} T = 5215.6 \ \mu\text{-°R} \tag{8-11}$$

Figure 8-5b indicates the relative radiation spectra from a blackbody at 3000°F and a corresponding ideal gray body with emissivity equal to 0.6. Also shown is a curve indicating an approximate behavior for a real surface, which may differ considerably from that of either an ideal blackbody or an ideal gray body. For analysis purposes surfaces are usually considered as gray bodies, with emissivities taken as the integrated average value.

The shift in the maximum point of the radiation curve explains the change in color of a body as it is heated. Since the band of wavelengths visible to the eye lies between about 0.3 and 0.7 $\mu$, only a very small portion

**Fig. 8-5** (a) Blackbody emissive power as a function of wavelength and temperature. (b) Comparison of emissive power of ideal blackbodies and gray bodies with that of a real surface.

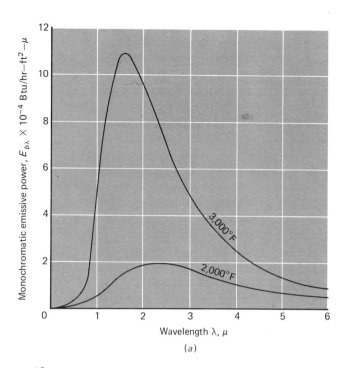

(a)

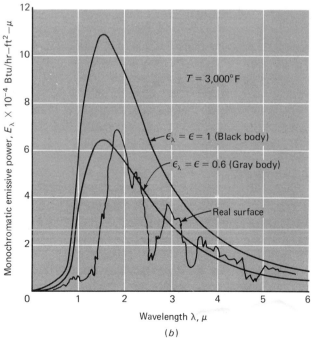

(b)

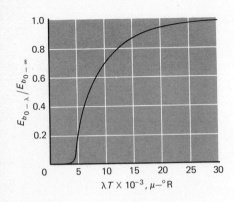

of the radiant-energy spectrum at low temperatures is detected by the eye. As the body is heated, the maximum intensity is shifted to the shorter wavelengths, and the first visible sign of the increase in temperature of the body is a dark-red color. With further increase in temperature, the color appears as a bright red, then bright yellow, and finally white. The material also appears much brighter at higher temperatures since a larger portion of the total radiation falls within the visible range.

We are frequently interested in the amount of energy radiated from a blackbody in a certain specified wavelength range. The fraction of the total energy radiated between 0 and $\lambda$ is given by

$$\frac{E_{b_{0-\lambda}}}{E_{b_{0-\infty}}} = \frac{\int_0^\lambda E_{b\lambda}\,d\lambda}{\int_0^\infty E_{b\lambda}\,d\lambda} \tag{8-12}$$

Equation (8-10) may be rewritten by dividing both sides by $T^5$, so that

$$\frac{E_{b\lambda}}{T^5} = \frac{C_1}{(\lambda T)^5 (e^{c_2/\lambda T} - 1)} \tag{8-13}$$

Now, for any specified temperature, the integrals in Eq. (8-12) may be expressed in terms of the single variable $\lambda T$. The results have been tabulated by Dunkle [2]. The ratio in Eq. (8-12) is plotted in Fig. 8-6 and tabulated in Table 8-1. If the radiant energy emitted between wavelengths $\lambda_1$ and $\lambda_2$ is desired,

$$E_{b\lambda_1-\lambda_2} = E_{b_{0-\infty}} \left( \frac{E_{b_{0-\lambda_2}}}{E_{b_{0-\infty}}} - \frac{E_{b_{0-\lambda_1}}}{E_{b_{0-\infty}}} \right) \tag{8-14}$$

where $E_{b_{0-\infty}}$ is the total radiation emitted over all wavelengths,

$$E_{b_{0-\infty}} = \sigma T^4 \tag{8-15}$$

## Table 8-1    Radiation Functions

| $\lambda T$ | $\dfrac{E_{b_0-\lambda T}}{\sigma T^4}$ | $\lambda T$ | $\dfrac{E_{b_0-\lambda T}}{\sigma T^4}$ | $\lambda T$ | $\dfrac{E_{b_0-\lambda T}}{\sigma T^4}$ |
|---|---|---|---|---|---|
| 1000 | 0 | 7,200 | 0.4809 | 13,400 | 0.8317 |
| 1200 | 0 | 7,400 | 0.5007 | 13,600 | 0.8370 |
| 1400 | 0 | 7,600 | 0.5199 | 13,800 | 0.8421 |
| 1600 | 0.0001 | 7,800 | 0.5381 | 14,000 | 0.8470 |
| 1800 | 0.0003 | 8,000 | 0.5558 | 14,200 | 0.8517 |
| 2000 | 0.0009 | 8,200 | 0.5727 | 14,400 | 0.8563 |
| 2200 | 0.0025 | 8,400 | 0.5890 | 14,600 | 0.8606 |
| 2400 | 0.0053 | 8,600 | 0.6045 | 14,800 | 0.8648 |
| 2600 | 0.0098 | 8,800 | 0.6195 | 15,000 | 0.8688 |
| 2800 | 0.0164 | 9,000 | 0.6337 | 16,000 | 0.8868 |
| 3000 | 0.0254 | 9,200 | 0.6474 | 17,000 | 0.9017 |
| 3200 | 0.0368 | 9,400 | 0.6606 | 18,000 | 0.9142 |
| 3400 | 0.0506 | 9,600 | 0.6731 | 19,000 | 0.9247 |
| 3600 | 0.0667 | 9,800 | 0.6851 | 20,000 | 0.9335 |
| 3800 | 0.0850 | 10,000 | 0.6966 | 21,000 | 0.9411 |
| 4000 | 0.1051 | 10,200 | 0.7076 | 22,000 | 0.9475 |
| 4200 | 0.1267 | 10,400 | 0.7181 | 23,000 | 0.9531 |
| 4400 | 0.1496 | 10,600 | 0.7282 | 24,000 | 0.9589 |
| 4600 | 0.1734 | 10,800 | 0.7378 | 25,000 | 0.9621 |
| 4800 | 0.1979 | 11,000 | 0.7474 | 26,000 | 0.9657 |
| 5000 | 0.2229 | 11,200 | 0.7559 | 27,000 | 0.9689 |
| 5200 | 0.2481 | 11,400 | 0.7643 | 28,000 | 0.9718 |
| 5400 | 0.2733 | 11,600 | 0.7724 | 29,000 | 0.9742 |
| 5600 | 0.2983 | 11,800 | 0.7802 | 30,000 | 0.9765 |
| 5800 | 0.3230 | 12,000 | 0.7876 | 40,000 | 0.9881 |
| 6000 | 0.3474 | 12,200 | 0.7947 | 50,000 | 0.9941 |
| 6200 | 0.3712 | 12,400 | 0.8015 | 60,000 | 0.9963 |
| 6400 | 0.3945 | 12,600 | 0.8081 | 70,000 | 0.9981 |
| 6600 | 0.4171 | 12,800 | 0.8144 | 80,000 | 0.9987 |
| 6800 | 0.4391 | 13,000 | 0.8204 | 90,000 | 0.9990 |
| 7000 | 0.4604 | 13,200 | 0.8262 | 100,000 | 1.9992 |
|  |  |  |  |  | 1.0000 |

and is obtained by integrating the Planck distribution formula of Eq. (8-10) over all wavelengths.

The concept of a blackbody is an idealization; i.e., a perfect blackbody does not exist—all surfaces reflect radiation to some extent, however slight. A blackbody may be approximated very accurately, however, in the following way:  A cavity is constructed, as shown in Fig. 8-7, so that it is very

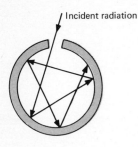

Incident radiation

**Fig. 8-7   Method of constructing a blackbody enclosure.**

large compared with the size of the opening in the side. An incident ray of energy is reflected many times on the inside before finally escaping from the side opening. With each reflection there is a fraction of the energy absorbed corresponding to the absorptivity of the inside of the cavity. After the many absorptions, practically all the incident radiation at the side opening is absorbed. It should be noted that the cavity of Fig. 8-7 behaves approximately as a blackbody emitter as well as an absorber.

### Example 8-1

A glass plate 1 ft square is used to view radiation from a furnace. The transmissivity of the glass is 0.5 from 0.2 to 3.5 $\mu$. The emissivity may be assumed to be 0.3 up to 3.5 $\mu$ and 0.9 above that. The transmissivity of the glass is zero, except in the range from 0.2 to 3.5 $\mu$. Assuming that the furnace is a blackbody at 3500°F, calculate the energy absorbed in the glass and the energy transmitted.

### Solution

$$T = 3500°F = 3960°R$$
$$\lambda_1 T = (0.2)(3960) = 792 \ \mu\text{-}°R$$
$$\lambda_2 T = (3.5)(3960) = 13,830 \ \mu\text{-}°R$$

From Table 8-1

$$\frac{E_{b_0-\lambda_1}}{E_{b_0-\infty}} \approx 0$$

$$\frac{E_{b_0-\lambda_2}}{E_{b_0-\infty}} = 0.8428$$

$$E_{b_0-\infty} = \sigma T^4 = (0.1714 \times 10^{-8})(3960)^4$$
$$= 4.2 \times 10^5 \ \text{Btu/hr-ft}^2$$

Total incident radiation 0.2 $\mu < \lambda < 3.5 \ \mu = (4.2 \times 10^5)(0.8428 - 0)$
$$= 3.54 \times 10^5 \ \text{Btu/hr-ft}^2$$

Total radiation transmitted $= (0.5)(3.54 \times 10^5) = 1.77 \times 10^5 \ \text{Btu/hr-ft}^2$

Radiation absorbed

$$= (0.3)(3.54 \times 10^5) = 1.062 \times 10^5 \text{ Btu/hr-ft}^2 \qquad \text{for } 0 < \lambda < 3.5 \ \mu$$
$$= (0.9)(1 - 0.8428)(4.2 \times 10^5)$$
$$= 0.594 \times 10^5 \text{ Btu/hr-ft}^2 \qquad \text{for } 3.5 < \lambda < \infty$$

Total radiation absorbed $= (1.062 + 0.594)(10^5)$
$$= 1.656 \times 10^5 \text{ Btu/hr-ft} \qquad (5.22 \times 10^5 \text{ W/m}^2)$$

## 8-4□THE RADIATION SHAPE FACTOR

Consider two black surfaces $A_1$ and $A_2$, as shown in Fig. 8-8. We wish to obtain a general expression for the energy exchange between these surfaces when they are maintained at different temperatures. The problem becomes essentially one of determining the amount of energy which leaves one surface and reaches the other. To solve this problem the *radiation shape factors* are defined as

$F_{1\text{-}2}$ = fraction of energy leaving surface 1 which reaches surface 2

$F_{2\text{-}1}$ = fraction of energy leaving surface 2 which reaches surface 1

$F_{m\text{-}n}$ = fraction of energy leaving surface $m$ which reaches surface $n$

Other names for the radiation shape factor are *view factor*, *angle factor*, and *configuration factor*. The energy leaving surface 1 and arriving at surface 2 is

$$E_{b1}A_1F_{12}$$

and that energy leaving surface 2 and arriving at surface 1 is

$$E_{b2}A_2F_{21}$$

Since the surfaces are black, all the incident radiation will be absorbed, and the net energy exchange is

$$E_{b1}A_1F_{12} - E_{b2}A_2F_{21} = Q_{1\text{-}2}$$

If both surfaces are at the same temperature, there can be no heat exchange, that is, $Q_{1\text{-}2} = 0$. Also

$$E_{b1} = E_{b2}$$

so that $\qquad\qquad A_1F_{12} = A_2F_{21} \qquad\qquad\qquad (8\text{-}16)$

The net heat exchange is therefore

$$Q_{1\text{-}2} = A_1F_{12}(E_{b1} - E_{b2}) = A_2F_{21}(E_{b1} - E_{b2}) \qquad (8\text{-}17)$$

Equation (8-16) is known as a reciprocity theorem. The problem now is to determine the value of $F_{12}$ (or $F_{21}$). To do this we consider the elements of area $dA_1$ and $dA_2$ in Fig. 8-8. The angles $\phi_1$ and $\phi_2$ are measured between a normal to the surface and the line drawn between the area elements $r$. The projection of $dA_1$ on the line between centers is

$$dA_1 \cos \phi_1$$

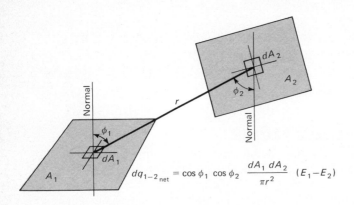

**Fig. 8-8 Sketch showing area elements used in deriving radiation shape factor.**

$$dq_{1-2\,\text{net}} = \cos \phi_1 \cos \phi_2 \, \frac{dA_1 \, dA_2}{\pi r^2} \, (E_1 - E_2)$$

This may be seen more clearly in the elevation drawing shown in Fig. 8-9. We assume that the surfaces are diffuse, i.e., that the intensity of the radiation is the same in all directions. The intensity is the radiation emitted per unit area and per unit solid angle in a certain specified direction. So, in order to obtain the energy emitted by the element of area $dA_1$ in a certain direction, we must multiply the intensity by the projection of $dA_1$ in the specified direction. Thus the energy leaving $dA_1$ in the direction given by the angle $\phi_1$ is

$$I_b \, dA_1 \cos \phi_1 \tag{a}$$

where $I_b$ is the blackbody intensity. The radiation arriving at some area element $dA_n$ at a distance $r$ from $A_1$ would be

$$I_b \, dA_1 \cos \phi_1 \frac{dA_n}{r^2} \tag{b}$$

where $dA_n$ is constructed normal to the radius vector. The quantity $dA_n/r^2$ represents the solid angle subtended by the area $dA_n$. The intensity may be

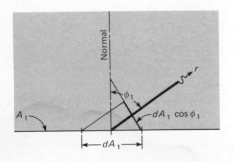

**Fig. 8-9 Elevation view of area shown in Fig. 8-8.**

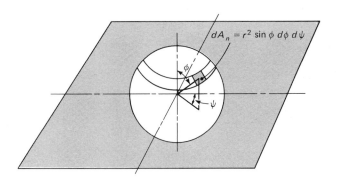

obtained in terms of the emissive power by integrating expression $(b)$ over a hemisphere enclosing the element of area $dA_1$. In a spherical coordinate system like that in Fig. 8-10

$$dA_n = r \sin \phi \, d\psi \, r \, d\phi$$

Then
$$E_b \, dA_1 = I_b \, dA_1 \int_0^{2\pi} \int_0^{\pi/2} \sin \phi \cos \phi \, d\phi \, d\psi$$
$$= \pi I_b \, dA_1$$

so that
$$E_b = \pi I_b \tag{8-18}$$

We may now return to the energy-exchange problem indicated in Fig. 8-8. The area element $dA_n$ is given by

$$dA_n = \cos \phi_2 \, dA_2$$

so that the energy leaving $dA_1$ which arrives at $dA_2$ is

$$dq_{1\text{-}2} = E_{b1} \cos \phi_1 \cos \phi_2 \frac{dA_1 \, dA_2}{\pi r^2}$$

That energy leaving $dA_2$ and arriving at $dA_1$ is

$$dq_{2\text{-}1} = E_{b2} \cos \phi_2 \cos \phi_1 \frac{dA_2 \, dA_1}{\pi r^2}$$

and the net energy exchange is

$$q_{\text{net}_{1\text{-}2}} = (E_{b1} - E_{b2}) \int_{A_2} \int_{A_1} \cos \phi_1 \cos \phi_2 \frac{dA_1 \, dA_2}{\pi r^2} \tag{8-19}$$

The integral is either $A_1 F_{12}$ or $A_2 F_{21}$ according to Eq. (8-17). To evaluate the integral, the specific geometry of the surfaces $A_1$ and $A_2$ must be known. We shall work out an elementary problem and then present the results for more complicated geometries in graphical form.

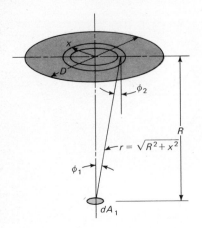

**Fig. 8-11  Radiation from a small-area element to a disk.**

**Fig. 8-12  Radiation shape factor for radiation between parallel rectangles.**

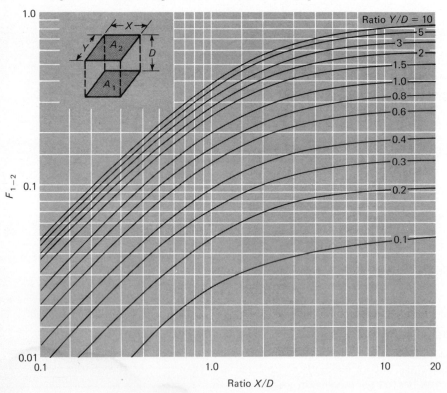

**Fig. 8-13** Radiation shape
factor for radiation between
parallel disks.

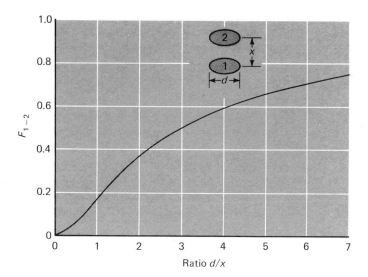

**Fig. 8-13** Radiation shape factor for radiation between parallel disks.

Consider the radiation from the small area $dA_1$ to the flat disk $A_2$, as shown in Fig. 8-11. The element of area $dA_2$ is chosen as the circular ring of radius $x$. Thus

$$dA_2 = 2\pi x \, dx$$

We note that $\phi_1 = \phi_2$ and apply Eq. (8-19), integrating over the area $A_2$.

$$dA_1 F_{dA_1\text{-}A_2} = dA_1 \int_{A_2} \cos^2 \phi_1 \frac{2\pi x \, dx}{\pi r^2}$$

Making the substitutions

$$r = (R^2 + x^2)^{\frac{1}{2}}$$

and

$$\cos \phi_1 = \frac{R}{(R^2 + x^2)^{\frac{1}{2}}}$$

we have

$$dA_1 F_{dA_1\text{-}A_2} = dA_1 \int_0^{D/2} \frac{2R^2 x \, dx}{(R^2 + x^2)^2}$$

Performing the integration,

$$dA_1 F_{dA_1\text{-}A_2} = - \, dA_1 \left( \frac{R^2}{R^2 + x^2} \right)_0^{D/2} = dA_1 \frac{D^2}{4R^2 + D^2}$$

so that

$$F_{dA_1\text{-}A_2} = \frac{D^2}{4R^2 + D^2} \tag{8-20}$$

The calculation of shape factors may be extended to more complex geometries. For our purposes only the results of calculation for parallel rectangles and disks and perpendicular rectangles with a common edge will be presented [3]. The shape factors for these situations are shown in Figs. 8-12 to 8-14, and the reader may consult Ref. 4 for the derivation of these shape factors.

**Fig. 8-14**  Radiation shape factor for radiation between perpendicular rectangles with a common edge.

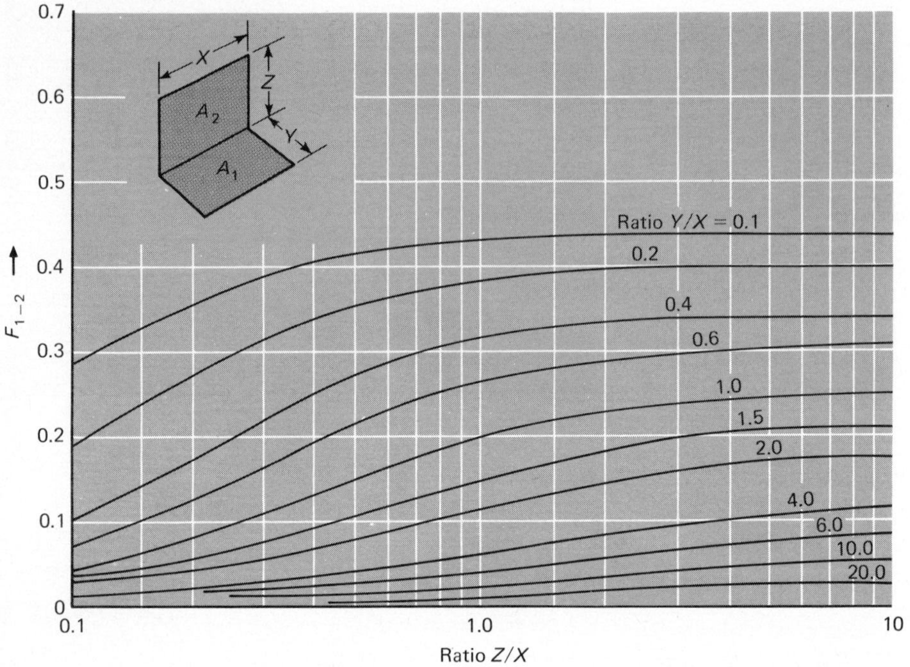

## Real Surface Behavior

Real surfaces exhibit interesting deviations from the ideal surfaces described in the preceding paragraphs. Real surfaces, for example, are not perfectly diffuse, and hence the intensity of emitted radiation is not constant over all directions. The directional-emittance characteristics of several types of surfaces are shown in Fig. 8-15. These curves illustrate the characteristically different behavior of electrical conductors and nonconductors. Nonconductors emit more energy in a direction normal to the surface, while conductors emit more energy in a direction having a large azimuth angle. This behavior may be satisfactorily explained with basic electromagnetic wave theory, and is discussed by Jakob [16]. As a result of this basic behavior of conductors and nonconductors, we may anticipate the appearance of a sphere which is heated to incandescent temperatures, as shown in Fig. 8-16. An electrical conducting sphere will appear bright around the rim since more energy is emitted at large angles $\phi$. A sphere constructed of a nonconducting material will have the opposite behavior and will appear bright in the center and dark around the edge.

**Fig. 8-15** Directional emissivity of materials according to Schmidt and Eckert [9]. (a) Wet ice; (b) wood; (c) glass; (d) paper; (e) clay; (f) copper oxide; (g) aluminum oxide.

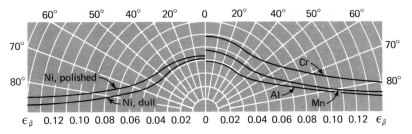

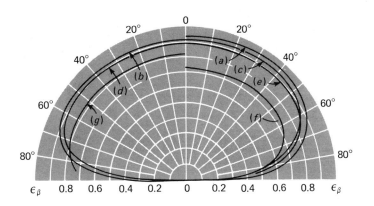

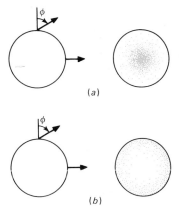

**Fig. 8-16** Effect of directional emittance on appearance of an incandescent sphere. (a) Electrical conductor; (b) electrical nonconductor.

Reflectance and absorptance of thermal radiation from real surfaces are a function not only of the surface itself, but also of the surroundings. These properties are dependent on the direction of the incident radiation and the wavelength. But the distribution of the intensity of incident radiation with wavelength may be a very complicated function of the temperatures and surface characteristics of all the surfaces which incorporate the surroundings. Let us denote the total incident radiation on a surface per unit time, per unit area, and per unit wavelength as $G_\lambda$. Then the total absorptivity will be given as the ratio of the total energy absorbed to the total energy incident on the surface, or

$$\alpha = \frac{\int_0^\infty \alpha_\lambda G_\lambda \, d\lambda}{\int_0^\infty G_\lambda \, d\lambda} \tag{8-21}$$

If we are fortunate enough to have a gray body such that $\epsilon_\lambda = \epsilon = $ const, this relation simplifies considerably. It may be shown that Kirchhoff's law [Eq. (8-7)] may be written for monochromatic radiation as

$$\epsilon_\lambda = \alpha_\lambda \tag{8-22}$$

Therefore, for a gray body, $\alpha_\lambda = $ const, and Eq. (8-21) expresses the result that the total absorptivity is also constant and independent of the wavelength distribution of incident radiation. Furthermore, since the emissivity and absorptivity are constant over all wavelengths for a gray body, they must be independent of temperature as well. Unhappily, real surfaces are not always "gray" in nature, and significant errors may ensue by assuming gray-body behavior. On the other hand, analysis of radiation exchange using real-surface behavior is so complicated that the ease and simplification of the gray-body assumption is justified by the practical utility it affords. Wiebelt [11] presents comparisons of heat-transfer calculations based on both gray and non-gray analyses.

## Example 8-2

Two parallel black plates 5 by 10 ft are spaced 5 ft apart. One plate is maintained at 1000°F, and the other at 500°F. What is the net radiant heat exchange between the two plates?

**Solution**   The ratios for use with Fig. 8-12 are

$$\frac{Y}{D} = \frac{5}{5} = 1.0$$

$$\frac{X}{D} = \frac{10}{5} = 2.0$$

Fig. 8-17 Sketch showing some
relations between shape factors.

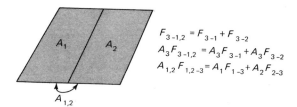

Fig. 8-17 Sketch showing some relations between shape factors.

$$F_{3-1,2} = F_{3-1} + F_{3-2}$$
$$A_3 F_{3-1,2} = A_3 F_{3-1} + A_3 F_{3-2}$$
$$A_{1,2} F_{1,2-3} = A_1 F_{1-3} + A_2 F_{2-3}$$

so that $F_{12} = 0.285$. The heat transfer is calculated from

$$
\begin{aligned}
q &= A_1 F_{12}(E_{b_1} - E_{b_2}) = \sigma A_1 F_{12}(T_1{}^4 - T_2{}^4) \\
&= (0.1714 \times 10^{-8})(50)(0.285)[(1460)^4 - (960)^4] \\
&= 90{,}200 \text{ Btu/hr}
\end{aligned}
$$

## 8-5□RELATIONS BETWEEN SHAPE FACTORS

Some useful relations between shape factors may be obtained by considering the system shown in Fig. 8-17. Suppose that the shape factor for radiation from $A_3$ to the combined area $A_{1,2}$ is desired. This shape factor must be given very simply as

$$F_{3-1,2} = F_{3-1} + F_{3-2} \qquad (8\text{-}23)$$

i.e., the total shape factor is the sum of its parts. We could also write Eq. (8-23) as

$$A_3 F_{3-1,2} = A_3 F_{3-1} + A_3 F_{3-2} \qquad (8\text{-}24)$$

and making use of the reciprocity relations

$$
\begin{aligned}
A_3 F_{3-1,2} &= A_{1,2} F_{1,2-3} \\
A_3 F_{3-1} &= A_1 F_{1-3} \\
A_3 F_{3-2} &= A_2 F_{2-3}
\end{aligned}
$$

the expression could be rewritten

$$A_{1,2} F_{1,2-3} = A_1 F_{1-3} + A_2 F_{2-3} \qquad (8\text{-}25)$$

which simply states that the total radiation arriving at surface 3 is the sum of the radiations from surfaces 1 and 2. Suppose we wish to determine the

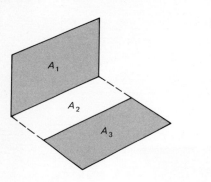

Fig. 8-18

shape factor $F_{1\text{-}3}$ for the surfaces in Fig. 8-18 in terms of known shape factors for perpendicular rectangles with a common edge. We may write

$$F_{1\text{-}2,3} = F_{1\text{-}2} + F_{1\text{-}3}$$

in accordance with Eq. (8-23). Both $F_{1\text{-}2,3}$ and $F_{1\text{-}2}$ may be determined from Fig. 8-14, so that $F_{1\text{-}3}$ is easily calculated when the dimensions are known. Now consider the somewhat more complicated situation shown in Fig. 8-19. An expression for the shape factor $F_{1\text{-}4}$ is desired in terms of known shape factors for perpendicular rectangles with a common edge. We write

$$A_{1,2}F_{1,2\text{-}3,4} = A_1 F_{1\text{-}3,4} + A_2 F_{2\text{-}3,4} \qquad (a)$$

in accordance with Eq. (8-25). Both $F_{1,2\text{-}3,4}$ and $F_{2\text{-}3,4}$ can be obtained from Fig. 8-14. $F_{1\text{-}3,4}$ may be expressed

$$A_1 F_{1\text{-}3,4} = A_1 F_{1\text{-}3} + A_1 F_{1\text{-}4} \qquad (b)$$

Also,    $A_{1,2}F_{1,2\text{-}3} = A_1 F_{1\text{-}3} + A_2 F_{2\text{-}3} \qquad (c)$

Fig. 8-19

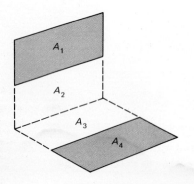

**Fig. 8-20 Generalized perpendicular rectangle arrangement.**

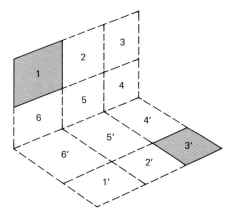

Solving for $A_1F_{1\text{-}3}$ from $(c)$, inserting this in $(b)$, and then inserting the resultant expression for $A_1F_{1\text{-}3,4}$ in $(a)$ gives

$$A_{1,2}F_{1,2\text{-}3,4} = A_{1,2}F_{1,2\text{-}3} - A_2F_{2\text{-}3} + A_1F_{1\text{-}4} + A_2F_{2\text{-}3,4} \qquad (d)$$

Notice that all shape factors except $F_{1\text{-}4}$ may be determined from Fig. 8-14. Thus

$$F_{1\text{-}4} = \frac{1}{A_1}\,(A_{1,2}F_{1,2\text{-}3,4} + A_2F_{2\text{-}3} - A_{1,2}F_{1,2\text{-}3} - A_2F_{2\text{-}3,4}) \qquad (8\text{-}26)$$

In the foregoing discussion the tacit assumption has been made that the various bodies do not see themselves; i.e.,

$$F_{11} = F_{22} = F_{33} = 0 \quad \cdots$$

To be perfectly general, we must include the possibility of concave curved surfaces, which may then see themselves. The general relation is therefore

$$\sum_{j=1}^{n} F_{ij} = 1.0 \qquad (8\text{-}27)$$

where $F_{ij}$ is the fraction of the total energy leaving surface $i$ which arrives at surface $j$. A certain amount of care is required when analyzing radiation exchange between curved surfaces.

Hamilton and Morgan [5] have presented generalized relations for parallel and perpendicular rectangles in terms of shape factors which may be obtained from Figs. 8-12 and 8-14. The two situations of interest are shown in Figs. 8-20 and 8-21. For the perpendicular rectangles of Fig. 8-20 it can be shown that the following reciprocity relations apply [5]:

$$A_1F_{13'} = A_3F_{31'} = A_{3'}F_{3'1} = A_{1'}F_{1'3} \qquad (8\text{-}28)$$

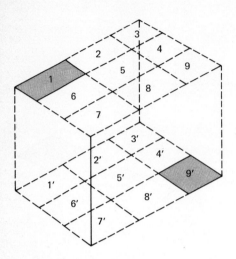

**Fig. 8-21  Generalized parallel rectangle arrangement.**

Making use of these reciprocity relations, the radiation shape factor $F_{13'}$ may be expressed by

$$A_1F_{13'} = \tfrac{1}{2}[K_{(1,2,3,4,5,6)^2} - K_{(2,3,4,5)^2} - K_{(1,2,5,6)^2}$$
$$+ K_{(4,5,6)^2} - K_{(4,5,6)-(1',2',3',4',5',6')}$$
$$- K_{(1,2,3,4,5,6)-(4',5',6')} + K_{(1,2,5,6)-(5'6')}$$
$$+ K_{(2,3,4,5)-(4',5')} + K_{(5,6)-(1',2',5',6')}$$
$$+ K_{(4,5)-(2',3',4',5')} + K_{(2,5)^2} - K_{(2,5)-5'}$$
$$- K_{(5,6)^2} - K_{(4,5)^2} - K_{5-(2',5')} + K_{52}] \quad (8\text{-}29)$$

where the $K$ terms are defined by

$$K_{m\text{-}n} = A_mF_{m\text{-}n} \quad (8\text{-}30)$$
$$K_{(m)^2} = A_mF_{m\text{-}m'} \quad (8\text{-}31)$$

The generalized parallel rectangle arrangement is depicted in Fig. 8-21. The reciprocity relations which apply to this situation are given in Ref. 5 as

$$A_1F_{19'} = A_3F_{37'} = A_9F_{91'} = A_7F_{73'} \quad (8\text{-}32)$$

Making use of these relations, it is possible to derive the shape factor $F_{19'}$ as

$$A_1F_{19'} = \tfrac{1}{4}[K_{(1,2,3,4,5,6,7,8,9)^2} - K_{(1,2,5,6,7,8)^2}$$
$$- K_{(2,3,4,5,8,9)^2} - K_{(1,2,3,4,5,6)^2} + K_{(1,2,5,6)^2}$$
$$+ K_{(2,3,4,5)^2} + K_{(4,5,8,9)^2} - K_{(4,5)^2} - K_{(5,8)^2}$$
$$- K_{(5,6)^2} - K_{(4,5,6,7,8,9)^2} + K_{(5,6,7,8)^2}$$
$$+ K_{(4,5,6)^2} + K_{(2,5,8)^2} - K_{(2,5)^2} + K_{(5)^2}] \quad (8\text{-}33)$$

The nomenclature for the $K$ terms is the same as given in Eqs. (8-30) and (8-31).

## 8-6□HEAT EXCHANGE BETWEEN NONBLACKBODIES

The calculation of the radiation heat transfer between black surfaces is relatively easy since all the radiant energy which strikes a surface is absorbed. The main problem is one of determining the geometric shape factor, but once this is accomplished, the calculation of the heat exchange is very simple. When nonblackbodies are involved, the situation is much more complex, for all the energy striking a surface will not be absorbed; part will be reflected back to another heat-transfer surface, and part may be reflected out of the system entirely. The problem can become complicated because the radiant energy can be reflected back and forth between the heat-transfer surfaces several times. The analysis of the problem must take into consideration these multiple reflections if correct conclusions are to be drawn.

We shall assume that all surfaces considered in our analysis are diffuse and uniform in temperature and that the reflective and emissive properties are constant over all the surface. Two new terms may be defined:

$G$ = irradiation = total radiation incident upon a surface per unit time and per unit area

$J$ = radiosity = total radiation which leaves a surface per unit time and per unit area

In addition to the assumptions stated above, we shall also assume that the radiosity and irradiation are uniform over each surface. This assumption is not strictly correct, even for ideal gray diffuse surfaces, but the problems become exceedingly complex when this analytical restriction is not imposed. Sparrow and Cess [10] give a discussion of such problems. The radiosity is the sum of the energy emitted and the energy reflected when no energy is transmitted, or

$$J = \epsilon E_b + \rho G \tag{8-34}$$

where $\epsilon$ is the emissivity and $E_b$ is the blackbody emissive power. Since the transmissivity is assumed to be zero, the reflectivity may be expressed

$$\rho = 1 - \alpha = 1 - \epsilon$$
so that
$$J = \epsilon E_b + (1 - \epsilon)G \tag{8-35}$$

The net energy leaving the surface is the difference between the radiosity and the irradiation.

$$\frac{q}{A} = J - G = \epsilon E_b + (1 - \epsilon)G - G$$

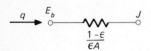

**Fig. 8-22**   Element representing "surface resistance" in radiation-network method.

Solving for $G$ in terms of $J$ from Eq. (8-35),

$$q = \frac{\epsilon A}{1 - \epsilon}(E_b - J)$$

or
$$q = \frac{E_b - J}{(1 - \epsilon)/\epsilon A} \tag{8-36}$$

At this point we introduce a very useful interpretation for Eq. (8-36). If the denominator of the right side is considered as the surface resistance to radiation heat transfer, the numerator as a potential difference, and the heat flow as the "current," then a network element could be drawn as in Fig. 8-22 to represent the physical situation. This is the first step in the network method of analysis originated by Oppenheim [20].

Now consider the exchange of radiant energy by two surfaces $A_1$ and $A_2$. Of that total radiation which leaves surface 1, the amount that reaches surface 2 is

$$J_1 A_1 F_{12}$$

and of that total energy leaving surface 2, the amount that reaches surface 1 is

$$J_2 A_2 F_{21}$$

The net interchange between the two surfaces is

$$q_{1\text{-}2} = J_1 A_1 F_{12} - J_2 A_2 F_{21}$$
But
$$A_1 F_{12} = A_2 F_{21}$$
so that
$$q_{1\text{-}2} = (J_1 - J_2)A_1 F_{12} = (J_1 - J_2)A_2 F_{21}$$
or
$$q_{1\text{-}2} = \frac{J_1 - J_2}{1/A_1 F_{12}} \tag{8-37}$$

We may thus construct a network element which represents Eq. (8-37), as shown in Fig. 8-23. The two network elements shown in Figs. 8-22 and 8-23 represent the essentials of the radiation-network method. To construct a network for a particular radiation heat-transfer problem we need only

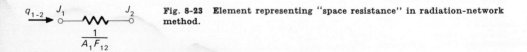

**Fig. 8-23**   Element representing "space resistance" in radiation-network method.

**Fig. 8-24** Radiation network for two surfaces which see each other and nothing else.

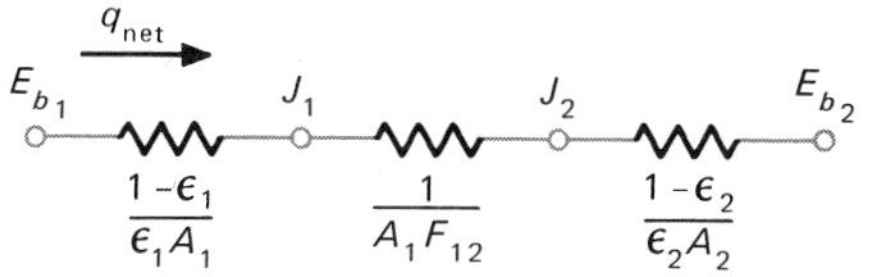

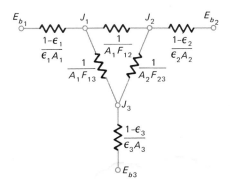

connect a "surface resistance" $(1 - \epsilon)/\epsilon A$ to each surface and a "space resistance" $1/A_m F_{m\text{-}n}$ between the radiosity potentials. For example, two surfaces which exchange heat with each other *and nothing else* would be represented by the network shown in Fig. 8-24. In this case the net heat transfer would be the overall potential difference divided by the sum of the resistances.

$$q_{net} = \frac{E_{b_1} - E_{b_2}}{(1 - \epsilon_1)/\epsilon_1 A_1 + 1/A_1 F_{12} + (1 - \epsilon_2)/\epsilon_2 A_2}$$
$$= \frac{\sigma(T_1{}^4 - T_2{}^4)}{(1 - \epsilon_1)/\epsilon_1 A_1 + 1/A_1 F_{12} + (1 - \epsilon_2)/\epsilon_2 A_2} \quad (8\text{-}38)$$

A three-body problem is shown in Fig. 8-25. In this case each of the bodies exchanges heat with the other two. The heat exchange between body 1 and body 2 would be

$$q_{1\text{-}2} = \frac{J_1 - J_2}{1/A_1 F_{12}}$$

and that between body 1 and body 3,

$$q_{1\text{-}3} = \frac{J_1 - J_3}{1/A_1 F_{13}}$$

To determine the heat flows in a problem of this type, the values of the radiosities must be calculated. This may be accomplished by performing

**Fig. 8-25** Radiation network for three surfaces which see each other and nothing else.

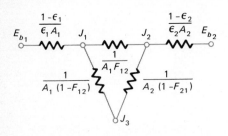

standard methods of analysis used in dc circuit theory. The most convenient method is an application of Kirchhoff's current law to the circuit, which states that the sum of the currents entering a node is zero. Example 8-3 illustrates the use of the method for the three-body problem.

A problem which may be easily solved with the network method is that of two surfaces exchanging heat with one another but connected by a third surface which does not exchange heat, i.e., one which is perfectly insulated. This third surface nevertheless influences the heat-transfer process because it absorbs and re-radiates energy to the other two surfaces which exchange heat. The network for this system is shown in Fig. 8-26. Notice that node $J_3$ is not connected to a radiation surface resistance because surface 3 does not exchange energy. Notice also that the values for the space resistances have been written

$$F_{13} = 1 - F_{12}$$
$$F_{23} = 1 - F_{21}$$

since surface 3 completely surrounds the other two surfaces. The network in Fig. 8-26 is a simple series-parallel system, and may be solved for the heat flow as

$$q_{net} = \frac{\sigma A_1(T_1{}^4 - T_2{}^4)}{\dfrac{A_1 + A_2 - 2A_1F_{12}}{A_2 - A_1(F_{12})^2} + \left(\dfrac{1}{\epsilon_1} - 1\right) + \dfrac{A_1}{A_2}\left(\dfrac{1}{\epsilon_2} - 1\right)} \tag{8-39}$$

where the reciprocity relation

$$A_1F_{12} = A_2F_{21}$$

has been used to simplify the expression. It is to be noted that Eq. (8-39) applies only to surfaces which do not see themselves.

This network, and others which follow, assume that the only heat exchange is by radiation. Conduction and convection are neglected for now.

### Example 8-3

Two parallel plates 5 by 10 ft are spaced 5 ft apart. One plate is maintained at 1000°F, and the other at 500°F. The emissivities of the plates are

0.2 and 0.5, respectively. The plates are located in a very large room, the walls of which are maintained at a temperature of 140°F. The plates exchange heat with each other and with the room. Find the total heat lost by the hotter plate.

**Solution**  The radiation network for this problem is shown in Fig. 8-25. From the data of the problem

$$T_1 = 1000°F = 1460°R$$
$$T_2 = 500°F = 960°R \qquad \epsilon_1 = 0.2$$
$$T_3 = 140°F = 600°R \qquad \epsilon_2 = 0.5$$

Since $A_3$, the area of the room, is very large, the resistance $(1 - \epsilon_3)/\epsilon_3 A_3$ may be taken as zero. The shape factor was given in Example 8-2.

Also
$$F_{1\text{-}2} = 0.285 = F_{2\text{-}1}$$
$$F_{1\text{-}3} = 1 - 0.285 = 0.715$$
$$F_{2\text{-}3} = 1 - 0.285 = 0.715$$

The resistances in the network are

$$\frac{1 - \epsilon_1}{\epsilon_1 A_1} = \frac{1 - 0.2}{(0.2)(50)} = 0.08$$

$$\frac{1 - \epsilon_2}{\epsilon_2 A_2} = \frac{1 - 0.5}{(0.5)(50)} = 0.02$$

$$\frac{1}{A_1 F_{12}} = \frac{1}{(50)(0.285)} = 0.0702$$

$$\frac{1}{A_1 F_{13}} = \frac{1}{(50)(0.715)} = 0.0280$$

$$\frac{1}{A_2 F_{23}} = \frac{1}{(50)(0.715)} = 0.0280$$

Taking the resistance $(1 - \epsilon_3)/\epsilon_3 A_3$ as zero, we have the network

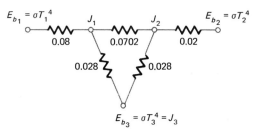

The network is solved by setting the sum of the heat currents entering nodes $J_1$ and $J_2$ to zero.

Node $J_1$:
$$\frac{E_{b1} - J_1}{0.08} + \frac{J_2 - J_1}{0.0702} + \frac{E_{b3} - J_1}{0.028} = 0 \qquad (a)$$

Node $J_2$:
$$\frac{J_1 - J_2}{0.0702} + \frac{E_{b3} - J_2}{0.028} + \frac{E_{b2} - J_2}{0.02} = 0 \qquad (b)$$

Now
$$E_{b1} = \sigma T_1{}^4 = 7780$$
$$E_{b2} = \sigma T_2{}^4 = 1455$$
$$E_{b3} = \sigma T_3{}^4 = 222$$

Inserting the values of $E_{b1}$, $E_{b2}$, and $E_{b3}$ into equations $(a)$ and $(b)$, we have two equations and two unknowns $J_1$ and $J_2$. Solving these equations simultaneously gives

$$J_1 = 1960 \text{ Btu/hr-ft}^2 \qquad (6{,}181 \text{ W/m}^2)$$
$$J_2 = 1086 \text{ Btu/hr-ft}^2 \qquad (3{,}425 \text{ W/m}^2)$$

The total heat lost by plate 1 is

$$q_1 = \frac{E_{b1} - J_1}{(1 - \epsilon_1)/\epsilon_1 A_1} = \frac{7780 - 1960}{0.08} = 72{,}750 \text{ Btu/hr}$$

The total heat lost by plate 2 is

$$q_2 = \frac{E_{b2} - J_2}{(1 - \epsilon_2)/\epsilon_2 A_2} = \frac{1455 - 1086}{0.02} = 18{,}450 \text{ Btu/hr} \qquad (5{,}406 \text{ W})$$

The total heat received by the room is

$$q_3 = \frac{J_1 - J_3}{1/A_1 F_{13}} + \frac{J_2 - J_3}{1/A_2 F_{23}} = 91{,}200 \text{ Btu/hr} \qquad (26{,}720 \text{ W})$$

## 8-7□INFINITE PARALLEL PLANES

When two infinite parallel planes are considered, $A_1$ and $A_2$ are equal; and the radiation shape factor is unity since all the radiation leaving one plane reaches the other. The network is the same as in Fig. 8-24, and the heat flow per unit area may be obtained from Eq. (8-38) by letting $A_1 = A_2$ and $F_{12} = 1.0$. Thus

$$\frac{q}{A} = \frac{\sigma(T_1{}^4 - T_2{}^4)}{1/\epsilon_1 + 1/\epsilon_2 - 1} \qquad (8\text{-}40)$$

When two long concentric cylinders as shown in Fig. 8-27 exchange heat, we may again apply Eq. (8-38). Rewriting the equation and noting that $F_{12} = 1.0$,

$$q = \frac{\sigma A_1(T_1{}^4 - T_2{}^4)}{1/\epsilon_1 + (A_1/A_2)(1/\epsilon_2 - 1)} \qquad (8\text{-}41)$$

The area ratio $A_1/A_2$ could be replaced by the diameter ratio $d_1/d_2$ when cylindrical bodies are concerned.

Equation (8-41) is particularly important when applied to the limiting

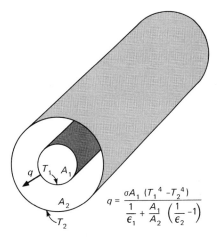

Fig. 8-27   Radiation exchange between two cylindrical surfaces.

$$q = \frac{\sigma A_1 (T_1{}^4 - T_2{}^4)}{\frac{1}{\epsilon_1} + \frac{A_1}{A_2}\left(\frac{1}{\epsilon_2} - 1\right)}$$

case of a convex object completely enclosed by a very large concave surface. In this instance $A_1/A_2 \rightarrow 0$ and the following simple relation results:

$$q = \sigma A_1 \epsilon_1 (T_1{}^4 - T_2{}^4) \qquad (8\text{-}41a)$$

This equation is readily applied to calculate the radiation energy loss from a hot object in a large room.

## 8-8□RADIATION SHIELDS

One way of reducing radiant heat transfer between two particular surfaces is to use materials which are highly reflective. An alternative method is to use radiation shields between the heat-exchange surfaces. These shields do not deliver or remove any heat from the overall system; they only place another resistance in the heat-flow path so that the overall heat transfer is retarded. Consider the two parallel infinite planes shown in Fig. 8-28a. We

Fig. 8-28   Radiation between parallel infinite planes with and without a radiation shield.

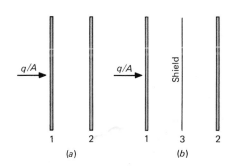

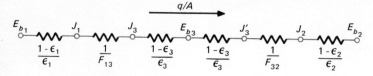

have shown that the heat exchange between these surfaces may be calculated with Eq. (8-40). Now consider the same two planes, but with a radiation shield placed between them as in Fig. 8-28b. The heat transfer will be calculated for this latter case and will be compared with the heat transfer without the shield.

Since the shield does not deliver or remove heat from the system, the heat transfer between plate 1 and the shield must be precisely the same as that between the shield and plate 2, and this is the overall heat transfer. Thus

$$\left(\frac{q}{A}\right)_{1\text{-}3} = \left(\frac{q}{A}\right)_{3\text{-}2} = \frac{q}{A}$$

$$\frac{q}{A} = \frac{\sigma(T_1^4 - T_3^4)}{1/\epsilon_1 + 1/\epsilon_3 - 1} = \frac{\sigma(T_3^4 - T_2^4)}{1/\epsilon_3 + 1/\epsilon_2 - 1} \tag{8-42}$$

The only unknown in Eq. (8-42) is the temperature of the shield $T_3$. Once this temperature is obtained, the heat transfer is easily calculated. If the emissivities of all three surfaces are equal, that is, $\epsilon_1 = \epsilon_2 = \epsilon_3$, we obtain the simple relation

$$T_3^4 = \tfrac{1}{2}(T_1^4 + T_2^4) \tag{8-43}$$

and the heat transfer is

$$\frac{q}{A} = \frac{\tfrac{1}{2}\sigma(T_1^4 - T_2^4)}{1/\epsilon_1 + 1/\epsilon_3 - 1}$$

But since $\epsilon_3 = \epsilon_2$, we observe that this heat flow is just one-half of that which would be experienced if there were no shield present. The radiation network corresponding to the situation in Fig. 8-28b is given in Fig. 8-29.

Multiple-radiation-shield problems may be treated in the same manner as that outlined above. When the emissivities of all surfaces are different, the overall heat transfer may be calculated most easily by using a series radiation network with the appropriate number of elements, similar to the one in Fig. 8-29. If the emissivities of all surfaces are equal, a rather simple relation may be derived for the heat transfer when the surfaces may be considered as infinite parallel planes. Let the number of shields be $n$. Considering the radiation network for the system, all the "surface resistances" would be the same since the emissivities are equal. There would be two of these resistances for each shield and one for each heat-transfer surface. There would be $n + 1$ "space resistances," and these would all be unity since the radiation shape factors are unity for the infinite parallel planes. The total resistance in the network would thus be

$$R(n \text{ shields}) = (2n + 2) \frac{1 - \epsilon}{\epsilon} + (n + 1)(1) = (n + 1) \left( \frac{2}{\epsilon} - 1 \right)$$

The resistance when no shield is present is

$$R(\text{no shield}) = \frac{1}{\epsilon} + \frac{1}{\epsilon} - 1 = \frac{2}{\epsilon} - 1$$

We note that the resistance with the shields in place is $n + 1$ times as large as when the shields are absent. Thus

$$\left( \frac{q}{A} \right)_{\substack{\text{with} \\ \text{shields}}} = \frac{1}{n + 1} \left( \frac{q}{A} \right)_{\substack{\text{without} \\ \text{shields}}} \tag{8-44}$$

if the temperatures of the heat-transfer surfaces are maintained the same in both cases. The radiation-network method may also be applied to shield problems involving cylindrical systems. In these cases the proper area relations must be used.

Notice that the analyses above, dealing with infinite parallel planes, have been carried out on a per-unit area basis since all areas are the same.

## Example 8-4

Two very large parallel planes with emissivities 0.3 and 0.8 exchange heat. Find the percentage reduction in heat transfer when a polished-aluminum radiation shield ($\epsilon = 0.04$) is placed between them.

**Solution**    The heat transfer without the shield is given by

$$\frac{q}{A} = \frac{\sigma(T_1{}^4 - T_2{}^4)}{1/\epsilon_1 + 1/\epsilon_2 - 1} = 0.279\sigma(T_1{}^4 - T_2{}^4)$$

The radiation network for the problem with the shield in place is shown in Fig. 8-29. The resistances are

$$\frac{1 - \epsilon_1}{\epsilon_1} = \frac{1 - 0.3}{0.3} = 2.333$$

$$\frac{1 - \epsilon_3}{\epsilon_3} = \frac{1 - 0.04}{0.04} = 24.0$$

$$\frac{1 - \epsilon_2}{\epsilon_2} = \frac{1 - 0.8}{0.8} = 0.25$$

The total resistance is

$$2.333 + (2)(24.0) + (2)(1) + 0.25 = 52.583$$

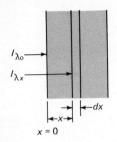

Fig. 8-30  Absorption in a gas layer.

and the heat transfer is

$$\frac{q}{A} = \frac{\sigma(T_1{}^4 - T_2{}^4)}{52.583} = 0.01712\sigma(T_1{}^4 - T_2{}^4)$$

so that the heat transfer is *reduced* by 93.6 percent.

### 3-9□GAS RADIATION

Radiation exchange between a gas and a heat-transfer surface is considerably more complex than the situations described in the preceding sections. Unlike most solid bodies, gases are in many cases transparent to radiation. When they absorb and emit radiation, they usually do so only in certain narrow wavelength bands. Some gases, such as $N_2$, $O_2$, and others of nonpolar symmetrical molecular structure, are essentially transparent at low temperatures, while $CO_2$, $H_2O$, and various hydrocarbon gases radiate to an appreciable extent.

The absorption of radiation in gas layers may be described analytically in the following way, considering the system shown in Fig. 8-30: A monochromatic beam of radiation having an intensity $I_\lambda$ impinges on the gas layer of thickness $dx$. The decrease in intensity resulting from absorption in the layer is assumed to be proportional to the thickness of the layer and the intensity of radiation at that point. Thus

$$dI_\lambda = -a_\lambda I_\lambda \, dx \tag{8-45}$$

where the proportionality constant $a_\lambda$ is called the monochromatic absorption coefficient. Integrating this equation,

$$\int_{I_{\lambda_0}}^{I_{\lambda_x}} \frac{dI_\lambda}{I_\lambda} = \int_0^x -a_\lambda \, dx$$

or

$$\frac{I_{\lambda_x}}{I_{\lambda_0}} = e^{-a_\lambda x} \tag{8-46}$$

Equation (8-46) is called Beer's law and represents the familiar exponential-decay formula experienced in many types of radiation analyses dealing with absorption. In accordance with our definitions in Sec. 8-3, the monochromatic transmissivity will be given as

$$\tau_\lambda = e^{-a_\lambda x} \tag{8-47}$$

If the gas is nonreflecting, then

$$\tau_\lambda + \alpha_\lambda = 1$$

and
$$\alpha_\lambda = 1 - e^{-a_\lambda x} \tag{8-48}$$

As mentioned above, gases frequently absorb only in narrow wavelength bands, as indicated for water vapor in Fig. 8-31. These curves also indicate the effect of thickness of the gas layer on monochromatic absorptivity.

The calculation of gas-radiation properties is quite complicated, and the reader should consult Refs. 6, 7, 10, and 11 for detailed information.

## 8-10□RADIATION HEAT EXCHANGE THROUGH AN ABSORBING AND TRANSMITTING MEDIUM

The foregoing discussions have shown the methods that may be used to calculate radiation heat transfer between surfaces separated by a completely transparent medium. The radiation-network method is used to great advantage in these types of problems.

Many practical problems involve radiation heat transfer through a medium which is both absorbing and transmitting. The various glass substances are one example of this type of medium; gases are another example. The problem of calculating heat transfer in these situations is attacked by performing an analysis similar to those in the preceding sections.

To begin with, let us consider a simple case, that of two nontransmitting surfaces which see each other and nothing else. In addition, we let the space between these surfaces be occupied by a transmitting and absorbing medium. The practical problem might be that of two large planes separated by either an absorbing gas or a transparent sheet of glass or plastic. The situation is shown schematically in Fig. 8-32. The transmitting medium will be designated by the subscript $m$. We make the assumption that the medium is nonreflecting and that Kirchhoff's identity applies, so that

$$\alpha_m + \tau_m = 1 = \epsilon_m + \tau_m \tag{8-49}$$

The assumption that the medium is nonreflecting is a valid one when gases are considered. For glass or plastic plates this is not necessarily true, and reflectivities of the order of 0.1 are common for many glass substances. In addition, the transmissive properties of glasses are usually limited to a narrow wavelength band between about 0.2 and 4 $\mu$. Thus the analysis which follows is highly idealized and serves mainly to furnish a starting point for

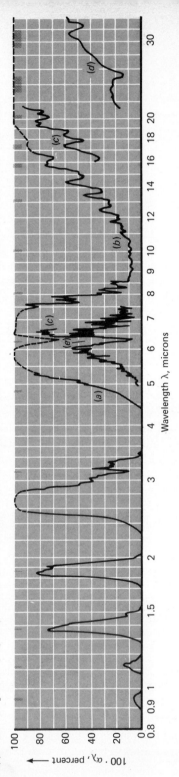

**Fig. 8-31** Monochromatic absorptivity for water vapor according to Ref. 8. For wavelengths from 0.8 to 4 $\mu$, steam temperature 260.6°F; thickness of layer 109 cm, wavelengths from 4 to 34 $\mu$. (a) Temperature 260.6°F, thickness of layer 109 cm. (b) Temperature 260.6°F, thickness of layer 104 cm. (c) Temperature 260.6°F, thickness of layer 32.4 cm. (d) Temperature 177.8°F, thickness of layer 32.4 cm., air-steam mixture corresponding to a steam layer approximately 4 cm thick. (e) Room temperature, layer of moist air 200 cm thick, corresponding to a layer of steam at atmospheric pressure approximately 7 cm thick.

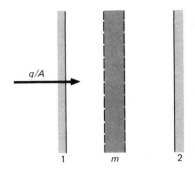

the solution of problems in which transmission of radiation must be considered. Other complications with gases are mentioned later in the discussion. When both reflection and transmission must be taken into account,
the analysis techniques to be discussed in Sec. 8-12 must be employed.

Returning to the analysis, we note that the medium can emit and
transmit radiation from one surface to the other. Our task is to determine
the network elements to use in describing these two types of exchange processes. The transmitted energy may be analyzed as follows:   The energy
leaving surface 1, which is transmitted through the medium and arrives at
surface 2, is

$$J_1 A_1 F_{12} \tau_m$$

and that which leaves surface 2 and arrives at surface 1 is

$$J_2 A_2 F_{21} \tau_m$$

The net exchange in the transmission process is therefore

$$q_{1\text{-}2\text{transmitted}} = A_1 F_{12} \tau_m (J_1 - J_2) = A_1 F_{12}(1 - \epsilon_m)(J_1 - J_2)$$

or     $$q_{1\text{-}2\text{transmitted}} = \frac{J_1 - J_2}{1/A_1 F_{12}(1 - \epsilon_m)} \qquad (8\text{-}50)$$

and the network element which may be used to describe this process is
shown in Fig. 8-33.

Now consider the exchange process between surface 1 and the transmitting medium. Since we have assumed that this medium is nonreflecting,

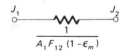

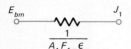

$$\frac{1}{A_1 F_{1m} \epsilon_m}$$

**Fig. 8-34** Network element for radiation exchange between medium and surface.

the energy leaving the medium (other than the transmitted energy, which we have already considered) is precisely that energy which is emitted by the medium.

$$J_m = \epsilon_m E_{bm}$$

And of the energy leaving the medium, the amount which reaches surface 1 is

$$A_m F_{m1} J_m = A_m F_{m1} \epsilon_m E_{bm}$$

Of that energy which leaves surface 1, the quantity which reaches the transparent medium is

$$J_1 A_1 F_{1m} \alpha_m = J_1 A_1 F_{1m} \epsilon_m$$

At this point we note that absorption in the medium means that the incident radiation has "reached" the medium. Consistent with the above relations, the net energy exchange between the medium and surface 1 is the difference between the amount emitted by the medium toward surface 1 and that absorbed which emanated from surface 1. Thus

$$q_{m-1\text{net}} = A_m F_{m1} \epsilon_m E_{bm} - J_1 A_1 F_{1m} \epsilon_m$$

Using the reciprocity relation

$$A_1 F_{1m} = A_m F_{m1}$$

we have

$$q_{m-1\text{net}} = \frac{E_{bm} - J_1}{1/A_1 F_{1m} \epsilon_m} \tag{8-51}$$

This heat-exchange process is represented by the network element shown in Fig. 8-34. The total network for the physical situation of Fig. 8-32 is shown in Fig. 8-35.

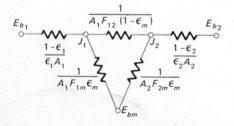

**Fig. 8-35** Total radiation network for system of Fig. 8-32.

If the transparent medium is maintained at some fixed temperature, then the potential $E_{bm}$ is fixed according to

$$E_{bm} = \sigma T_m{}^4$$

On the other hand, if no net energy is delivered to the medium, then $E_{bm}$ becomes a floating node, and its potential is determined by the other network elements.

In reality, the radiation shape factors $F_{1\text{-}2}$, $F_{1\text{-}m}$, and $F_{2\text{-}m}$ are unity for this example, so that the expression for the heat flow could be simplified to some extent; however, these shape factors are included in the network resistances for the sake of generality in the analysis.

When the practical problem of heat exchange between "gray" surfaces through an absorbing gas is encountered, the major difficulty is that of determining the transmissivity and emissivity of the gas. These properties are functions, not only of the temperature of the gas, but also of the thickness of the gas layer; i.e., thin gas layers transmit more radiation than thick layers. The usual practical problem almost always involves more than two heat-transfer surfaces, as in the simple example given above. As a result, the transmissivities between the various heat-transfer surfaces can be quite different, depending on their geometric orientation. Since the temperature of the gas will vary, the transmissive and emissive properties will vary with their location in the gas. One way of handling this situation is to divide the gas body into layers and set up a radiation network accordingly, letting the potentials of the various nodes "float," and thus arriving at the gas-temperature distribution. Even with this procedure, an iterative method must eventually be employed because the radiation properties of the gas are functions of the unknown "floating potentials." Naturally, if the temperature of the gas is uniform, the solution is much easier.

We shall not present the solution of a complex gas-radiation problem since the tedious effort required for such a solution is beyond the scope of our present discussion; however, it is worthwhile to analyze a two-layer transmitting system in order to indicate the general scheme of reasoning which might be applied to more complex problems.

Consider the physical situation shown in Fig. 8-36. Two radiating and absorbing surfaces are separated by two layers of transmitting and absorbing media. These two layers might represent two sheets of transparent media, such as glass, or they might represent the division of a separating gas into two parts for purposes of analysis. We designate the two transmitting and absorbing layers with the subscripts $m$ and $n$. The energy exchange between surface 1 and $m$ is given by

$$q_{1\text{-}m} = A_1 F_{1m} \epsilon_m J_1 - A_m F_{m1} \epsilon_m E_{bm} = \frac{J_1 - E_{bm}}{1/A_1 F_{1m} \epsilon_m} \qquad (8\text{-}52)$$

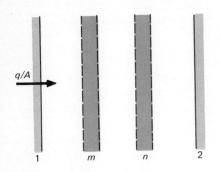

Fig. 8-36  Radiation system consisting of two trans-mitting layers between two planes.

and that between surface 2 and $n$ is

$$q_{2-n} = A_2F_{2n}\epsilon_nJ_2 - A_nF_{n2}\epsilon_nE_{bn} = \frac{J_2 - E_{bn}}{1/A_2F_{2n}\epsilon_n} \tag{8-53}$$

Of that energy leaving surface 1, the amount arriving at surface 2 is

$$q_{1-2} = A_1F_{12}J_1\tau_m\tau_n = A_1F_{12}J_1(1 - \epsilon_m)(1 - \epsilon_n)$$

and of that energy leaving surface 2, the amount arriving at surface 1 is

$$q_{2-1} = A_2F_{21}J_2\tau_n\tau_m = A_2F_{12}J_2(1 - \epsilon_n)(1 - \epsilon_m)$$

so that the net energy exchange by transmission between surfaces 1 and 2 is

$$q_{1-2\text{transmitted}} = A_1F_{12}(1 - \epsilon_m)(1 - \epsilon_n)(J_1 - J_2) = \frac{J_1 - J_2}{1/A_1F_{12}(1 - \epsilon_m)(1 - \epsilon_n)} \tag{8-54}$$

and the network element representing this transmission is shown in Fig. 8-37. Of that energy leaving surface 1, the amount which is absorbed in $n$ is

$$q_{1-n} = A_1F_{1n}J_1\tau_m\epsilon_n = A_1F_{1n}J_1(1 - \epsilon_m)\epsilon_n$$

Also,
$$q_{n-1} = A_nF_{n1}J_n\tau_m = A_nF_{n1}\epsilon_nE_{bn}(1 - \epsilon_m)$$
since
$$J_n = \epsilon_nE_{bn}$$

The net exchange between surface 1 and $n$ is therefore

$$q_{1-n_{\text{net}}} = A_1F_{1n}(1 - \epsilon_m)\epsilon_n(J_1 - E_{bn}) = \frac{J_1 - E_{bn}}{1/A_1F_{1n}(1 - \epsilon_m)\epsilon_n} \tag{8-55}$$

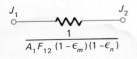

Fig. 8-37  Network element for transmitted radiation between planes.

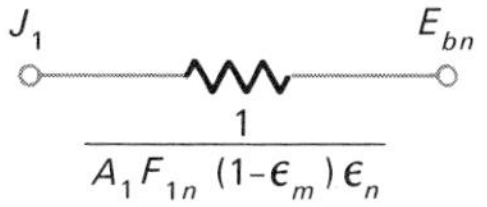

**Fig. 8-38   Network element for transmitted radiation from medium** $n$
**to plane 1.**

$$J_1 \qquad\qquad E_{bn}$$

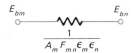

$$\frac{1}{A_1 F_{1n} (1-\epsilon_m)\epsilon_n}$$

and the network element representing this situation is shown in Fig. 8-38. In like manner, the net exchange between surface 2 and $m$ is

$$q_{2\text{-}m_{\text{net}}} = \frac{J_2 - E_{bm}}{1/A_2 F_{2m}(1 - \epsilon_n)\epsilon_m} \tag{8-56}$$

Of that radiation leaving $m$, the amount absorbed in $n$ is

$$q_{m\text{-}n} = J_m A_m F_{mn}\alpha_n = A_m F_{mn}\epsilon_m\epsilon_n E_{bm}$$

and

$$q_{n\text{-}m} = A_n F_{nm}\epsilon_n\epsilon_m E_{bn}$$

so that the net energy exchange between $m$ and $n$ is

$$q_{m\text{-}n_{\text{net}}} = A_m F_{mn}\epsilon_m\epsilon_n(E_{bm} - E_{bn}) = \frac{E_{bm} - E_{bn}}{1/A_m F_{mn}\epsilon_m\epsilon_n} \tag{8-57}$$

and the network element representing this energy transfer is given in Fig. 8-39.

The final network for the entire heat-transfer process is shown in Fig. 8-40, with the surface resistances added. If the two transmitting layers $m$ and $n$ are maintained at given temperatures, the solution to the network is relatively easy to obtain because only two unknown potentials $J_1$ and $J_2$ need be determined to establish the various heat-flow quantities. In this case the two transmitting layers will either absorb or lose a certain quantity of energy, depending on the temperature at which they are maintained.

When no net energy is delivered to the transmitting layers, the nodes $E_{bm}$ and $E_{bn}$ must be left "floating" in the analysis; and for this particular system four nodal equations would be required for a solution of the problem.

## 8-11□RADIATION EXCHANGE WITH SPECULAR SURFACES

All the preceding discussions have considered radiation exchange between diffuse surfaces. In fact, the radiation shape factors defined by Eq. (8-19) hold only for diffuse radiation because the radiation was assumed to have no preferred direction in the derivation of this relation. In this section we wish to extend the analysis to take into account some simple geometries

**Fig. 8-39   Network element for radiation exchange between two transparent layers.**

$$E_{bm} \qquad\qquad E_{bn}$$

$$\frac{1}{A_m F_{mn}\epsilon_m\epsilon_n}$$

**Fig. 8-40    Total radiation network for system of Fig. 8-36.**

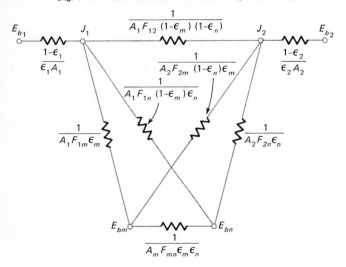

containing surfaces that may have a specular type of reflection. No real surface is completely diffuse or completely specular. We shall assume, however, that all the surfaces to be considered *emit* radiation diffusely but that they may *reflect* radiation partly in a specular manner and partly in a diffuse manner. We therefore take the reflectivity to be the sum of a specular component and a diffuse component.

$$\rho = \rho_s + \rho_D \tag{8-58}$$

It is still assumed that Kirchhoff's identity applies so that

$$\epsilon = \alpha = 1 - \rho \tag{8-59}$$

The net heat lost by a surface is the difference between the energy emitted and absorbed.

$$q = A(\epsilon E_b - \alpha G) \tag{8-60}$$

We define the *diffuse radiosity* $J_D$ as the total *diffuse* energy leaving the surface per unit area and per unit time, or

$$J_D = \epsilon E_b + \rho_D G \tag{8-61}$$

Solving for the irradiation $G$ from Eq. (8-61) and inserting in Eq. (8-60) gives

$$q = \frac{\epsilon A}{\rho_D} [E_b(\epsilon + \rho_D) - J_D]$$

**Fig. 8-41    Network element representing Eq. (8-62).**

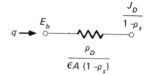

or written in a different form,

$$q = \frac{E_b - J_D/(1 - \rho_s)}{\rho_D/[\epsilon A (1 - \rho_s)]} \qquad (8\text{-}62)$$

where $1 - \rho_s$ has been substituted for $\epsilon + \rho_D$. It is easy to see that Eq. (8-62) may be represented with the network element shown in Fig. 8-41. A quick inspection will show that this network element reduces to that in Fig. 8-22 for the case of a surface which reflects in only a diffuse manner, i.e., for $\rho_s = 0$.

Now let us compute the radiation exchange between two specular-diffuse surfaces. For the moment, we shall assume that the surfaces are oriented as shown in Fig. 8-42. In this arrangement any diffuse radiation leaving surface 1 which is specularly reflected by 2 will not be reflected directly back to 1. This is an important point, for in eliminating such reflections we are considering only the *direct* diffuse exchange between the two surfaces. In subsequent paragraphs we shall show how the specular reflections must be analyzed. For the surfaces in Fig. 8-42 the *diffuse* exchanges are given by

$$q_{1 \to 2} = J_{1D} A_1 F_{12}(1 - \rho_{2s}) \qquad (8\text{-}63)$$
$$q_{2 \to 1} = J_{2D} A_2 F_{21}(1 - \rho_{1s}) \qquad (8\text{-}64)$$

Equation (8-63) expresses the diffuse radiation leaving 1 which arrives at 2 *and* which may contribute to a diffuse radiosity of surface 2. The factor $1 - \rho_s$ represents the fraction absorbed plus the fraction reflected diffusely. The inclusion of this factor is most important because we are considering only diffuse direct exchange, and thus must leave out the specular-reflection

**Fig. 8-42**

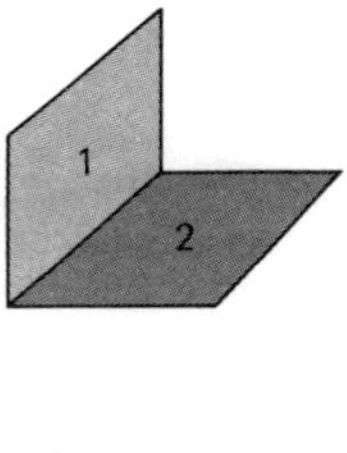

Fig. 8-43    Network element representing Eq. (8-65).

$q_{12} \rightarrow$

$$\frac{J_{1D}}{1-\rho_{1S}} \qquad\qquad \frac{J_{2D}}{1-\rho_{2S}}$$

$$\frac{1}{A_1 F_{12}(1-\rho_{1S})(1-\rho_{2S})}$$

contribution for now. The net exchange is given by the difference between (8-63) and (8-64), according to Ref. 21.

$$q_{12} = \frac{J_{1D}/(1-\rho_{1s}) - J_{2D}/(1-\rho_{2s})}{1/[A_1 F_{12}(1-\rho_{1s})(1-\rho_{2s})]} \tag{8-65}$$

The network element representing Eq. (8-65) is shown in Fig. 8-43.

To analyze specular reflections we utilize a technique presented by Sparrow et al. [12, 13]. Consider the four long surface enclosures shown in Fig. 8-44. Surfaces 1, 2, and 4 reflect diffusely, while surface 3 has both a specular and a diffuse component of reflection. The dashed lines represent mirror images of the surfaces 1, 2, and 4 in surface 3. (A specular reflection produces a mirror image.) The nomenclature 2(3) designates the mirror image of surface 2 in mirror 3.

Now consider the radiation leaving 2 which arrives at 1. There is a direct diffuse radiation of

$$(q_{2\rightarrow1})_{\substack{\text{direct}\\ \text{diffuse}}} = J_2 A_2 F_{21} \tag{8-66}$$

Part of the diffuse radiation from 2 is specularly reflected in 3 and strikes 1. This specularly reflected radiation acts like *diffuse* energy coming from the image surface 2(3). Thus we may write

$$(q_{2\rightarrow1})_{\substack{\text{specular}\\ \text{reflected}}} = J_2 A_{2(3)} F_{2(3)1} \rho_{3s} \tag{8-67}$$

The radiation shape factor $F_{2(3)1}$ is the one between surface 2(3) and surface 1. The reflectivity $\rho_{3s}$ is inserted because only this fraction of the radiation gets to 1. Of course, $A_2 = A_{2(3)}$. We now have

$$q_{2\rightarrow1} = J_2 A_2 (F_{21} + \rho_{3s} F_{2(3)1}) \tag{8-68}$$

Fig. 8-44    System with one specular-diffuse surface.

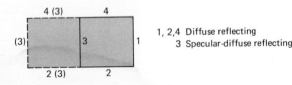

1, 2, 4  Diffuse reflecting
    3  Specular-diffuse reflecting

Fig. 8-45  Network element for Eq. (8-70).

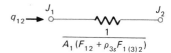

Using similar reasoning,

$$q_{1\rightarrow 2} = J_1 A_1 (F_{12} + \rho_{3s} F_{1(3)2}) \qquad (8\text{-}69)$$

Combining Eqs. (8-68) and (8-69) and making use of the reciprocity relation $A_1 F_{12} = A_2 F_{21}$ gives

$$q_{12} = \frac{J_1 - J_2}{1/[A_1(F_{12} + \rho_{3s} F_{1(3)2})]} \qquad (8\text{-}70)$$

The network element represented by Eq. (8-70) is shown in Fig. 8-45.

Analogous network elements may be developed for radiation between the other surfaces in Fig. 8-44, so that the final complete network becomes as shown in Fig. 8-46. It is to be noted that the elements connecting to $J_{3D}$ are simple modifications of the one shown in Fig. 8-43 since $\rho_{1s} = \rho_{2s} = \rho_{4s} = 0$. An interesting observation can be made about this network for the case where $\rho_{3D} = 0$. In this instance surface 3 is completely specular and

$$J_{3D} = \epsilon_3 E_{b3}$$

Fig. 8-46  Complete radiation network for system in Fig. 8-44.

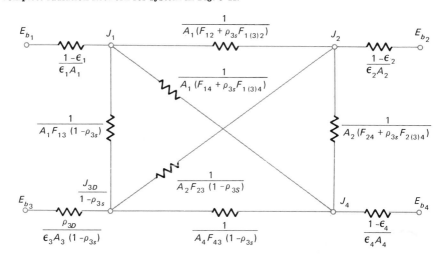

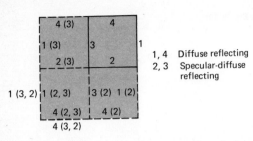

**Fig. 8-47   System with two specular-diffuse surfaces.**

1, 4    Diffuse reflecting
2, 3    Specular-diffuse
          reflecting

so that we are left with only three unknowns, $J_1$, $J_2$, and $J_4$, when surface 3 is completely specular-reflecting.

Now let us complicate the problem a step further by letting the enclosure have two specular-diffuse surfaces as shown in Fig. 8-47. In this case multiple images may be formed as shown. Surface 1(3,2) represents the image of 1 after it is viewed first through 3 and then through 2. In other words, it is the image of surface 1(3) in mirror 2. At the same location is surface 1(2,3), which is the image of surface 1(2) in mirror 3.

This problem is complicated because multiple specular reflections must be considered. Consider the exchange between surfaces 1 and 4. Diffuse energy leaving 1 can arrive at 4 in five possible ways:

| | |
|---|---|
| Direct: | $J_1 A_1 F_{14}$ |
| Reflection in 2 only: | $J_1 A_1 F_{1(2)4} \rho_{2s}$ |
| Reflection in 3 only: | $J_1 A_1 F_{1(3)4} \rho_{3s}$ |

Reflection first in 2 and then in 3:

$$J_1 A_1 \rho_{3s} \rho_{2s} F_{1(2,3)4}$$

$$(8\text{-}71)$$

Reflection first in 3 and then in 2:

$$J_1 A_1 \rho_{2s} \rho_{3s} F_{1(3,2)4}$$

The last shape factor, $F_{1(3,2)4}$, is zero because surface 1(3,2) cannot "see" surface 4 when looking *through* mirror 2. On the other hand, $F_{1(2,3)4}$ is not zero because surface 1(2,3) can see surface 4 when looking through mirror 3. The sum of the above terms is given as

$$q_{1 \to 4} = J_1 A_1 (F_{14} + \rho_{2s} F_{1(2)4} + \rho_{3s} F_{1(3)4} + \rho_{3s} \rho_{2s} F_{1(2,3)4}) \qquad (8\text{-}72)$$

In a similar manner,

$$q_{4 \to 1} = J_4 A_4 (F_{41} + \rho_{2s} F_{4(2)1} + \rho_{3s} F_{4(3)1} + \rho_{3s} \rho_{2s} F_{4(3,2)1}) \qquad (8\text{-}73)$$

Subtracting these two equations and applying the usual reciprocity relations gives the network element shown in Fig. 8-48.

Now consider the diffuse exchange between surfaces 1 and 3. Of the

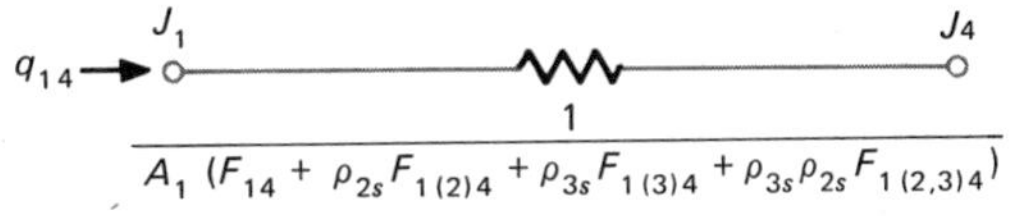

**Fig. 8-48**  Network element representing
exchange  between  surfaces  1  and  4  of
**Fig. 8-47.**

$$q_{14} \longrightarrow \overset{J_1}{\circ} \quad\text{———}\mkern-8mu\bigvee\mkern-8mu\text{———}\quad \overset{J_4}{\circ}$$

$$\frac{1}{A_1 \left(F_{14} + \rho_{2s}F_{1(2)4} + \rho_{3s}F_{1(3)4} + \rho_{3s}\rho_{2s}F_{1(2,3)4}\right)}$$

energy leaving 1, the amount which contributes to the diffuse radiosity of
surface 3 is

$$q_{1\rightarrow 3} = J_1 A_1 F_{13}(1 - \rho_{3s}) + J_1 A_1 \rho_{2s} F_{1(2)3}(1 - \rho_{3s}) \qquad (8\text{-}74)$$

The first term represents the direct exchange, and the second term represents the exchange after one specular reflection in mirror 2. As before, the
factor $1 - \rho_{3s}$ is included to leave out of consideration the specular reflection
from 3. This reflection, of course, is taken into account in other terms. The
*diffuse* energy going from 3 to 1 is

$$q_{3\rightarrow 1} = J_{3D} A_3 F_{31} + J_{3D} A_3 \rho_{2s} F_{3(2)1} \qquad (8\text{-}75)$$

The first term is the direct radiation, and the second term is that which is
specularly reflected in mirror 2. Combining Eqs. (8-74) and (8-75) gives the
network element shown in Fig. 8-49.

The above two elements are typical for the enclosure of Fig. 8-47, and
the other elements may be drawn by analogy. Thus the final complete
network is given in Fig. 8-50.

If both surfaces 2 and 3 are pure specular reflectors, that is,

$$\rho_{2D} = \rho_{3D} = 0$$

we have

$$J_{2D} = \epsilon_2 E_{b2}$$
$$J_{3D} = \epsilon_3 E_{b3}$$

and the network involves only two unknowns, $J_1$ and $J_4$, under these
circumstances.

We could complicate the calculation further by installing the specular
surfaces opposite each other. In this case there would be an infinite number
of images, and a series solution would have to be obtained; however, the
series for such problems usually converge rather rapidly. The reader should
consult Ref. 13 for further information on this aspect of radiation exchange
between specular surfaces.

**Fig. 8-49**  Network element representing exchange between surfaces 1 and 3 in Fig. 8-47.

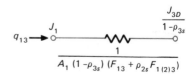

$$q_{13} \longrightarrow \overset{J_1}{\circ} \quad\text{———}\mkern-8mu\bigvee\mkern-8mu\text{———}\quad \overset{\frac{J_{3D}}{1-\rho_{3s}}}{\circ}$$

$$\frac{1}{A_1 (1 -\rho_{3s}) (F_{13} + \rho_{2s}F_{1(2)3})}$$

**Fig. 8-50**  Complete radiation network for system in **Fig. 8-47.**

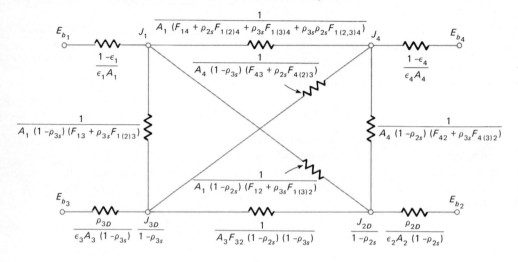

# 8-12□RADIATION EXCHANGE WITH TRANSMITTING, REFLECTING, AND ABSORBING MEDIA

We now consider a simple extension of the presentations in Secs. 8-10 and 8-11 to analyze a medium where reflection, transmission, and absorption modes are all important. As in Sec. 8-10, we shall analyze a system consisting of two parallel diffuse planes with a medium in between which may absorb, transmit, and reflect radiation. For generality we assume that the surface of the transmitting medium may have both a specular and a diffuse component of reflection. The system is shown in Fig. 8-51.

For the transmitting medium $m$ we have

$$\alpha_m + \rho_{mD} + \rho_{ms} + \tau_m = 1 \tag{8-76}$$

Also,

$$\epsilon_m = \alpha_m$$

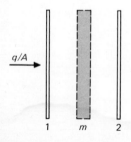

$q/A$

1    $m$    2

**Fig. 8-51**  Physical system for analysis of transmitting and reflecting layers.

**Fig. 8-52   Network element representing Eq. (8-79).**

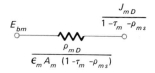

The diffuse radiosity of a particular surface of the medium is defined by

$$J_{mD} = \epsilon_m E_{bm} + \rho_{mD} G \tag{8-77}$$

where $G$ is the irradiation on the particular surface. Note that $J_{mD}$ no longer represents the total diffuse energy leaving a surface. Now it represents only emission and diffuse reflection. The transmitted energy will be analyzed with additional terms. As before, the heat exchange is written

$$q = A(\epsilon E_b - \alpha G) \tag{8-78}$$

Solving for $G$ from (8-77) and making use of (8-76) gives

$$q = \frac{E_{bm} - J_{mD}/(1 - \tau_m - \rho_{ms})}{\rho_{mD}/[\epsilon_m A_m(1 - \tau_m - \rho_{ms})]} \tag{8-79}$$

The network element representing Eq. (8-79) is shown in Fig. 8-52. This element is quite similar to the one shown in Fig. 8-41, except that here we must take the transmissivity into account.

The transmitted heat exchange between surfaces 1 and 2 is the same as in Sec. 8-10; i.e.,

$$q = \frac{J_1 - J_2}{1/A_1 F_{12} \tau_m} \tag{8-80}$$

The heat exchange between surface 1 and $m$ is computed in the following way:   Of that energy leaving surface 1, the amount which arrives at $m$ and contributes to the diffuse radiosity of $m$ is

$$q_{1 \to m} = J_1 A_1 F_{1m}(1 - \tau_m - \rho_{ms}) \tag{8-81}$$

The diffuse energy leaving $m$ which arrives at 1 is

$$q_{m \to 1} = J_{mD} A_m F_{m1} \tag{8-82}$$

Subtracting (8-82) from (8-81) and using the reciprocity relation

$$A_1 F_{1m} = A_m F_{m1}$$

gives

$$q_{1m} = \frac{J_1 - J_{mD}/(1 - \tau_m - \rho_{ms})}{1/A_1 F_{1m}(1 - \tau_m - \rho_{ms})} \tag{8-83}$$

Fig. 8-53   Complete radiation network for system in Fig. 8-51.

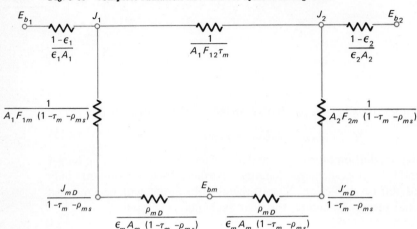

Fig. 8-53   Complete radiation network for system in Fig. 8-51.

The network element corresponding to Eq. (8-83) is quite similar to the one shown in Fig. 8-43. An equation similar to (8-83) can be written for the radiation exchange between surface 2 and $m$. Finally, the complete network may be drawn as in Fig. 8-53. It is to be noted that $J_{mD}$ represents the diffuse radiosity of the left side of $m$, while $J'_{mD}$ represents the diffuse radiosity of the right side of $m$.

If $m$ is maintained at a fixed temperature, then $J_1$ and $J_2$ must be obtained as a solution to nodal equations for the network. On the other hand, if no net energy is delivered to $m$, then $E_{bm}$ is a floating node, and the network reduces to a simple series-parallel arrangement. In this latter case the temperature of $m$ must be obtained by solving the network for $E_{bm}$.

We may extend the analysis a few steps further by distinguishing between specular and diffuse transmission. A specular transmission is one where the incident radiation goes "straight through" the material, while a diffuse transmission is encountered when the incident radiation is scattered in passing through the material, so that it emerges from the other side with a random spatial orientation. As with reflected energy, the assumption is made that the transmissivity may be represented with a specular and a diffuse component.

$$\tau = \tau_s + \tau_D \tag{8-84}$$

The diffuse radiosity is still defined as in Eq. (8-77), and the net energy exchange with a transmitting surface is given by Eq. (8-79). The analysis of

**Fig. 8-54** Radiation network for infinite parallel planes separated by a transmitting specular-diffuse plane.

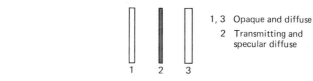

1, 3  Opaque and diffuse

2  Transmitting and specular diffuse

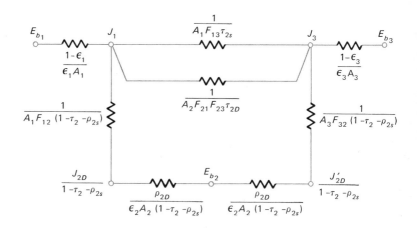

transmitted energy exchange with other surfaces must be handled somewhat differently, however.

Consider, for example, the arrangement in Fig. 8-54. The two diffuse opaque surfaces are separated by a specular-diffuse transmitting and reflecting plane. For this example all planes are assumed to be infinite in extent. The *specular*-transmitted exchange between surfaces 1 and 3 may be calculated immediately with

$$(q_{13})_{\substack{\text{specular} \\ \text{transmitted}}} = \frac{J_1 - J_3}{1/A_1 F_{13}\tau_{2s}} \tag{8-85}$$

The *diffuse*-transmitted exchange between 1 and 3 is a bit more complicated. The energy leaving 1 which is transmitted diffusely through 2 is

$$J_1 A_1 F_{12}\tau_{2D}$$

Of this amount transmitted through 2 the amount which arrives at 3 is

$$(q_{13})_{\substack{\text{diffuse} \\ \text{transmitted}}} = J_1 A_1 F_{12}\tau_{2D}F_{23} \tag{8-86}$$

**Fig. 8-55** Radiation network for system of finite specular-diffuse planes in an enclosure.

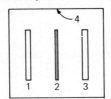

1, 3, 4    Opaque and diffuse
2    Transmitting and specular diffuse

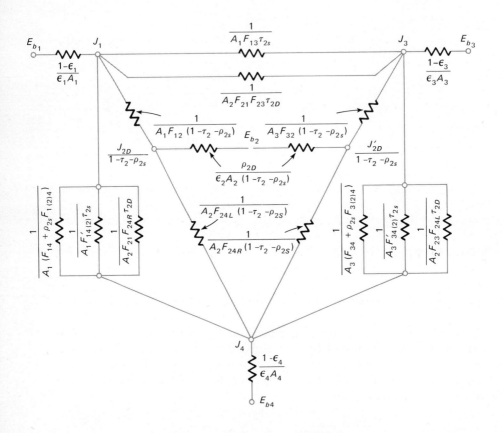

Similarly, the amount leaving 3 which is diffusely transmitted to 1 is

$$(q_{31})_{\substack{\text{diffuse} \\ \text{transmitted}}} = J_3 A_3 F_{32} \tau_{2D} F_{21} \tag{8-87}$$

Now, making use of the reciprocity relations $A_1 F_{12} = A_2 F_{21}$ and $A_3 F_{32} = A_2 F_{23}$, subtraction of (8-87) from (8-86) gives

$$(q_{13})_{\substack{\text{net diffuse} \\ \text{transmitted}}} = \frac{J_1 - J_3}{1/A_2 F_{21} F_{23} \tau_{2D}} \tag{8-88}$$

Making use of Eqs. (8-85) and (8-88) gives the complete network for the system as shown in Fig. 8-54. Of course, all the radiation shape factors in the above network are unity, but they have been included for the sake of generality. In this network $J_{2D}$ refers to the diffuse radiosity on the left side of 2, while $J'_{2D}$ is the diffuse radiosity on the right side of this surface.

It is easy to extend the network method to the case of two finite parallel planes separated by a transmitting plane inside a large enclosure. Such an arrangement is shown in Fig. 8-55. In this network the notation $F_{24R}$ means the radiation shape factor for radiation leaving the right side of surface 2, while $F_{24L}$ refers to the radiation leaving the left side of surface 2. The notation $F'_{14(2)}$ means the fraction of energy leaving 1 which arrives at the enclosure 4 after specular transmission through 2. The notation $F'_{34(2)}$ has a similar meaning with respect to surface 3. The factors $F_{1(2)4}$ and $F_{3(2)4}$ designate the radiation shape factors between the image surfaces 1(2) and 3(2) and the enclosure 4.

The above networks require information on the specular and diffuse properties of materials which are to be analyzed. Unfortunately, such information is quite meager at this time. Some data are available in Refs. 17, 18, and 19.

It should be rather obvious by now that all the equations that are obtained from the radiation network could be formulated without recourse to this analogy. For example, an energy balance on a particular opaque surface could be written

Net heat lost by surface = energy emitted − energy absorbed

or on a unit area basis with the usual gray body assumptions,

$$\frac{q}{A} = \epsilon E_b - \alpha G$$

Considering the $i$th surface, the total irradiation is the sum of all irradiations $G_j$ from the other $j$ surfaces. Thus, for $\epsilon = \alpha$,

$$\frac{q_i}{A_i} = \epsilon_i \left[ E_{b_i} - \sum_j G_j \right] \tag{8-89}$$

But, the irradiations can be expressed by

$$A_j J_j F_{ji} = G_j A_i \qquad (8\text{-}90)$$

From reciprocity, we have

$$A_j F_{ji} = A_i F_{ij}$$

so that we can combine the equations to give

$$\frac{q_i}{A_i} = \epsilon_i \left[ E_{bi} - \sum_j F_{ij} J_j \right] \qquad (8\text{-}91)$$

The heat transfer at each surface is then evaluated in terms of the radio-isties $J_j$. These parameters are obtained by recalling that the heat-transfer can also be expressed as

$$\frac{q_i}{A_i} = J_i - G = J_i - \sum_j F_{ij} J_j \qquad (8\text{-}92)$$

Combining Eqs. (8-91) and (8-92) gives

$$J_i - (1 - \epsilon_i) \sum_j F_{ij} J_j = \epsilon_i E_{bi} \qquad (8\text{-}93)$$

In the equations above it must be noted that the summations must be performed over *all* surfaces in the enclosure. For a three-surface enclosure, with $i = 1$, the summation would then become

$$\sum_j F_{ij} J_j = F_{11} J_1 + F_{12} J_2 + F_{13} J_3$$

Of course, if surface 1 is convex, $F_{11} = 0$ and some simplification could be effected.

Equation (8-93) may now be written for each surface in the enclosure and the set solved for the $J_i$'s. Other formulations may be developed for enclosures with specular surfaces and the reader is referred to Ref. 10 for additional information. Problems at the end of the chapter offer an opportunity to explore this formulation further.

## 8-13□SOLAR RADIATION

Solar radiation is a form of thermal radiation having a particular wavelength distribution. Its intensity is strongly dependent on atmospheric conditions, time of year, and the angle of incidence for the sun's rays on the surface of the earth. At the outer limit of the atmosphere the total solar irradiation when the earth is at its mean distance from the sun is 442.4 Btu/hr-ft². This number is called the solar *constant* and is subject to modification upon collection of more precise experimental data.

Not all the energy expressed by the solar constant reaches the surface of the earth, because of strong absorption by carbon dioxide and water vapor in the atmosphere. The solar radiation incident on the earth's surface

**Fig. 8-56** Spectral distribution of solar radiation as functions of atmospheric conditions and angle of incidence, according to Threlkeld and Jordan [15].

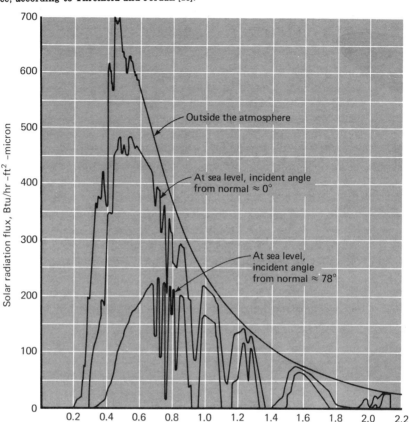

is also dependent on the atmospheric content of dust and other pollutants. The maximum solar energy reaches the surface of the earth when the rays are directly incident on the surface since (1) a larger view area is presented to the incoming solar flux, and (2) the solar rays travel a smaller distance through the atmosphere so that there is less absorption than there would be for an incident angle tilted from the normal. Figure 8-56 indicates the atmospheric absorption effects according to Ref. 15. These data are for a sea-level location on clear days in a moderately dusty atmosphere with moderate water-vapor content.

It is quite apparent from Fig. 8-56 that solar radiation which arrives at the surface of the earth does not behave like the radiation from an ideal gray body, while outside the atmosphere the distribution of energy follows more of an ideal pattern. To determine an equivalent blackbody temperature for the solar radiation we might employ the wavelength at which the maximum in the spectrum occurs (about 0.5 $\mu$ according to Fig. 8-56) and

Wien's displacement law [Eq. (8-11)]. This estimate gives

$$T \approx \frac{5215.6}{0.5} = 10{,}431.2°\text{R} \qquad (5{,}795 \text{ K})$$

The equivalent solar temperature for thermal radiation is therefore about 10,000°R.

If all materials exhibited gray-body behavior, solar-radiation analysis would not present a particularly unusual problem; however, since solar radiation is concentrated at short wavelengths, as opposed to much longer wavelengths for most "earth-bound" thermal radiation, a particular material may exhibit entirely different absorptance and transmittance properties for the two types of radiation. The classic example of this behavior is a greenhouse. Ordinary glass transmits radiation very readily at wavelengths below 2 $\mu$; thus it transmits the large part of solar radiation incident upon it. This glass, however, is essentially opaque to long-wavelength radiation above 3 or 4 $\mu$. Practically all the low-temperature radiation emitted by objects in the greenhouse is of such a long-wavelength character that it remains trapped in the greenhouse. Thus the glass allows much more radiation to come in than can escape, thereby producing the familiar heating effect. The solar radiation absorbed by objects in the greenhouse must eventually be dissipated to the surroundings by convection from the outside walls of the greenhouse.

Similar behavior is observed for the absorptance and reflectance of solar as opposed to low-temperature radiation from opaque metal or painted surfaces. In many instances, the total absorptivity for solar radiation can be quite different from the absorptivity for blackbody radiation at some moderate temperature like 100°F. Table 8-2 gives a brief comparison of the

**Table 8-2   Comparisons of Absorptivities of Various Surfaces to Solar and Low-temperature Thermal Radiation as Compiled from Ref. 14**

| Surface | Absorptivity for solar radiation | Absorptivity for low-temperature radiation $\sim$100°F |
|---|---|---|
| Aluminum, polished | 0.15 | 0.04 |
| Cast iron | 0.94 | 0.21 |
| White marble | 0.46 | 0.95 |
| White paper | 0.28 | 0.95 |
| Asphalt | 0.90 | 0.90 |
| Flat black lacquer | 0.96 | 0.95 |
| White paints, various types of pigments | 0.12–0.16 | 0.90–0.95 |

absorptivities for some typical surfaces for both solar and low-temperature radiation as compiled from Ref. 14. As will be noted, rather striking differences can occur.

This brief discussion of solar radiation is not intended to be comprehensive. Rather, it has the purpose of alerting the reader to some of the property information (like that of Ref. 14) when making calculations for solar radiation. Additional information on radiation transmission and absorption in the atmosphere is given in Chap. 13 along with the discussions pertaining to environmental heat transfer.

## 8-14□THE RADIATION HEAT - TRANSFER COEFFICIENT

In the development of convection heat transfer in the preceding chapters we found it convenient to define a heat-transfer coefficient by

$$q_{conv} = h_{conv}A(T_w - T_\infty)$$

Since radiation heat-transfer problems are often very closely associated with convection problems, and the total heat transfer by both convection and radiation is often the objective of an analysis, it is worthwhile to put both processes on a common basis by defining a radiation heat-transfer coefficient $h_r$ as

$$q_{rad} = h_r A_1(T_1 - T_2)$$

where $T_1$ and $T_2$ are the temperatures of the two bodies exchanging heat by radiation. The total heat transfer is then the sum of the convection and radiation,

$$q = (h_c + h_r)A_1(T_w - T_\infty) \tag{8-94}$$

if we assume that the second radiation-exchange surface is an enclosure and is at the same temperature as the fluid. For example, the heat loss by free convection and radiation from a hot steam pipe passing through a room could be calculated from Eq. (8-94).

In many instances the convection heat-transfer coefficient is not strongly dependent on temperature. However, this is not so with the radiation heat-transfer coefficient. The value of $h_r$, corresponding to Eq. (8-41), could be calculated from

$$\frac{q}{A_1} = \frac{\sigma(T_1^4 - T_2^4)}{1/\epsilon_1 + (A_1/A_2)(1/\epsilon_2 - 1)} = h_r(T_1 - T_2)$$

$$h_r = \frac{\sigma(T_1^2 + T_2^2)(T_1 + T_2)}{1/\epsilon_1 + (A_1/A_2)(1/\epsilon_2 - 1)} \tag{8-95}$$

Obviously, the radiation coefficient is a very strong function of temperature.

## REVIEW QUESTIONS

1.  How does thermal radiation differ from other types of electromagnetic radiation?
2.  What is the Stefan-Boltzmann law?
3.  Distinguish between specular and diffuse surfaces.
4.  Define radiation intensity.
5.  What is Kirchhoff's identity? When does it apply?
6.  What is a gray body?
7.  What is meant by the radiation shape factor?
8.  Define irradiation and radiosity.
9.  What is Beer's law?
10. Why do surfaces absorb differently for solar or earthbound radiation?
11. Explain the greenhouse effect.

## PROBLEMS

8-1.  Find the radiation shape factors $F_{1-2}$ for the situations shown.

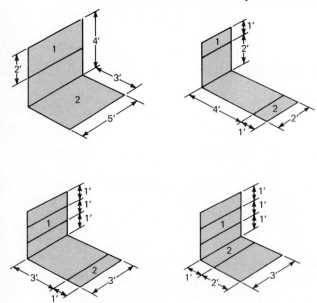

8-2.  Fused quartz transmits 90 percent of the incident thermal radiation between 0.2 and 4 $\mu$. Suppose a certain heat source is viewed through a quartz window. What heat flux in Btu per hour will be transmitted through the material from blackbody radiation sources at (a) 1500°F, (b) 1000°F, (c) 500°F, and (d) 150°F?

8-3. Repeat Prob. 8-2 for synthetic sapphire, which has a transmissivity of 0.85 between 0.2 and 5.5 $\mu$.

8-4. Repeat Prob. 8-2 for cesium iodide, which has a transmissivity of approximately 0.92 between 0.3 and 52 $\mu$.

8-5. A plate whose emissivity is 0.5 is 1 by 1 ft square. It is attached to the side of a spaceship so that it is perfectly insulated from the inside of the ship. Assuming that outer space is a blackbody at zero absolute, determine the equilibrium temperature for the plate at a point in space where the radiant heat flux from the sun is 500 Btu/hr-ft². Assume gray body behavior.

8-6. A conical hole is machined in a block of metal whose emissivity is 0.5. The hole is 1 in. in diameter at the surface and 2 in. deep. If the metal block is heated to 1000°F, calculate the radiation emitted by the hole. Calculate the value of an apparent emissivity of the hole, defined as the ratio of the actual energy emitted by the hole to that energy which would be emitted by a black surface having an area equal to that of the opening and a temperature equal to that of the inside surfaces.

8-7. A hole 1 in. in diameter is drilled in a 3-in. metal plate which is maintained at 500°F. The hole is lined with a thin foil having an emissivity of 0.07. A heated surface at 800°F having an emissivity of 0.5 is placed over the hole on one side of the plate, and the hole is left open on the other side of the plate. The 800°F surface is insulated from the plate insofar as conduction is concerned. Calculate the energy emitted from the open hole.

8-8. A cylindrical hole of depth $x$ and diameter $d$ is drilled in a block of metal having an emissivity of $\epsilon$. Using the definition given in Prob. 8-6, plot the apparent emissivity of the hole as a function of $x/d$ and $\epsilon$.

8-9. A cylindrical hole of diameter $d$ is drilled in a metal plate of thickness $x$. Assuming that the radiation emitted through the hole on each side of the plate is due only to the temperature of the plate, plot the apparent emissivity of the hole as a function of $x/d$ and the emissivity of the plate material $\epsilon$.

8-10. An artificial satellite 3 ft in diameter circles the earth at an altitude of 250 miles. Assuming that the diameter of the earth is 8,000 miles and the outer surface of the satellite is polished aluminum, calculate the radiation equilibrium temperature of the satellite when it is on the "dark" side of the earth. Take the earth as a blackbody at 60°F, and outer space as a blackbody at 0°R. The geometric shape factor from the satellite to the earth may be taken as the ratio of the solid angle subtended by the earth to the total solid angle for radiation from the satellite. When the satellite is on the "bright" side of the earth, it is irradiated with a heat flux of approximately 450 Btu/hr-ft² from the sun. Recalculate the equilibrium temperature of the satellite under these conditions. Assume that the satellite receives radiation from the sun as a disk and radiates to space as a sphere.

8-11. A long rod heater with $\epsilon = 0.8$ is maintained at 1800°F and is placed near a half-cylinder reflector as shown. The diameter of the rod is

3 in., and the diameter of the reflector is 20 in. The reflector is insulated, and the combined heater reflector is placed in a large room whose walls are maintained at 60°F. Calculate the radiant heat loss per unit length of the heater rod. How does this compare with the energy which would be radiated by the rod if it were used without the reflector?

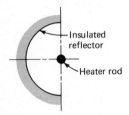

8-12.    A heated rod protrudes from a spaceship. The rod loses heat to outer space by radiation. Assuming that the emissivity of the rod is $\epsilon$ and that none of the radiation leaving the rod is reflected back, set up the differential equation for the temperature distribution in the rod. Also set up the boundary conditions which the differential equation must satisfy. The length of the rod is $L$, its cross-sectional area is $A$, its perimeter is $P$, and its base temperature is $T_0$. Assume that outer space is a blackbody at 0° abs.

8-13.    A room 10 by 10 by 10 ft has one sidewall maintained at 500°F, while the floor is maintained at 100°F. The other four surfaces are perfectly insulated. Assume that all surfaces are black. Calculate the net heat transfer between the hot wall and the cool floor.

8-14.    A square room 10 by 10 ft has a floor heated to 80°F, a ceiling at 55°F, and walls that are assumed perfectly insulated. The height of the room is 8 ft. The emissivity of all surfaces is 0.8. Using the network method, find the net interchange between floor and ceiling and the wall temperature.

8-15.    It is desired to transmit energy from one spaceship to another. A 5-ft-square plate is available on each ship to accomplish this. The ships are guided so that the plates are parallel and 1 ft apart. One plate is maintained at 1450°F, and the other at 540°F. The emissivities are 0.5 and 0.8, respectively. Find

(a) The net heat transferred between the spaceships in Btu per hour.

(b) The total heat lost by the hot plate in Btu per hour. Assume that outer space is a blackbody at 0°R.

8-16.    Two perfectly black parallel planes 4 by 4 ft are separated by a distance of 4 ft. One plane is maintained at 1000°F, and the other at 500°F. The planes are located in a large room whose walls are at 70°F. What is the net heat transfer between the planes?

8-17.    Two parallel planes 3 by 2 ft are separated by a distance of 2 ft. One plane is maintained at a temperature of 1000°F and has an emissivity of 0.6. The other plane is insulated. The planes are placed in a large room which is maintained at 50°F. Calculate the temperature of the insulated plane and the energy lost by the heated plane.

8-18.    Three infinite parallel plates are arranged as shown. Plate 1 is maintained at 2000°R, and plate 3 is maintained at 100°R. $\epsilon_1 = 0.2$, $\epsilon_2 = 0.5$, and $\epsilon_3 = 0.8$. Plate 2 receives no heat from external sources. What is the temperature of plate 2?

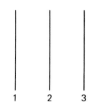

8-19.    Two large parallel planes having emissivities of 0.3 and 0.5 are maintained at temperatures of 1500 and 700°F, respectively. A radiation shield having an emissivity of 0.05 on both sides is placed between the two planes. Calculate (a) the heat-transfer rate per unit area if the shield were not present, (b) the heat-transfer rate per unit area with the shield present, and (c) the temperature of the shield.

8-20.    Two parallel planes 4 by 4 ft are separated by a distance of 4 ft. The emissivities of the planes are 0.4 and 0.6, and the temperatures are 1400 and 600°F, respectively. A 4- by 4-ft radiation shield having an emissivity of 0.05 on both sides is located equidistant between the two planes. The combined arrangement is placed in a large room which is maintained at 100°F. Calculate

(a) The heat-transfer rate from each of the two planes if the shield were not present.

(b) The heat-transfer rate from each of the two planes with the shield present.

(c) The temperature of the shield.

8-21.    A vertical plate 2 ft high and 1 ft wide is maintained at a temperature of 200°F in a room where the air is 70°F and 14.7 psia. The walls of the room are also at 70°F. Assume that $\epsilon = 0.8$ for the plate. How much heat is lost by the plate?

8-22.    A horizontal pipe 20 ft long and 5 in. in diameter is maintained at a temperature of 300°F in a large room where the air is 70°F and 14.7 psia. The walls of the room are at 100°F. Assume that $\epsilon = 0.7$ for the pipe. How much heat is lost by the pipe?

8-23.    A thermocouple is placed in a large heated duct to measure the temperature of the gas flowing through the duct. The duct walls are at 800°F, and the thermocouple indicates a temperature of 340°F. The heat-transfer coefficient from the gas to the thermocouple is 26 Btu/hr-ft²-°F.

The emissivity of the thermocouple material is 0.43. What is the temperature of the gas?

8-24.  A long pipe 2 in. in diameter passes through a room and is exposed to air at atmospheric pressure and temperature of 70°F. The pipe surface temperature is 200°F. Assuming that the emissivity of the pipe is 0.6, calculate the heat loss per foot of length of pipe.

8-25.  An annular space is filled with a gas whose emissivity and transmissivity are 0.3 and 0.7, respectively. The inside and outside diameters of the annular space are 1 and 2 ft, and the emissivities of the surfaces are 0.5 and 0.3, respectively. The inside surface is maintained at 1400°F, while the outside surface is maintained at 700°F. Calculate the net heat transfer per unit length from the hot surface to the cooler surface. What is the temperature of the gas? Neglect convection heat transfer.

8-26.  For the conditions in Prob. 8-25, plot the net heat transfer per unit length of the annulus as a function of the gas emissivity, assuming that

$$\epsilon_m + \tau_m = 1$$

8-27.  Repeat Prob. 8-25 for two infinite parallel planes with the same temperatures and emissivities. Calculate the heat-transfer rates per unit area of the parallel planes.

8-28.  The gas of Prob. 8-25 is forced through the annular space at a velocity of 20 ft/sec and is maintained at a temperature of 2000°F. The properties of the gas are

$\rho = 0.1 \ \text{lb}_m/\text{ft}^3$
$c_p = 0.4 \ \text{Btu/lb}_m\text{-°F}$
$\mu = 0.13 \ \text{lb}_m/\text{hr-ft}$
$k = 0.06 \ \text{Btu/hr-ft-°F}$

Assuming the same temperatures and emissivities of the surfaces as in Prob. 8-25, estimate the heating or cooling required for the inner and outer surfaces to maintain them at these temperatures. Assume that the convection heat-transfer coefficient may be estimated with the Dittus-Boelter equation (6-1).

8-29.  Suppose a flat plate is exposed to a high-speed environment. We define the radiation equilibrium temperature of the plate as that temperature it would attain if insulated so that the energy received by aerodynamic heating is just equal to the heat lost by radiation to the surroundings; i.e.,

$$hA(T_w - T_{aw}) = -\sigma A\epsilon(T_w{}^4 - T_s{}^4)$$

where the surroundings are supposed to be infinite and at the temperature $T_s$, and $\epsilon$ is the emissivity of the plate surface. Assuming an emissivity of 0.8 for the surface, calculate the radiation equilibrium temperature for the flow conditions of Example 5-7. Assume that the effective radiation temperature for the surroundings is −40°F.

8-30.   On a clear night the effective radiation temperature of the sky may be taken as $-100°F$. Assuming that there is no wind and the convection heat-transfer coefficient from the air to the dew which has collected on the grass is 5.0 Btu/hr-ft²-°F, estimate the minimum temperature which the air must have to prevent formation of frost. Neglect evaporation of the dew, and assume that the grass is insulated from the ground insofar as conduction is concerned. Take the emissivity as unity for the water.

8-31.   One way of constructing a blackbody cavity is to drill a hole in a metal plate. As a result of multiple reflections on the inside of the hole, the interior walls appear to have a higher emissivity than a flat surface in free space would have. A strict analysis of the cavity must take into account the fact that the irradiation is nonuniform over the interior surface. Thus the specific location at which a radiant-flux measuring device sights on this surface must be known in order to say how "black" the surface may be. Consider a $\frac{1}{2}$-in.-diameter hole 1 in. deep. Divide the interior surface into three sections, as shown. Assume, as an approximation, that the irradiation is uniform over each of these three surfaces and that the temperature and emissivity are uniform inside the entire cavity. A radiometer will detect the total energy leaving a surface (radiosity). Calculate the ratio $J/E_b$ for each of the three surfaces, assuming $\epsilon = 0.6$ and no appreciable radiation from the exterior surroundings.

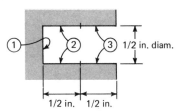

8-32.   Repeat Prob. 8-31, assuming a cavity temperature of 250°F and a surrounding temperature of 100°F.

8-33.   A thermocouple enclosed in a $\frac{1}{8}$-in. stainless-steel sheath ($\epsilon = 0.6$) is inserted horizontally into a furnace to measure the air temperature inside. The walls of the furnace are at 1200°F, and the true air temperature is 1050°F. What temperature will be indicated by the thermocouple? Assume free convection from the thermocouple.

8-34.   The thermocouple of Prob. 8-33 is placed horizontally in an air-conditioned room. The walls of the room are at 90°F, and the air temperature in the room is 69°F. What temperature is indicated by the thermocouple? What would be the effect on the reading if the thermocouple were enclosed by a polished-aluminum radiation shield?

8-35.   Two parallel disks 4 in. in diameter are separated by a distance of 1 in. One disk is maintained at 1000°F, is completely diffuse-reflecting, and has an emissivity of 0.3. The other disk is maintained at 500°F but is a specular-diffuse reflector such that $\rho_D = 0.2$, $\rho_s = 0.4$. The surroundings are maintained at 70°F. Calculate the heat lost by the inside surface of each disk.

8-36.    Two parallel infinite plates are maintained at 1500 and 100°F with emissivities of 0.5 and 0.8, respectively. To reduce the heat-transfer rate a radiation shield is placed between the two plates. Both sides of the shield are specular-diffuse-reflecting and have $\rho_D = 0.4$, $\rho_s = 0.4$. Calculate the heat-transfer rate with and without the shield. Compare this result with that obtained when the shield is completely diffuse-reflecting with $\rho = 0.8$.

8-37.    Rework Prob. 8-17, assuming that the 1000°F plane reflects in only a specular manner. The insulated plane is diffuse.

8-38.    Draw the radiation network for a specular-diffuse surface losing heat to a large enclosure. Obtain an expression for the heat transfer under these circumstances. How does this heat transfer compare with that which would be lost by a completely diffuse surface with the same emissivity?

8-39.    A 1- by 2-ft plate with $\epsilon = 0.6$ is placed in a large room and heated to 700°F. Only one side of the plate exchanges heat with the room. A highly reflecting plate ($\rho_s = 0.7$, $\rho_D = 0.1$) of the same size is placed perpendicular to the heated plate so that the 2-ft sides are in contact. Both sides of the reflector can exchange heat with the room. The room temperature is 100°F. Calculate the energy lost by the hot plate both with and without the reflector. What is the temperature of the reflector? Neglect convection.

8-40.    A 2- by 2- by 1-in. cavity is constructed of stainless steel ($\epsilon = 0.6$) and heated to 500°F. Over the top is placed a special ground-glass window ($\rho_s = 0.1$, $\rho_D = 0.1$, $\tau_D = 0.3$, $\tau_s = 0.3$, $\epsilon = 0.2$) 2 by 2 in. Calculate the heat lost to a very large room at 70°F, and compare with the energy which would be lost to the room if the glass window were not in place.

8-41.    Repeat Prob. 8-40 for the case of a window which is all diffuse-reflecting and all specular-transmitting; that is, $\rho_D = 0.2$, $\tau_s = 0.6$, $\epsilon = 0.2$.

8-42.    The cavity of Prob. 8-40 has a fused-quartz window placed over it, and the cavity is assumed to be perfectly insulated with respect to conduction and convection loss to the surroundings. The cavity is exposed to a solar irradiation flux of 400 Btu/hr-ft². Assuming that the quartz is nonreflecting and $\tau = 0.9$, calculate the equilibrium temperature of the inside surface of the cavity. Recall that the transmission range for quartz is 0.2 to 4 $\mu$. Neglect convection loss from the window. The surroundings may be assumed to be at 70°F.

8-43.    Apply Eqs. (8-91) and (8-93) to the problem described by Eq. (8-41). Apply the equations directly to obtain Eq. (8-41a) for a convex object completely enclosed by a very large concave surface.

8-44.    Rework Example 8-3 using the formulation of Eqs. (8-91) and (8-93).

8-45.   Rework Prob. 8-14 using the formulation of Eqs. (8-91) and (8-93). Recall that $J = E_b$ for the insulated surface.

8-46.   Rework Prob. 8-5 for a polished aluminum plate having the radiation characteristics given in Table 8-2.

8-47.   A slab of white marble is exposed to a solar radiation flux of 340 Btu/hr-ft². Assuming the effective radiation temperature of the sky is −100°F, calculate the radiation equilibrium temperature of the slab using the properties given in Table 8-2. For this calculation neglect all conduction and convection losses.

8-48.   The plate of Example 5-4 is 1 ft by 1 ft square and is sprayed with a white paint having a solar absorptivity of 0.16 and a low temperature absorptivity of 0.90. The plate is exposed to a solar radiation flux of 350 Btu/hr-ft² and allowed to reach equilibrium with the convection surroundings. Assuming that the underside of the plate is insulated, calculate the equilibrium temperature of the plate.

8-49.   Calculate the energy emitted between 4 and 15 $\mu$ by a gray body at 100°F with $\epsilon = 0.6$.

8-50.   An oven with a radiant heater for drying painted metal parts on a moving conveyor belt is designed as shown.

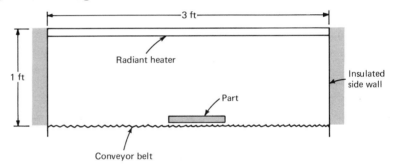

The length of the heating section is 10 ft and the heater temperature is 800°F. The side walls are insulating and it is determined experimentally that the conveyor belt and parts attain a temperature of 250°F. The belt-part combination has an effective emissivity of 0.8, and the radiant heater surface has $\epsilon = 0.7$. Calculate the energy supplied to the heater. Be sure to consider the radiation that is lost from the ends of the channel. Take the surroundings as a blackbody at 80°F.

## REFERENCES

1.   Sears, F. W.: "Introduction to Thermodynamics, Kinetic Theory, and Statistical Mechanics," pp. 123–124, Addison-Wesley Publishing Company, Inc., Reading, Mass., 1953.
2.   Dunkle, R. V.: Thermal Radiation Tables and Applications, *Trans. ASME*, vol. 76, p. 549, 1954.
3.   Mackey, C. O., L. T. Wright, Jr., R. E. Clark, and N. R. Gay: Radiant Heating and Cooling, part I, *Cornell Univ. Eng. Exp. Sta. Bull.*, vol. 32, 1943.

4.  Chapman, A. J.: "Heat Transfer," pp. 319–323, The Macmillan Company, New York, 1960.

5.  Hamilton, D. C., and W. R. Morgan: Radiant Interchange Configuration Factors, *NACA Tech. Notes*, no. 2836, 1952.

6.  Eckert, E. R. G., and R. M. Drake: "Heat and Mass Transfer," 2d ed., pp. 381–393, McGraw-Hill Book Company, New York, 1959.

7.  McAdams, W. H.: "Heat Transmission," 3d ed., chap. 2, McGraw-Hill Book Company, New York, 1954.

8.  Schmidt, E.: Messung der Gesamtstrahlung das Wasserdampfes bei Temperaturen bis 1000C, *Forsch. Gebiete Ingenieurw.*, vol. 3, p. 57, 1932.

9.  Schmidt, E., and E. R. G. Eckert: Uber die Richtungsverteilung der Warmestrahlung von Oberflachen, *Forsch. Gebiete Ingenieurw.*, vol. 6, p. 175, 1935.

10. Sparrow, E. M., and R. D. Cess: "Radiation Heat Transfer," Wadsworth Publishing Co., Inc., Englewood Cliffs, N.J., 1966.

11. Wiebelt, J. A.: "Engineering Radiation Heat Transfer," Holt, Rinehart and Winston, Inc., New York, 1966.

12. Eckert, E. R. G., and E. M. Sparrow: Radiative Heat Exchange between Surfaces with Specular Reflection, *Intern. J. Heat Mass Transfer*, vol. 3, pp. 42–54, 1961.

13. Sparrow, E. M., E. R. G. Eckert, and V. K. Jonsson: An Enclosure Theory for Radiative Exchange between Specular and Diffusely Reflecting Surfaces, *J. Heat Transfer*, ser. C, vol. 84, pp. 294–299, 1962.

14. Gubareff, G. G., J. E. Janssen, and R. H. Torborg: "Thermal Radiation Properties Survey," 2d ed., Minneapolis Honeywell Regulator Co., Minneapolis, Minn., 1960.

15. Threlkeld, J. L., and R. C. Jordan: Direct Solar Radiation Available on Clear Days, *ASHAE Trans.*, vol. 64, pp. 45–56, 1958.

16. Jakob, M.: "Heat Transfer," vol. 2, John Wiley & Sons, Inc., New York, 1957.

17. Birkebak, R. D., and E. R. G. Eckert: Effects of Roughness of Metal Surfaces on Angular Distribution of Monochromatic Radiation, *J. Heat Transfer*, vol. 87, p. 85, 1965.

18. Torrance, K. E., and E. M. Sparrow: Off Specular Peaks in the Directional Distribution of Reflected Thermal Radiation, *J. Heat Transfer*, vol. 88, p. 223, 1966.

19. Hering, R. G., and T. F. Smith: Surface Roughness Effects on Radiant Transfer between Surfaces, *Intern. J. Heat Mass Transfer*, vol. 13, p. 725, 1970.

20. Oppenheim, A. K.: Radiation Analysis by the Network Method, *Trans. ASME*, vol. 78, pp. 725–735, 1956.

21. Holman, J. P.: Radiation Networks for Specular-Diffuse Transmitting and Reflecting Surfaces, *ASME Paper* 66 WA/HT-9, Dec. 1966.

# CHAPTER

# 9

# CONDENSATION AND BOILING HEAT TRANSFER

## 9-1□INTRODUCTION

Our preceding discussions of convection heat transfer have considered homogeneous single-phase systems. Of equal importance are the convection processes associated with a change of phase of a fluid. The two most important examples are condensation and boiling phenomena, although heat transfer with solid-gas-phase changes has recently become important because of its application to ablation problems associated with reentry conditions on a ballistic missile.

In many types of power or refrigeration cycles one is interested in changing a vapor to a liquid, or a liquid to a vapor, depending on the particular part of the cycle under study. These changes are accomplished by boiling or condensation, and the engineer must understand the processes involved in order to design the appropriate heat-transfer equipment. High heat-transfer rates are usually involved in boiling and condensation, and this fact has also led designers of compact heat exchangers to utilize the phenomena for heating or cooling purposes not necessarily associated with power cycles.

## 9-2□CONDENSATION HEAT - TRANSFER PHENOMENA

Consider the vertical flat plate exposed to a condensable vapor shown in Fig. 9-1. If the temperature of the plate is below the saturation temperature of the vapor, condensate will form on the surface and under the action of gravity will flow down the plate. If the liquid wets the surface, a smooth film is formed, and the process is called *film condensation*. If the liquid does not wet the surface, droplets are formed which fall down the surface in some random fashion. This process is called *dropwise condensation*. In the film-condensation process the surface is blanketed by the film, which grows in thickness as it moves down the plate. A temperature gradient exists in the film, and the film represents a thermal resistance to heat transfer. In dropwise condensation a large portion of the area of the plate is directly exposed

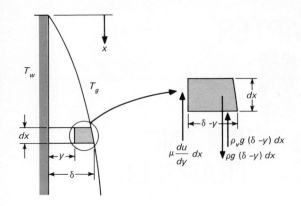

**Fig. 9-1  Film condensation on a vertical flat plate.**

to the vapor; there is no film barrier to heat flow, and higher heat-transfer rates are experienced. In fact, heat-transfer rates in dropwise condensation may be as much as ten times higher than in film condensation.

Because of the higher heat-transfer rates, dropwise condensation would be preferred to film condensation, but it is extremely difficult to maintain since most surfaces become "wetted" after exposure to a condensing vapor over an extended period of time. Various surface coatings and vapor additives have been used in attempts to maintain dropwise condensation, but these methods have not met with general success to date. Some of the pioneer work on drop condensation was conducted by Schmidt [26] and a good summary of the overall problem is presented in Ref. 27.

Film condensation on a vertical plate may be analyzed in a manner first proposed by Nusselt [1]. Consider the coordinate system shown in Fig. 9-1. The plate temperature is maintained at $T_w$, and the vapor temperature at the edge of the film is the saturation temperature $T_g$. The film thickness is represented by $\delta$, and we choose the coordinate system with the positive direction of $x$ measured downward, as shown. It is assumed that the viscous shear of the vapor on the film is negligible at $y = \delta$. It is further assumed that a linear temperature distribution exists between wall and vapor conditions. The weight of the fluid element of thickness $dx$ between $y$ and $\delta$ is balanced by the viscous-shear force at $y$ and the buoyancy force due to the displaced vapor. Thus

$$\rho g(\delta - y)\, dx = \mu \frac{du}{dy}\, dx + \rho_v g(\delta - y)\, dx \qquad (9\text{-}1)$$

Integrating and using the boundary condition that $u = 0$ at $y = 0$ gives

$$u = \frac{(\rho - \rho_v)g}{\mu}\left(\delta y - \tfrac{1}{2}y^2\right) \qquad (9\text{-}2)$$

The mass flow of condensate through any $x$ position of the film is thus given by

$$\text{Mass flow} = \dot{m} = \int_0^\delta \rho \left[ \frac{(\rho - \rho_v)g}{\mu} \left( \delta y - \tfrac{1}{2}y^2 \right) \right] dy$$

$$= \frac{\rho(\rho - \rho_v)g\delta^3}{3\mu} \tag{9-3}$$

when unit depth is assumed. The heat transfer at the wall in the area $dx$ is

$$q_x = -k \, dx \left. \frac{\partial T}{\partial y} \right)_{y=0} = k \, dx \, \frac{T_g - T_w}{\delta} \tag{9-4}$$

since a linear temperature profile was assumed. As the flow proceeds from $x$ to $x + dx$, the film grows from $\delta$ to $\delta + d\delta$ as a result of the influx of additional condensate. The amount of condensate added between $x$ and $x + dx$ is

$$\frac{d}{dx} \left[ \frac{\rho(\rho - \rho_v)g\delta^3}{3\mu} \right] dx = \frac{d}{d\delta} \left[ \frac{\rho(\rho - \rho_v)g\delta^3}{3\mu} \right] \frac{d\delta}{dx} \, dx$$

$$= \frac{\rho(\rho - \rho_v)g\delta^2 \, d\delta}{\mu}$$

The heat removed by the wall must equal this incremental mass flow times the latent heat of condensation of the vapor. Thus

$$\frac{\rho(\rho - \rho_v)g\delta^2 \, d\delta}{\mu} h_{fg} = k \, dx \, \frac{T_g - T_w}{\delta} \tag{9-5}$$

Equation (9-5) may be integrated with the boundary condition $\delta = 0$ at $x = 0$ to give

$$\delta = \left[ \frac{4\mu kx(T_g - T_w)}{gh_{fg}\rho(\rho - \rho_v)} \right]^{\frac{1}{4}} \tag{9-6}$$

The heat-transfer coefficient is now written

$$h \, dx \, (T_w - T_g) = -k \, dx \, \frac{T_g - T_w}{\delta}$$

or

$$h = \frac{k}{\delta}$$

so that

$$h_x = \left[ \frac{\rho(\rho - \rho_v)gh_{fg}k^3}{4\mu x(T_g - T_w)} \right]^{\frac{1}{4}} \tag{9-7}$$

Expressed in dimensionless form in terms of the Nusselt number,

$$\text{Nu}_x = \frac{hx}{k} = \left[ \frac{\rho(\rho - \rho_v)gh_{fg}x^3}{4\mu k(T_g - T_w)} \right]^{\frac{1}{4}} \tag{9-8}$$

The average value of the heat-transfer coefficient is obtained by integrating over the length of the plate.

$$\bar{h} = \frac{1}{L} \int_0^L h_x \, dx = \tfrac{4}{3}h_{x=L} \tag{9-9}$$

For laminar film condensation on horizontal tubes Nusselt obtained the relation

$$\bar{h} = 0.725 \left[ \frac{\rho(\rho - \rho_v)gh_{fg}k^3}{\mu d(T_g - T_w)} \right]^{\frac{1}{4}} \tag{9-10}$$

where $d$ is the diameter of the tube. The properties to be used in Eqs. (9-8) and (9-10) should be evaluated at the film temperature, taken as the arithmetic average between wall and saturation temperatures. Experimental measurements have shown that the theoretical analysis of Nusselt for laminar film condensation is conservative and that heat-transfer coefficients approximately 20 percent higher than that predicted by Eq. (9-8) may be expected in practice.

The "error" in the analysis results primarily from the assumption of a linear temperature profile. A more precise treatment would analyze the film in much the same manner as the boundary-layer problems of Chap. 5. With such an analysis it is possible to match experimental data very well. For an outline of such analysis techniques the reader is referred to Gebhart [20]. The final recommended relation for the average heat-transfer coefficient on a vertical plate is

$$\mathrm{Nu}_L = \frac{\bar{h}L}{k} = 1.60 \left[ \frac{\rho(\rho - \rho_v)gh_{fg}L^3}{4\mu k(T_g - T_w)} \right]^{\frac{1}{4}} \tag{9-9a}$$

When condensation occurs on a horizontal tube bank with $n$ tubes placed directly over each other in the vertical direction, the heat-transfer coefficient may be calculated by replacing the diameter in Eq. (9-10) with $nd$.

When a plate on which condensation occurs is sufficiently large or there is a sufficient amount of condensate flow, turbulence may appear in the condensate film. This turbulence results in higher heat-transfer rates. As in forced-convection flow problems, the criterion for determining whether the flow is laminar or turbulent is the Reynolds number, and for the condensation system it is defined as

$$\mathrm{Re}_f = \frac{D_H \rho V}{\mu_f} = \frac{4A\rho V}{P\mu_f}$$

where $D_H$ = hydraulic diameter
   $A$ = flow area
   $P$ = shear, or "wetted," perimeter
   $V$ = average velocity in the flow

But                                $\dot{m} = \rho A V$

so that                    $\mathrm{Re}_f = \dfrac{4\dot{m}}{P\mu_f}$ \hfill (9-11)

where $\dot{m}$ is the mass flow through the particular section of the condensate film. For a vertical plate of unit depth, $P = 1$, and for a vertical tube,

$P = \pi d$. The critical Reynolds number is approximately 1800, and turbulent correlations for heat transfer must be used at Reynolds numbers greater than this value. The Reynolds number is sometimes expressed in terms of the mass flow per unit depth of plate $\Gamma$, so that

$$\mathrm{Re}_f = \frac{4\Gamma}{\mu_f} \tag{9-12}$$

Thus the average heat transfer for laminar condensation may be written in terms of the Reynolds number as

$$\bar{h}\left[\frac{\mu_f^2}{k_f^3\rho_f(\rho_f - \rho_v)g}\right]^{\frac{1}{3}} = 1.47\ \mathrm{Re}_f^{-\frac{1}{3}} \qquad \text{for } \mathrm{Re}_f < 1800 \tag{9-13}$$

The average heat-transfer coefficient for turbulent film condensation is given by Kirkbride [2] as

$$\bar{h}\left[\frac{\mu_f^2}{k_f^3\rho_f(\rho_f - \rho_v)g}\right]^{\frac{1}{3}} = 0.0077\ \mathrm{Re}_f^{0.4} \qquad \text{for } \mathrm{Re}_f > 1800 \tag{9-14}$$

In calculating the Reynolds number the mass flow may be related to the total heat transfer and the heat-transfer coefficient by

$$q = \bar{h}A(T_{\text{sat}} - T_w) = \dot{m}h_{fg}$$

where $A$ is the total surface area for heat transfer. Thus

$$\dot{m} = \frac{q}{h_{fg}} = \frac{\bar{h}A(T_{\text{sat}} - T_w)}{h_{fg}}$$

$$\mathrm{Re}_f = \frac{4\bar{h}A(T_{\text{sat}} - T_w)}{h_{fg}P\mu_f} \tag{9-15}$$

But $\qquad\qquad A = LW \qquad \text{and} \qquad P = W$

where $L$ and $W$ are the height and width of the plate, respectively, so that

$$\mathrm{Re}_f = \frac{4\bar{h}L(T_{\text{sat}} - T_w)}{h_{fg}\mu_f} \tag{9-16}$$

## Example 9-1

A 1-ft-square vertical plate is exposed to steam at atmospheric pressure ($212°F$, $h_{fg} = 970.3$ Btu/lb$_m$). The plate temperature is $208°F$. Calculate the heat transfer and the mass of condensate condensed per hour.

**Solution**  The Reynolds number must be checked to determine if the condensate film is laminar or turbulent. The properties are evaluated at the film temperature.

$T_f = (212 + 208)/2 = 210°F$
$\rho_f = 59.9\ \text{lb}_m/\text{ft}^3 \qquad (960\ \text{kg/m}^3)$
$\mu_f = 0.693\ \text{lb}_m/\text{hr-ft}$
$k_f = 0.393\ \text{Btu/hr-ft-°F}$

For this problem the density of the vapor is very small in comparison with that of the liquid, and we are justified in making the substitution

$$\rho_f(\rho_f - \rho_v) \approx \rho_f^2$$

In trying to calculate the Reynolds number we find that it is dependent on the mass flow of condensate. But this is dependent on the heat-transfer coefficient, which is dependent on the Reynolds number. This dependence of the heat coefficient on the Reynolds number takes the form of either Eq. (9-13) or (9-14). To solve the problem we assume either laminar or turbulent flow, calculate the heat-transfer coefficient, and then check the Reynolds number to see if our assumption was correct. Let us assume laminar film condensation. Then, from Eq. (9-9a),

$$\bar{h} = 1.60 \left[ \frac{\rho_f^2 g h_{fg} k_f^3}{4\mu_f L (T_g - T_w)} \right]^{\frac{1}{4}}$$

$$= 1.60 \left[ \frac{(59.9)^2(32.2)(3600)^2(970.3)(0.393)^3}{(4)(0.693)(1)(212 - 208)} \right]^{\frac{1}{4}}$$

$$= 2686 \text{ Btu/hr-ft}^2\text{-}°\text{F} \qquad (15297 \text{ W/m}^2\text{-}°\text{C})$$

Checking the Reynolds number with Eq. (9-16), we have

$$\text{Re}_f = \frac{4\bar{h}L(T_{\text{sat}} - T_w)}{h_{fg}\mu_f}$$

$$= \frac{(4)(2686)(1)(4)}{(970.3)(0.693)} = 63.9$$

so that the laminar assumption was correct. The heat transfer is now calculated from

$$q = \bar{h}A(T_{\text{sat}} - T_w) = (2686)(1)(4) = 10744 \text{ Btu/hr} \qquad (2947 \text{ W})$$

The total mass flow of condensate is calculated as

$$\dot{m} = \frac{q}{h_{fg}} = \frac{10744}{970.3} = 11.07 \text{ lb}_m/\text{hr} \qquad (34.69 \text{ kg/h})$$

If the vapor to be condensed is superheated, the preceding equations may be used to calculate the heat-transfer coefficient, provided the heat flow is calculated on the basis of the temperature difference between the surface and the saturation temperature corresponding to the system pressure. When a noncondensable gas is present along with the vapor, there may be an impediment of the heat transfer since the vapor must diffuse through the gas before it can condense on the surface. The reader should consult Refs. 3 and 4 for more information on this subject.

Fig. 9-2   Heat-flux data from an electrically heated platinum wire, from Farber and Scorah [9].

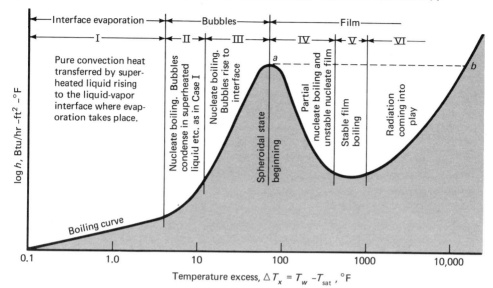

## 9-3□BOILING HEAT TRANSFER

When a surface is exposed to a liquid and is maintained at a temperature above the saturation temperature of the liquid, boiling may occur, and the heat flux will depend on the difference in temperature between the surface and the saturation temperature. When the heated surface is submerged below a free surface of liquid, the process is referred to as *pool boiling*. If the temperature of the liquid is below the saturation temperature, the process is called *subcooled*, or *local*, *boiling*. If the liquid is maintained at saturation temperature, the process is known as *saturated*, or *bulk*, *boiling*.

The different regimes of boiling are indicated in Fig. 9-2, where heat-flux data from an electrically heated platinum wire submerged in water are plotted against temperature excess $T_w - T_{sat}$. In region I free-convection currents are responsible for motion of the fluid near the surface. In this region the liquid near the heated surface is superheated slightly, and it subsequently evaporates when it rises to the surface. In region II bubbles begin to form on the surface of the wire and are dissipated in the liquid after breaking away from the surface. This region indicates the beginning of *nucleate boiling*. As the temperature excess is increased further, bubbles form more rapidly and rise to the surface of the liquid, where they are dissipated. This is indicated in region III. Eventually, bubbles are formed so rapidly that they blanket the heating surface and prevent the inflow of fresh liquid from taking their place. At this point the bubbles coalesce and form a vapor film which covers the surface. The heat must be conducted through this film before it can reach the liquid and effect the boiling process. The thermal resistance of this film causes a reduction in heat flux, and this phenomenon is illustrated in region IV, the *film-boiling* region. This region represents a transition from nucleate boiling to film boiling and is unstable. Stable film boiling is eventually encountered in region V. The surface temperatures required to maintain stable film boiling are high, and once this condition is attained, a significant portion of the heat lost by the surface may be the result of thermal radiation, as indicated in region VI.

An electrically heated wire is unstable at point $a$ since a small increase in $\Delta T_x$ at this point results in a decrease in the boiling heat flux. But the wire still must dissipate the same heat flux, or its temperature will rise, resulting in operation farther down on the boiling curve. Eventually, equilibrium may be reestablished only at point $b$ in the film-boiling region. This temperature usually exceeds the melting temperature of the wire, so that burnout results. If the electric-energy input is quickly reduced when the system attains point $a$, it may be possible to observe the partial nucleate boiling and unstable film region.

In nucleate boiling, bubbles are created by the expansion of entrapped gas or vapor at small cavities in the surface. The bubbles grow to a certain size, depending on the surface tension at the liquid-vapor interface and the temperature and pressure. Depending on the temperature excess, the bubbles may collapse on the surface, may expand and detach from the surface to be dissipated in the body of the liquid, or at sufficiently high temperatures may rise to the surface of the liquid before being dissipated. When local boiling conditions are observed, the primary mechanism of heat transfer is thought to be the intense agitation at the heat-transfer surface which creates the high heat-transfer rates observed in boiling. In saturated, or bulk, boiling the bubbles may break away from the surface because of the buoyancy action and move into the body of the liquid. In this case the heat-transfer rate is influenced by both the agitation caused by the bubbles and the vapor transport of energy into the body of the liquid.

Experiments have shown that the bubbles are not always in thermodynamic equilibrium with the surrounding liquid; i.e., the vapor inside the bubble is not necessarily at the same temperature as the liquid. Considering a spherical bubble as shown in Fig. 9-3, the pressure forces of the liquid and vapor must be balanced by the surface-tension force at the vapor-liquid interface. The pressure force acts on an area of $\pi r^2$, and the surface tension acts on the interface length of $2\pi r$. Making the force balance,

$$\pi r^2(p_v - p_l) = 2\pi r\sigma$$

or
$$p_v - p_l = \frac{2\sigma}{r} \qquad (9\text{-}17)$$

Fig. 9-3  Force balance on a vapor
bubble.

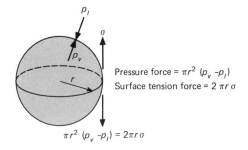

Pressure force $= \pi r^2 (p_v - p_l)$
Surface tension force $= 2\pi r \sigma$

$$\pi r^2 (p_v - p_l) = 2\pi r \sigma$$

where $p_v$ = vapor pressure inside bubble

$p_l$ = liquid pressure

$\sigma$ = surface tension of vapor-liquid interface

Now, suppose we consider a bubble in pressure equilibrium, i.e., one which is not growing or collapsing. Let us assume that the temperature of the vapor inside the bubble is the saturation temperature corresponding to the pressure $p_v$. If the liquid is at the saturation temperature corresponding to the pressure $p_l$, it is below the temperature inside the bubble. Consequently, heat must be conducted out of the bubble, the vapor inside must condense, and the bubble must collapse. This is the phenomenon which occurs when the bubbles collapse on the heating surface or in the body of the liquid. In order for the bubbles to grow and escape to the surface, they must receive heat from the liquid. This requires that the liquid be in a superheated condition so that the temperature of the liquid is greater than the vapor temperature inside the bubble. This is a metastable condition, but it is observed experimentally and accounts for the bubble growth after leaving the surface in some regions of nucleate boiling.

Figure 9-4 from Ref. 17 is a photograph illustrating several boiling regimes. The horizontal 0.25-in.-diameter copper rod is heated from the right side and immersed in isopropanol. As a result of the temperature gradient along the rod, it was possible to observe the different regimes simultaneously. At the left end of the rod the surface temperature is only slightly greater than the bulk fluid temperature, so that free-convection boiling is observed. Farther to the right higher surface temperatures are experienced and nucleate boiling is observed. Still farther to the right transition boiling takes place, and finally, film boiling is observed at the wall. Note the blanketing action of the vapor film on the right-hand portion of the rod.

The process of bubble growth is a complex one, but a simple qualitative explanation of the physical mechanism may be given in the following way: Bubble growth takes place when heat is conducted to the liquid-vapor inter-

Fig. 9-4  Photograph of 0.25-in.-diameter copper rod heated on the right side and immersed in isopropanol, from Haley and Westwater [17]. (Photograph courtesy of Professor Westwater.)

face from the liquid. Evaporation then takes place at the interface, thereby increasing the total vapor volume. Assuming that the liquid pressure remains constant, Eq. (9-17) requires that the pressure inside the bubble be reduced. Corresponding to a reduction in pressure inside the bubble will be a reduction in the vapor temperature and a larger temperature difference between the liquid and vapor if the bubble stays at its same spatial position in the liquid. However, the bubble will likely rise from the heated surface, and the farther away it moves, the lower the liquid temperature will be. Once the bubble moves into a region where the liquid temperature is below that of the vapor, heat will be conducted out, and the bubble will collapse. Hence the bubble growth process may reach a balance at some location in the liquid, or if the liquid is superheated enough, the bubbles may rise to the surface before being dissipated.

There is considerable controversy as to exactly how bubbles are initially formed on the heat-transfer surface. Surface conditions—both roughness and type of material—can play a central role in the bubble formation and growth drama. The mystery has not been completely solved, and remains a subject of intense research. Excellent summaries of the status of knowledge of boiling heat transfer have been presented by Rohsenow [18], Leppert and Pitts [19], and Tong [23], and the interested reader is referred to these discussions for more extensive information than is presented in this chapter. Heat-transfer problems in two-phase flow are discussed by Wallis [28].

Before presenting specific relations for calculating boiling heat transfer it is suggested that the reader review the discussion of the last few pages and correlate it with some simple experimental observations of boiling. For this purpose a careful study of the boiling process in a pan of water on the kitchen stove can be quite enlightening.

Rohsenow [5] correlated experimental data for nucleate pool boiling with the following relation:

$$\frac{C_l \, \Delta T_x}{h_{fg} \, \mathrm{Pr}_l^{1.7}} = C_{sf} \left[ \frac{q/A}{\mu_l h_{fg}} \sqrt{\frac{g_c \sigma}{g(\rho_l - \rho_v)}} \right]^{0.33} \qquad (9\text{-}18)$$

where $C_l$ = specific heat of saturated liquid, Btu/lb$_m$ = °F

$\Delta T_x$ = temperature excess = $T_w - T_{\mathrm{sat}}$, °F

$h_{fg}$ = enthalpy of vaporization, Btu/lb$_m$

$\mathrm{Pr}_l$ = Prandtl number of saturated liquid

$q/A$ = heat flux per unit area, Btu/hr-ft$^2$

$\mu_l$ = liquid viscosity, lb$_m$/hr-ft

$\sigma$ = surface tension of liquid-vapor interface, lb$_f$/ft

$g$ = gravitational acceleration

$\rho_l$ = density of saturated liquid, lb$_m$/ft$^3$

$\rho_v$ = density of saturated vapor, lb$_m$/ft$^3$

$C_{sf}$ = const, determined from experimental data

Values of the surface tension are given in Ref. 10, and a brief tabulation of the vapor-liquid surface tension for water is given in Table 9-2 (page 286).

The functional form of Eq. (9-18) was determined by analyzing the significant parameters in bubble growth and dissipation. Experimental data for nucleate boiling of water on a platinum wire are shown in Fig. 9-5, and a correlation of these data by the Rohsenow equation is shown in Fig. 9-6, indicating good agreement. The value of the constant $C_{sf}$ for the water-platinum combination is 0.013. Values for other fluid-surface combinations are given in Table 9-1. Equation (9-18) may be used for geometries other than horizontal wires, and in general it is found that geometry is not a strong factor in determining heat transfer for pool boiling. This would be expected because the heat transfer is primarily dependent on bubble formation and agitation, which is dependent on surface area, and not surface shape. Vachon, Nix, and Tanger [29] have determined values of the constants in the Rohsenow equation for a large number of surface-fluid combinations. There are several extenuating circumstances that influence the determination of the constants.

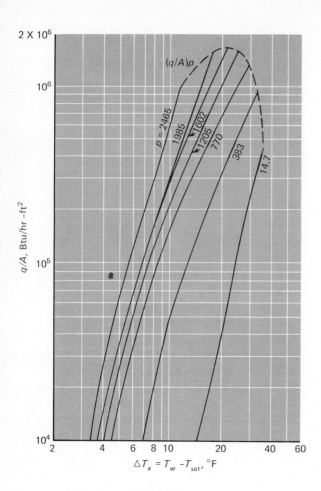

**Fig. 9-5  Heat-flux data for water boiling on a platinum wire, from McAdams [3].**

## Table 9-1  Values of the Coefficient $C_{sf}$ for Various Liquid-Surface Combinations

| Fluid-heating-surface combination | $C_{sf}$ |
| --- | --- |
| Water-copper [11]† | 0.013 |
| Water-platinum [12] | 0.013 |
| Water-brass [13] | 0.0060 |
| n-Butyl alcohol–copper [11] | 0.00305 |
| Isopropyl alcohol–copper [11] | 0.00225 |
| n-Pentane–chromium [14] | 0.015 |
| Benzene–chromium [14] | 0.010 |
| Ethyl alcohol–chromium [14] | 0.027 |

† Numbers in brackets refer to source of data.

Fig. 9-6   Correlation of pool boiling data by Eq. (9-18), from Rohsenow [5].

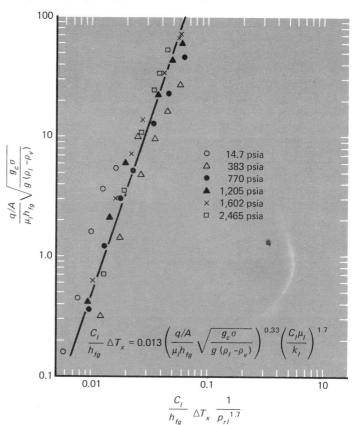

When a liquid is forced through a channel or over a surface maintained at a temperature greater than the saturation temperature of the liquid, forced-convection boiling may result. For forced-convection boiling in smooth tubes Rohsenow and Griffith [6] recommend that the forced-convection effect be computed with the Dittus-Boelter relation of Chap. 6 [Eq. (6-1)] and that this effect be added to the boiling heat flux computed from Eq. (9-18). Thus

$$\left(\frac{q}{A}\right)_{\text{total}} = \left(\frac{q}{A}\right)_{\text{boiling}} + \left(\frac{q}{A}\right)_{\substack{\text{forced} \\ \text{convection}}} \qquad (9\text{-}19)$$

For computing the forced-convection effect it is recommended that the coefficient 0.023 be replaced by 0.019 in the Dittus-Boelter equation. The

## Table 9-2 Vapor-Liquid Surface Tension for Water

| Surface tension $\sigma \times 10^4$, $lb_f/ft$ | Saturation temperature, °F |
|---|---|
| 51.8 | 32 |
| 50.2 | 60 |
| 47.8 | 100 |
| 45.2 | 140 |
| 41.2 | 200 |
| 40.3 | 212 |
| 31.6 | 320 |
| 21.9 | 440 |
| 11.1 | 560 |
| 1.0 | 680 |
| 0 | 705.4 |

temperature difference between wall and liquid bulk temperature is used to compute the forced-convection effect.

Forced convection boiling is not necessarily as simple as might be indicated by Eq. (9-19). This equation is generally applicable to forced-convection situations where the bulk liquid temperature is subcooled; in other words, for *local* forced-convection boiling. Once saturated or bulk boiling conditions are reached the situation changes rapidly. A fully developed nucleate boiling phenomenon is eventually encountered which is independent of the flow velocity or forced-convection effects. Various relations have been presented for calculating the heat flux in the fully developed boiling state. McAdams [21] suggests the following relation for low-pressure boiling water:

$$\frac{q}{A} = 0.074(\Delta T_x)^{3.86} \qquad 30 < p < 100 \text{ psia} \qquad (9\text{-}19a)$$

For higher pressures Levy [22] recommends the relation

$$\frac{q}{A} = \frac{p^{\frac{4}{3}}}{495} (\Delta T_x)^3 \qquad 100 < p < 2000 \text{ psia} \qquad (9\text{-}19b)$$

If boiling is maintained for a sufficiently long length of tube the majority of the flow area will be occupied by vapor. In this instance the vapor may flow rapidly in the central portion of the tube while a liquid film is vaporized along the outer surface. This situation is called *forced-convection vaporization* and is normally treated as a subject in two-phase flow and heat transfer. There are several complications which arise in this interesting subject, many of which are summarized by Tong [23] and Wallis [28].

The peak heat flux in flow boiling is a more complicated situation because the rapid generation of vapor produces a complex two-phase flow system which strongly influences the maximum heat flux which may be attained at the heat-transfer surface. Near the heater surface a thin layer of superheated liquid is formed, followed by a layer containing both bubbles and liquid. The core of the flow is occupied, for the most part, by vapor. The heat transfer at the wall is influenced by the boundary-layer development in that region and also by the rate at which diffusion of vapor and bubbles can proceed radially. Still further complications may arise from flow oscillations which are generated under certain conditions. Gambill [24] has suggested that the critical heat flux in flow boiling may be calculated by a superposition of the critical heat flux for pool boiling [Eq. (9-20)] and a forced-convection effect similar to the technique employed in Eq. (9-19). Levy [25] has considered the effects of vapor diffusion on the peak heat flux in flow boiling and Tong [23] presents a summary of available data on the subject.

### Example 9-2

A heated brass plate is submerged in a container of water at atmospheric pressure. The plate temperature is 242°F. Calculate the heat transfer per unit area of plate.

**Solution**    We could solve this problem by determining all the properties for use in Eq. (9-18) and subsequently determining the heat flux. An alternative method is to use the data of Fig. 9-5 in conjunction with Table 9-1. Upon writing Eq. (9-18), we find that if the heat flux for *one* particular water-surface combination is known, the heat flux for some other surface may easily be determined in terms of the constants $C_{sf}$ for the two surfaces since the fluid properties at any given temperature and pressure are the same regardless of the surface material. From Fig. 9-5 the heat flux for the water-platinum combination is

$$\frac{q}{A} = 3 \times 10^5 \text{ Btu/hr-ft}^2$$

since
$$T_w - T_{sat} = 242 - 212 = 30°F$$

From Table 9-1
$$C_{sf} = 0.013 \qquad \text{for water-platinum}$$
$$C_{sf} = 0.006 \qquad \text{for water-brass}$$

Accordingly,
$$\frac{(q/A)_{\substack{\text{water} \\ \text{brass}}}}{(q/A)_{\substack{\text{water} \\ \text{platinum}}}} = \left(\frac{C_{sf\substack{\text{water} \\ \text{platinum}}}}{C_{sf\substack{\text{water} \\ \text{brass}}}}\right)^3$$

and
$$\left(\frac{q}{A}\right)_{\substack{\text{water} \\ \text{brass}}} = (3 \times 10^5)\left(\frac{0.013}{0.006}\right)^3$$

$$= 3.4 \times 10^6 \text{ Btu/hr-ft}^2 \quad (1.072 \times 10^7 \text{ W/m}^2)$$

The peak heat flux for nucleate pool boiling is indicated as point $a$ in Fig. 9-2 and by a dashed line in Fig. 9-5. Zuber [7] has developed an analytical expression for the peak heat flux in nucleate boiling by considering the stability requirements of the interface between the vapor film and liquid. This relation is

$$\left(\frac{q}{A}\right)_{\max} = \frac{\pi}{24} h_{fg}\rho_v \left[\frac{\sigma g(\rho_l - \rho_v)}{\rho_v^2}\right]^{\frac{1}{4}} \left(1 + \frac{\rho_v}{\rho_l}\right)^{\frac{1}{2}} \tag{9-20}$$

where $\sigma$ is the vapor-liquid surface tension. This relation is in good agreement with experimental data. In general, the type of surface material does not affect the peak heat flux, although surface cleanliness can be an influence, dirty surfaces causing increases of approximately 15 percent in the peak value.

An interesting peak heat-flux phenomenon is observed when liquid droplets impinge on hot surfaces. Experiments with water, acetone, alcohol, and some of the Freons indicate that the maximum heat transfer is observed for temperature excesses of about 300°F, for all the fluids. The peak flux is a function of the fluid properties and the normal component of the impact velocity. A correlation of experimental data is given in Ref. 30 as

$$\frac{Q_{\max}}{\rho_L d^3 \lambda} = 1.83 \times 10^{-3} \left(\frac{\rho_L^2 V^2 d}{\rho_{vf} \sigma g_c}\right)^{0.341} \tag{9-21}$$

where $Q_{\max}$ is the maximum heat transfer per drop, $\rho_L$ is the density of the liquid droplet, $V$ is the normal component of the impact velocity, $\rho_{vf}$ is the vapor density evaluated at the film temperature $(T_w + T_{\text{sat}})/2$, $\sigma$ is the surface tension, $d$ is the drop diameter, and $\lambda$ is a modified heat of vaporization defined by

$$\lambda = h_{fg} + C_{pv}\left(\frac{T_w - T_{\text{sat}}}{2}\right)$$

While not immediately apparent from this equation, the heat-transfer rates in droplet impingement are quite high, and as much as 50 percent of the droplet is evaporated during the short time interval of impact and bouncing. The case of zero impact velocity is of historical note and is called the *Leidenfrost phenomenon* [31]. This latter case can be observed by watching water droplets sizzle and dance about on a hot plate.

Lienhard [34] has presented a relation for the peak boiling heat flux on horizontal cylinders which is in good agreement with experimental data. The relation is

$$\frac{q''_{\max}}{q''_{\max_F}} = 0.89 + 2.27 \exp\left(-3.44 \sqrt{R'}\right) \qquad \text{for } 0.15 < R' \tag{9-22}$$

where $R'$ is a dimensionless radius defined by

$$R' = R \left[\frac{g(\rho_l - \rho_v)}{\sigma}\right]^{\frac{1}{2}}$$

and $q''_{\text{max}_F}$ is the peak heat flux on an infinite horizontal plate derived by Zuber, et al. [33] as

$$q''_{\text{max}_F} = 0.131 \sqrt{\rho_v} \, h_{fg} [\sigma g (\rho_l - \rho_v)]^{\frac{1}{4}} \tag{9-23}$$

$\sigma$ is the surface tension.

Bromley [8] suggests the following relation for calculation of heat-transfer coefficients in the stable film-boiling region on a horizontal tube:

$$h_b = 0.62 \left[ \frac{k_v{}^3 \rho_v (\rho_l - \rho_v) g (h_{fg} + 0.4 c_{pv} \, \Delta T_x)}{d \mu_v \, \Delta T_x} \right]^{\frac{1}{4}} \tag{9-24}$$

where $d$ is the tube diameter. This heat-transfer coefficient considers only the conduction through the film, and does not include the effects of radiation. The total heat-transfer coefficient may be calculated from the empirical relation

$$h = h_b \left( \frac{h_b}{h} \right)^{\frac{1}{3}} + h_r \tag{9-25}$$

where $h_r$ is the radiation heat-transfer coefficient and is calculated assuming an emissivity of unity for the liquid. Thus

$$h_r = \frac{\sigma \epsilon (T_w{}^4 - T_{\text{sat}}^4)}{T_w - T_{\text{sat}}} \tag{9-26}$$

where $\sigma$ is the Stefan-Boltzmann constant and $\epsilon$ is the emissivity of the surface. Note that Eq. (9-25) will require an iterative solution for the total heat-transfer coefficient.

The properties of the vapor in Eq. (9-24) are to be evaluated at the film temperature defined by

$$T_f = \tfrac{1}{2} (T_w + T_{\text{sat}})$$

while the enthalpy of vaporization $h_{fg}$ is to be evaluated at the saturation temperature.

## 9-4□SIMPLIFIED RELATIONS FOR BOILING HEAT TRANSFER WITH WATER

Many empirical relations have been developed to estimate the boiling heat-transfer coefficients for water. Some of the simplest relations are those presented by Jakob and Hawkins [15] for water boiling on the outside of submerged surfaces at atmospheric pressure (Table 9-3). These heat-transfer coefficients may be modified to take into account the influence of pressure by using the empirical relation

$$h_p = h_1 \left( \frac{p}{p_1} \right)^{0.4} \tag{9-27}$$

**Table 9-3    Simplified Relations for Boiling Heat-transfer Coefficients to Water at Atmospheric Pressure According to Jakob and Hawkins [15]**

| Type of surface | $\dfrac{q}{A}$ | $h$ |
|---|---|---|
| Horizontal | $\dfrac{q}{A} < 5000$ | $h = 151(\Delta T_x)^{\frac{1}{3}}$ |
|  | $5000 < \dfrac{q}{A} < 75{,}000$ | $h = 0.168(\Delta T_x)^3$ |
| Vertical | $\dfrac{q}{A} < 1000$ | $h = 87(\Delta T_x)^{\frac{1}{7}}$ |
|  | $1000 < \dfrac{q}{A} < 20{,}000$ | $h = 0.240(\Delta T_x)^3$ |

$h$ in Btu/hr-ft²-°F     $\dfrac{q}{A}$ in Btu/hr     $\Delta T_x = T_w - T_{sat}$, °F

where $h_p$ = heat-transfer coefficient at some pressure $p$

$h_1$ = heat-transfer coefficient at atmospheric pressure as determined from Table 9-3

$p$ = system pressure

$p_1$ = standard atmospheric pressure

For forced-convection local boiling inside tubes the following relation is recommended [16]:

$$h = 0.077(\Delta T_x)^3 e^{p/225} \qquad (9\text{-}28)$$

where $\Delta T_x$ is the temperature difference between the surface and saturated liquid in degrees Fahrenheit and $p$ is the pressure in pounds per square inch absolute. The heat-transfer coefficient has the units of Btu/hr-ft²-°F.

## REVIEW QUESTIONS

1. Why are higher heat-transfer rates experienced in dropwise condensation than in film condensation?

2. How is the Reynolds number defined for film condensation?

3. What is meant by subcooled and saturated boiling?

4. Distinguish between nucleate and film boiling.

5. How is forced-convection boiling calculated?

6. Why does radiation play a significant role in film boiling heat transfer?

## PROBLEMS

9-1.    Using Eq. (9-14) as a starting point, develop an expression for the average heat-transfer coefficient in turbulent condensation as a function of only the fluid properties, length of the plate, and temperature difference;

i.e., eliminate the Reynolds number from Eq. (9-14) to obtain a relation similar to Eq. (9-8) for laminar condensation.

9-2. Show that the condensation Reynolds number for laminar condensation on a vertical plate may be expressed as

$$\text{Re}_f = 3.77 \left[ \frac{L^3(T_g - T_w)^3 \rho_f(\rho_f - \rho_v)gk_f^3}{\mu_f^5 h_{fg}^3} \right]^{\frac{1}{4}}$$

9-3. Plot Eqs. (9-13) and (9-14) as

$$\log \left\{ \bar{h} \left[ \frac{\mu_f^2}{k_f^3 \rho_f(\rho_f - \rho_v)g} \right]^{\frac{1}{3}} \right\} \quad \text{versus} \quad \log \text{Re}_f$$

Discuss this plot.

9-4. A square array of four hundred $\frac{1}{4}$-in. tubes is used to condense steam at atmospheric pressure. The tube walls are maintained at 190°F by a coolant flowing on the inside of the tubes. Calculate the amount of steam condensed per hour per unit length of the tubes.

9-5. A vertical plate 1 ft wide and 4 ft high is maintained at 160°F and exposed to saturated steam at 14.7 psia. Calculate the heat transfer and the total mass of steam condensed per hour.

9-6. Saturated steam at 100 psia condenses on the outside of a horizontal 1-in.-diameter tube. The tube wall temperature is maintained at 280°F. Calculate the heat-transfer coefficient and the condensate flow per unit length of tube.

9-7. Develop an expression for the total condensate flow in a turbulent film in terms of the fluid properties, the temperature difference, and the dimensions of the plate.

9-8. An uninsulated, chilled water pipe carrying water at 35°F passes through a hot, humid factory area where the temperature is 95°F and the relative humidity is 80 percent because of steam-operated equipment in the factory. If the pipe is 2 in. in diameter and the exposed length is 25 ft, estimate the condensate which will drip off the pipe. For this estimate assume that the pipe is exposed to saturated vapor at the partial pressure of the water vapor in the air.

9-9. A certain pressure cooker is designed to operate at 20 psig. It is well known that an item of food will cook faster in such a device because of the higher steam temperature at the higher pressure. Consider a certain item of food as a horizontal cylinder 4 in. in diameter at a temperature of 95°F when placed in the cooker. Calculate the percentage increase in heat transfer to this cylinder for the 20-psig condition as compared with condensation on the cylinder at standard atmospheric pressure.

9-10. A heated vertical plate at a temperature of 225°F is immersed in a tank of water exposed to atmospheric pressure. The temperature of the water is 212°F, and boiling occurs at the surface of the plate. The area of the plate is 3.2 ft². What is the heat lost from the plate in Btu per hour?

9-11.   The surface tension of water at 212°F is 58.8 dynes/cm for the vapor in contact with the liquid. Assuming that the saturated vapor inside a bubble is at 213°F while the surrounding liquid is saturated at 212°F, calculate the size of the bubble.

9-12.   Assuming that the bubble in Prob. 9-11 moves through the liquid at a velocity of 15 ft/sec, estimate the time required to cool the bubble $\frac{1}{2}$°F by calculating the heat-transfer coefficient for flow over a sphere and using this in a lumped-capacity analysis as described in Chap. 4.

9-13.   A heated 1- by 1-ft-square copper plate serves as the bottom for a pan of water at 1 atm pressure. The temperature of the plate is maintained at 242°F. Estimate the heat transferred per hour by the plate.

9-14.   Water at 14.7 psia flows in a $\frac{1}{2}$-in.-diameter brass tube at a velocity of 4 ft/sec. The tube wall is maintained at 230°F, and the average bulk temperature of the water is 205°F. Calculate the heat-transfer rate per unit length of tube.

9-15.   A kettle with a flat bottom 1 ft in diameter is available. It is desired to boil 5 lb_m/hr of water at atmospheric pressure in this kettle. At what temperature must the bottom surface of the kettle be maintained to accomplish this?

9-16.   A steel bar $\frac{1}{2}$ in. in diameter and 2 in. long is removed from a 2200°F furnace and placed in a container of water at atmospheric pressure. Estimate the heat-transfer rate from the bar when first placed in the water.

9-17.   Compare the heat-transfer coefficients for nucleate boiling of water, as shown in Fig. 9-4, with the simplified relations given in Table 9-3.

9-18.   Using Eqs. (9-12) and (9-7), develop Eq. (9-13).

9-19.   Saturated steam at 100 psia condenses on the outside of a horizontal 1-in.-diameter tube. The tube wall temperature is maintained at 280°F. Calculate the heat-transfer coefficient and the condensate flow per unit length of tube. Take

$$T_{\text{sat}} = 328°F \qquad h_{f_g} = 889 \text{ Btu/lb}$$

9-20.   Estimate the peak heat flux for boiling water at normal atmospheric pressure.

9-21.   A 0.2-in.-diameter copper heater rod is submerged in water at 14.7 psia. The temperature excess is 20°F. Estimate the heat loss per unit length of the rod.

9-22.   Compare the heat-transfer coefficients for boiling water and condensing steam on a horizontal tube for normal atmospheric pressure.

9-23.   Heat-transfer coefficients for boiling are usually large compared with those for ordinary convection. Estimate the flow velocity which would be necessary to produce a value of $h$ for forced convection through a smooth

$\frac{1}{4}$-in.-diameter brass tube comparable with that which could be obtained by pool boiling with $\Delta T_x = 30°F$, $p = 100$ psia, and water as the fluid. See Prob. 9-19 for data on properties.

9-24. A platinum wire is submerged in saturated water at 770 psia. What is the heat flux for a temperature excess of 20°F?

9-25. A horizontal tube having a 0.500 in. OD is submerged in water at 14.7 psia and 212°F. Calculate the heat flux for surface temperatures of (a) 1000°F, (b) 1200°F, and (c) 1500°F. Assume $\epsilon = 0.8$, Eq. (9-24).

9-26. A condenser is to be designed to condense 10,000 $lb_m/hr$ of steam at atmospheric pressure. A square array of 0.500 OD tubes is to be used with the outside tube walls maintained at 200°F. The spacing of the tubes is to be 0.75 in. between centers and their length is three times the square dimension. How many tubes are required for the condenser and what are the outside dimensions?

9-27. Calculate the peak heat flux for boiling water at atmospheric pressure on a horizontal cylinder 0.500 in. OD. Use the Lienhard relation.

9-28. Compare the heat flux calculated from the simple relations of Table 9-3 with the curve for atmospheric pressure in Fig. 9-5. Make the comparisons for two or three values of the temperature excess.

## REFERENCES

1. Nusselt, W.: Die Oberflachenkondensation des Wasserdampfes, *VDI Z.*, vol. 60, p. 541, 1916.
2. Kirkbride, C. G.: Heat Transfer by Condensing Vapors on Vertical Tubes, *Trans. AIChE*, vol. 30, p. 170, 1934.
3. McAdams, W. H.: "Heat Transmission," 3d ed., McGraw-Hill Book Company, New York, 1954.
4. Kern, D. Q.: "Process Heat Transfer," McGraw-Hill Book Company, New York, 1950.
5. Rohsenow, W. M.: A Method of Correlating Heat Transfer Data for Surface Boiling Liquids, *Trans. ASME*, vol. 74, p. 969, 1952.
6. Rohsenow, W. M., and P. Griffith: Correlation of Maximum Heat Flux Data for Boiling of Saturated Liquids, AIChE-ASME Heat Transfer Symposium, Louisville, Ky., 1955.
7. Zuber, N.: On the Stability of Boiling Heat Transfer, *Trans. ASME*, vol. 80, p. 711, 1958.
8. Bromley, L. A.: Heat Transfer in Stable Film Boiling, *Chem. Eng. Progr.*, vol. 46, p. 221, 1950.
9. Farber, E. A., and E. L. Scorah: Heat Transfer to Water Boiling under Pressure, *Trans. ASME*, vol. 70, p. 369, 1948.
10. "Handbook of Chemistry and Physics," Chemical Rubber Publishing Company, Cleveland, Ohio, 1960.
11. Piret, E. L., and H. S. Isbin: Natural Circulation Evaporation Two-phase Heat Transfer, *Chem. Eng. Progr.*, vol. 50, p. 305, 1954.
12. Addoms, J. N.: Heat Transfer at High Rates to Water Boiling outside Cylinders, Sc.D. thesis, Massachusetts Institute of Technology, Cambridge, Mass., 1948.

13. Cryder, D. S., and A. C. Finalbargo: Heat Transmission from Metal Surfaces to Boiling Liquids: Effect of Temperature of the Liquid on Film Coefficient, *Trans. AIChE*, vol. 33, p. 346, 1937.

14. Cichelli, M. T., and C. F. Bonilla: Heat Transfer to Liquids Boiling under Pressure, *Trans. AIChE*, vol. 41, p. 755, 1945.

15. Jakob, M., and G. Hawkins: "Elements of Heat Transfer," 3d ed., John Wiley & Sons, Inc., New York, 1957.

16. Jakob, M.: "Heat Transfer," vol. 2, p. 584, John Wiley & Sons, Inc., New York, 1957.

17. Haley, K. W., and J. W. Westwater: Heat Transfer from a Fin to a Boiling Liquid, *Chem. Engr. Sci.*, vol. 20, p. 711, 1965.

18. Rohsenow, W. M. (ed.): "Developments in Heat Transfer," The M.I.T. Press, Cambridge, Mass., 1964.

19. Leppert, G., and C. C. Pitts: Boiling, *Advan. Heat Transfer*, vol. 1, 1964.

20. Gebhart, B.: "Heat Transfer," McGraw-Hill Book Company, New York, 1961.

21. McAdams, W. H., et al.: Heat Transfer at High Rates to Water with Surface Boiling, *Ind. Eng. Chem.*, vol. 41, pp. 1945–1955, 1949.

22. Levy, S.: Generalized Correlation of Boiling Heat Transfer, *J. Heat Transfer*, vol. 81C, pp. 37–42, 1959.

23. Tong, L. S.: "Boiling Heat Transfer and Two-phase Flow," John Wiley & Sons, Inc., New York, 1965.

24. Gambill, W. R.: Generalized Prediction of Burnout Heat Flux for Flowing, Subcooled, Wetting Liquids, AIChE Reprint 17, 5th National Heat Transfer Conference, Houston, 1962.

25. Levy, S.: Prediction of the Critical Heat Flux in Forced Convection Flow, *USAEC Rept.* 3961, 1962.

26. Schmidt, E., W. Schurig, and W. Sellschop: Versuche uber die Kondensation von Wasserdampf in Film—und Tropfenform, *Tech. Mech. Thermodyn. Bul.*, vol. 1, p. 53, 1930.

27. Citakoglu, E., and J. W. Rose: Dropwise Condensation—Some Factors Influencing the Validity of Heat Transfer Measurements, *Intern. J. Heat Mass Trans.*, vol. 11, p. 523, 1968.

28. Wallis, G. B.: One-Dimensional Two-Phase Flow, McGraw-Hill Book Company, New York, 1969.

29. Vachon, R. I., G. H. Nix, and G. E. Tanger: Evaluation of Constants for the Rohsenow Pool-Booling Correlation, *ASME Paper* 67-HT-33, August, 1967.

30. McGinnis, F. K., and J. P. Holman: Individual Droplet Heat Transfer Rates for Splattering on Hot Surfaces, *Intern. J. Heat Mass Trans.*, vol. 12, p. 95, 1969.

31. Bell, K. J.: The Leidenfrost Phenomenon—A Survey, *Chem. Eng. Progr. Symp. Ser.*, no. 79, p. 73, 1967.

32. Rohsenow, W. M.: Nucleation with Boiling Heat Transfer, *ASME Paper* 70-HT-18.

33. Zuber, N., M. Tribus, and J. W. Westwater: The Hydrodynamic Crises in Pool Boiling of Saturated and Subcooled Liquids, *Intern. Div. Heat Transfer*, pp. 230–235, 1963.

34. Sun, K. H., and J. H. Lienhard: The Peak Boiling Heat Flux on Horizontal Cylinders, *Intern. J. Heat Mass Trans.*, vol. 13, p. 1425, 1970.

# CHAPTER
# 10
# HEAT EXCHANGERS

## 10-1□INTRODUCTION

The application of the principles of heat transfer to the design of equipment to accomplish a certain engineering objective is of extreme importance, for in applying the principles to design, the individual is working toward the important goal of product development for economic gain. Eventually, economics plays a key role in the design and selection of heat-exchange equipment, and the engineer should bear this in mind when embarking on any new heat-transfer design problem. The weight and size of heat exchangers used in space or aeronautical applications are very important parameters, and in these cases cost considerations are frequently subordinated insofar as material and heat-exchanger construction costs are concerned; however, the weight and size are important cost factors in the overall application in these fields, and thus may still be considered as economic variables.

A particular application will dictate the rules which one must follow to obtain the best design commensurate with economic considerations, size, weight, etc. An analysis of all these factors is beyond the scope of our present discussion, but it is well to remember that they all must be considered in practice. Our discussion of heat exchangers will take the form of technical analysis; i.e., the methods of predicting heat-exchanger performance will be outlined, along with a discussion of the methods which may be used to estimate the heat-exchanger size and type necessary to accomplish a particular task. In this respect, we limit our discussion to heat exchangers where the primary modes of heat transfer are conduction and convection. This is not to imply that radiation is not important in heat-exchanger design, for in many space applications it is the predominant means available for effecting an energy transfer. The reader is referred to the discussions by Kreith [1] and Sparrow and Cess [7] for detailed consideration of radiation heat-exchanger design.

## 10-2□THE OVERALL HEAT - TRANSFER COEFFICIENT

Consider the plane wall shown in Fig. 10-1 exposed to a hot fluid $A$ on one side and a cooler fluid $B$ on the other side. The heat transfer is expressed by

$$q = h_1 A (T_A - T_1) = \frac{kA}{\Delta x} (T_1 - T_2) = h_2 A (T_2 - T_B)$$

**Fig. 10-1   Overall heat transfer through a plane wall.**

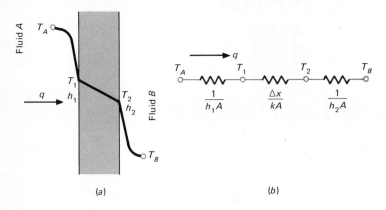

(a)                                    (b)

The heat-transfer process may be represented by the resistance network in Fig. 10-1b, and the overall heat transfer is calculated as the ratio of the overall temperature difference to the sum of the thermal resistances.

$$q = \frac{T_A - T_B}{1/h_1 A + \Delta x/kA + 1/h_2 A} \tag{10-1}$$

Observe that the value $1/hA$ is used to represent the convection resistance. The overall heat transfer by combined conduction and convection is frequently expressed in terms of an overall heat-transfer coefficient $U$, defined by the relation

$$q = UA \,\Delta T_{\text{overall}} \tag{10-2}$$

where $A$ is some suitable area for the heat flow. In accordance with Eq. (10-1), the overall heat-transfer coefficient would be

$$U = \frac{1}{1/h_1 + \Delta x/k + 1/h_2}$$

From the standpoint of heat-exchanger design the plane wall is of infrequent application; a more important case for consideration would be that of a double-pipe heat exchanger as shown in Fig. 10-2. In this application one fluid flows on the inside of the smaller tube, while the other fluid flows in the annular space between the two tubes. Note that the area for convection is not the same for both fluids in this case, these areas depending on the inside tube diameter and wall thickness. In this case the overall heat transfer would be expressed by

$$q = \frac{T_A - T_B}{\dfrac{1}{h_i A_i} + \dfrac{\ln (r_o/r_i)}{2\pi kL} + \dfrac{1}{h_o A_o}}$$

**Fig. 10-2  Double-pipe heat exchanger. (a) Schematic; (b) thermal-resistance network for overall heat transfer.**

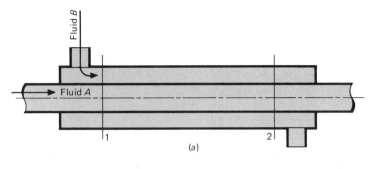

(a)

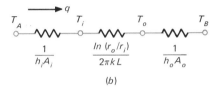

(b)

in accordance with the thermal network shown in Fig. 10-2b. The terms $A_i$ and $A_o$ represent the inside and outside surface areas of the inner tube. The overall heat-transfer coefficient may be based on either the inside or outside area of the tube at the discretion of the designer. Accordingly,

$$U_i = \frac{1}{\dfrac{1}{h_i} + \dfrac{A_i \ln (r_o/r_i)}{2\pi kL} + \dfrac{A_i}{A_o}\dfrac{1}{h_o}} \qquad (10\text{-}3)$$

$$U_o = \frac{1}{\dfrac{A_o}{A_i}\dfrac{1}{h_i} + \dfrac{A_o \ln (r_o/r_i)}{2\pi kL} + \dfrac{1}{h_o}} \qquad (10\text{-}4)$$

Calculations of the convection heat-transfer coefficients for use in the overall heat-transfer coefficient are made in accordance with the methods described in Chaps. 5 to 7.

## 10-3 □ FOULING FACTORS

After a period of time of operation the heat-transfer surfaces for a heat exchanger may become coated with various deposits present in the flow systems, or the surfaces may become corroded as a result of the interaction between the fluids and the material used for construction of the heat exchanger. In either event, this coating represents an additional resistance to the heat flow, and thus results in decreased performance. The overall effect is usually represented by a *fouling factor*, or fouling resistance, $R_f$, which must be included along with the other thermal resistances which make up the overall heat-transfer coefficient.

### Table 10-1    Table of Normal Fouling Factors

| Types of fluid | Fouling factor, hr-ft²-°F/Btu |
|---|---|
| Sea water below 125°F | 0.0005 |
| Sea water above 125°F | 0.001 |
| Treated boiler feedwater above 125°F | 0.001 |
| Fuel oil | 0.005 |
| Quenching oil | 0.004 |
| Alcohol vapors | 0.0005 |
| Steam, non-oil-bearing | 0.0005 |
| Industrial air | 0.002 |
| Refrigerating liquid | 0.001 |

Fouling factors must be obtained experimentally by determining the values of $U$ for both clean and dirty conditions in the heat exchanger. The fouling factor is thus defined as

$$R_f = \frac{1}{U_{\text{dirty}}} - \frac{1}{U_{\text{clean}}}$$

Recommended values of the fouling factor for various fluids are given in Table 10-1 in accordance with Ref. 2.

### 10-4□TYPES OF HEAT EXCHANGERS

One type of heat exchanger has already been mentioned, that of a double-pipe arrangement as shown in Fig. 10-2. Either counterflow or parallel flow may be used in this type of exchanger, with either the hot or cold fluid occupying the annular space and the other fluid occupying the inside of the inner pipe.

A type of heat exchanger widely used in the chemical-process industries is that of the shell-and-tube arrangement shown in Fig. 10-3. One fluid flows on the inside of the tubes, while the other fluid is forced through the shell and over the outside of the tubes. To ensure that the shell-side fluid will flow across the tubes and thus induce higher heat transfer, baffles are placed in the shell as shown in the figure. Depending on the head arrangement at the ends of the exchanger, one or more tube passes may be utilized. In Fig. 10-3a one tube pass is used, and the head arrangement for two tube passes is shown in Fig. 10-3b. A variety of baffle arrangements are used in

Fig. 10-3 (a) Shell-and-tube heat exchanger with one tube pass. (Young Radiator Company.) (b) Head arrangement for shell-and-tube heat exchanger with two tube passes. (Young Radiator Company.)

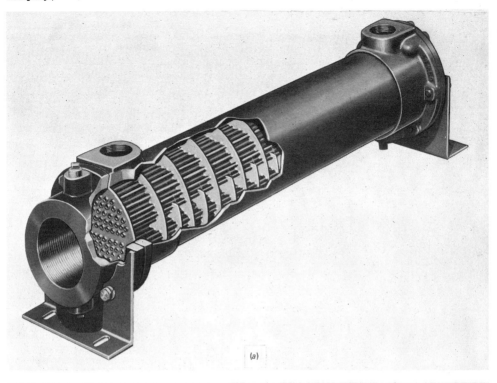

(a)

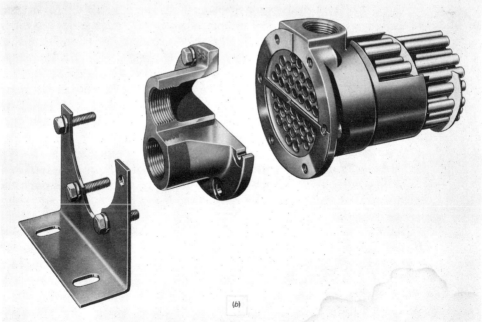

(b)

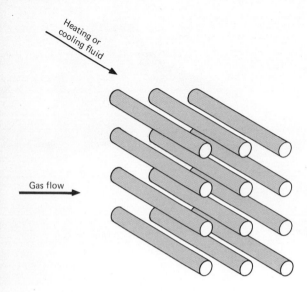

Heating or
cooling fluid

Gas flow

Fig. 10-4 Cross-flow heat exchanger, one fluid mixed, one fluid unmixed.

practice, and the reader is referred to Ref. 2 for more information on this matter.

Cross-flow heat exchangers are commonly used in air or gas heating and cooling applications. An example of such an exchanger is shown in Fig. 10-4, where a gas may be forced across a tube bundle, while another fluid is used inside the tubes for heating or cooling purposes. In this exchanger the gas flowing across the tubes is said to be a *mixed* stream, while the fluid in the tubes is said to be *unmixed*. The gas is mixed because it can move about freely in the exchanger as it exchanges heat. The other fluid is confined in separate tubular channels while in the exchanger so that it cannot mix with itself during the heat-transfer process.

A different type of cross-flow exchanger is shown in Fig. 10-5. In this case the gas flows across finned-tube bundles, and thus is unmixed since it is confined in separate channels between the fins as it passes through the exchanger. This exchanger is typical of the types used in air-conditioning applications.

If a fluid is unmixed, there can be a temperature gradient both parallel and normal to the flow direction, whereas when the fluid is mixed, there will be a tendency for the fluid temperature to equalize in the direction normal to the flow as a result of the mixing. An approximate temperature profile for the gas flowing in the exchanger of Fig. 10-5 is indicated in Fig. 10-6, assuming that the gas is being heated as it passes through the exchanger.

**Fig. 10-5**  Cross-flow heat exchanger, both fluids unmixed.

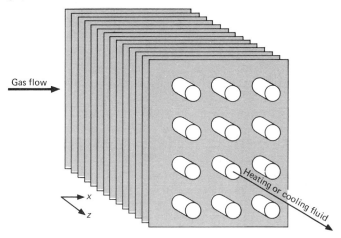

The fact that a fluid is mixed or unmixed influences the overall heat transfer in the exchanger because this heat transfer is dependent on the temperature difference between the hot and cold fluids.

## 10-5□THE LOG - MEAN TEMPERATURE DIFFERENCE

Consider the double-pipe heat exchanger shown in Fig. 10-2. The fluids may flow in either parallel flow or counterflow, and the temperature profiles

**Fig. 10-6**  Typical temperature profile for cross-flow heat exchanger of Fig. 10-5.

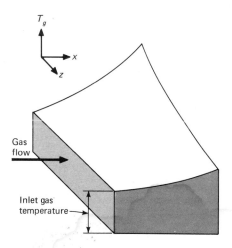

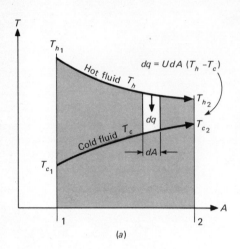

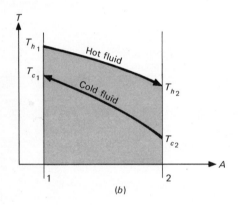

for these two cases are indicated in Fig. 10-7. We propose to calculate the
heat transfer in this double-pipe arrangement with

$$q = UA \, \Delta T_m \qquad (10\text{-}5)$$

where $U$ = overall heat-transfer coefficient

$A$ = surface area for heat transfer consistent with definition of $U$

$\Delta T_m$ = suitable mean temperature difference across heat exchanger

An inspection of Fig. 10-7 shows that the temperature difference between
the hot and cold fluids varies between inlet and outlet, and we must determine the average value for use in Eq. (10-5). For the parallel-flow heat
exchanger shown in Fig. 10-7 the heat transferred through an element of

area $dA$ may be written

$$dq = -\dot{m}_h c_h \, dT_h = \dot{m}_c c_c \, dT_c \qquad (10\text{-}6)$$

where the subscripts $h$ and $c$ designate the hot and cold fluids, respectively. The heat transfer could also be expressed

$$dq = U(T_h - T_c) \, dA \qquad (10\text{-}7)$$

From Eq. (10-6)

$$dT_h = \frac{-dq}{\dot{m}_h c_h}$$

$$dT_c = \frac{dq}{\dot{m}_c c_c}$$

where $\dot{m}$ represents the mass flow-rate and $c$ is the specific heat of the fluid. Thus

$$dT_h - dT_c = d(T_h - T_c) = -dq \left( \frac{1}{\dot{m}_h c_h} + \frac{1}{\dot{m}_c c_c} \right) \qquad (10\text{-}8)$$

Solving for $dq$ from Eq. (10-7) and substituting into Eq. (10-8) gives

$$\frac{d(T_h - T_c)}{T_h - T_c} = -U \left( \frac{1}{\dot{m}_h c_h} + \frac{1}{\dot{m}_c c_c} \right) dA \qquad (10\text{-}9)$$

This differential equation may now be integrated between conditions 1 and 2 as indicated in Fig. 10-7. The result is

$$\ln \frac{T_{h_2} - T_{c_2}}{T_{h_1} - T_{c_1}} = -UA \left( \frac{1}{\dot{m}_h c_h} + \frac{1}{\dot{m}_c c_c} \right) \qquad (10\text{-}10)$$

Returning to Eq. (10-6), the products $\dot{m}_c c_c$ and $\dot{m}_h c_h$ may be expressed in terms of the total heat transfer $q$ and the overall temperature differences of the hot and cold fluids. Thus

$$\dot{m}_h c_h = \frac{q}{T_{h_1} - T_{h_2}}$$

$$\dot{m}_c c_c = \frac{q}{T_{c_2} - T_{c_1}}$$

Substituting these relations into Eq. (10-10) gives

$$q = UA \frac{(T_{h_2} - T_{c_2}) - (T_{h_1} - T_{c_1})}{\ln [(T_{h_2} - T_{c_2})/(T_{h_1} - T_{c_1})]} \qquad (10\text{-}11)$$

Comparing Eq. (10-11) with Eq. (10-5), we find that the mean temperature difference is the grouping of terms in the brackets. Thus

$$\Delta T_m = \frac{(T_{h_2} - T_{c_2}) - (T_{h_1} - T_{c_1})}{\ln [(T_{h_2} - T_{c_2})/(T_{h_1} - T_{c_1})]} \qquad (10\text{-}12)$$

**Fig. 10-8** Correction-factor plot for exchanger with one shell pass and two, four, or any multiple of tube passes.

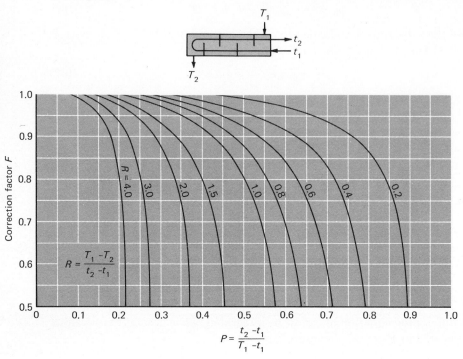

This temperature difference is called the *log-mean temperature difference,* abbreviated LMTD. Stated verbally, it is the temperature difference at one end of the heat exchanger less the temperature difference at the other end of the exchanger, divided by the natural logarithm of the ratio of these two temperature differences. It is left as an exercise for the reader to show that this relation may be used to calculate the log mean temperature differences for counterflow conditions.

The above derivation for LMTD involves two important assumptions: (1) the fluid specific heats do not vary with temperature, and (2) the convection heat-transfer coefficients are constant throughout the heat exchanger. The second assumption is usually the most serious one because of entrance effects, fluid viscosity, and thermal conductivity changes, etc. Numerical methods must normally be employed to correct for these effects. Section 10-7 describes one way of performing a variable properties analysis.

If a heat exchanger other than the double-pipe type is used, the heat transfer is calculated by using a correction factor applied to the LMTD for a counterflow double-pipe arrangement with the same hot and cold fluid

**Fig. 10-9** Correction-factor plot for exchanger with two shell passes and four, eight, or any multiple of tube passes.

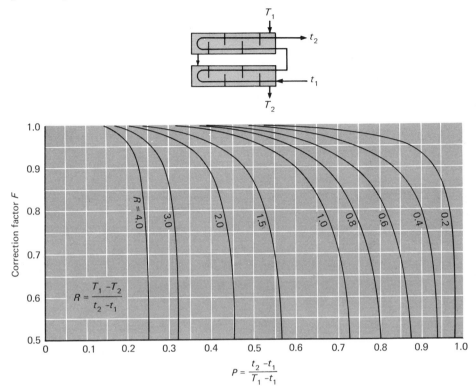

temperatures. The heat-transfer equation then takes the form

$$q = UAF \, \Delta T_m \qquad (10\text{-}13)$$

Values of the correction factor $F$ according to Ref. 4 are plotted in Figs. 10-8 to 10-11 for several different types of heat exchangers. Examples 10-1 to 10-3 illustrate the use of the log-mean temperature difference for calculation of heat-exchanger performance.

## Example 10-1

Water at the rate of 150 $\text{lb}_m/\text{min}$ is heated from 100 to 160°F by an oil having a specific heat of 0.45. The fluids are used in a counterflow double-pipe heat exchanger, and the oil enters the exchanger at 230°F and leaves at 150°F. The overall heat-transfer coefficient is 60 Btu/hr-ft²-°F. Calculate the heat-exchanger area.

**Fig. 10-10**   Correction-factor plot for single-pass cross-flow exchanger, both fluids unmixed.

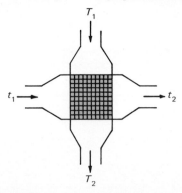

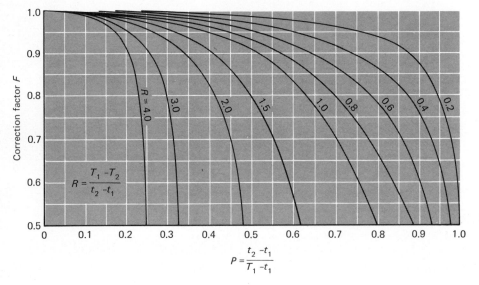

$$R = \frac{T_1 - T_2}{t_2 - t_1}$$

$$P = \frac{t_2 - t_1}{T_1 - t_1}$$

**Solution**   The total heat transfer is

$$q = \dot{m}_w c_{pw}(\Delta T_w) = (150)(1)(160 - 100)$$
$$= 9000 \text{ Btu/min} = 540{,}000 \text{ Btu/hr} \qquad (158{,}000 \text{ W})$$

The log-mean temperature difference is calculated as

$$\Delta T_m = \frac{(230 - 160) - (150 - 100)}{\ln[(230 - 160)/(150 - 100)]} = 59.6°\text{F}$$

Then, since

$$q = UA \, \Delta T_m$$

$$A = \frac{q}{U \, \Delta T_m} = \frac{5.4 \times 10^5}{(60)(59.6)} = 151 \text{ ft}^2 \qquad (14.03 \text{ m}^2)$$

**Fig. 10-11** Correction-factor plot for single-pass cross-flow exchanger, one fluid mixed, the other unmixed.

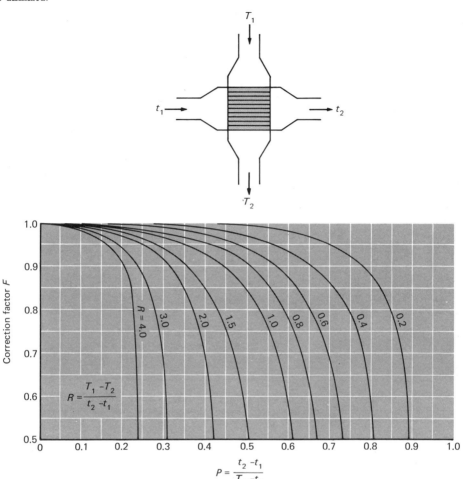

$$P = \frac{t_2 - t_1}{T_1 - t_1}$$

## Example 10-2

Instead of the double-pipe heat exchanger of Example 10-1, it is desired to use a shell-and-tube heat exchanger, the water making one shell pass and the oil making two tube passes. Calculate the area required for this exchanger, assuming that the overall heat-transfer coefficient remains at 60 Btu/hr-ft²-°F.

**Solution**   To solve this problem, we determine a correction factor from Fig. 10-8 to be used with the LMTD. The parameters according to the nomenclature of Fig. 10-8 are

$$T_1 = 100°F \qquad T_2 = 160°F \qquad t_2 = 150°F \qquad t_1 = 230°F$$
$$P = (150 - 230)/(100 - 230) = 0.615$$
$$R = (100 - 160)/(150 - 230) = 0.75$$

From Fig. 10-8, $F = 0.73$. Then

$$q = UAF \, \Delta T_m$$

so that
$$A = \frac{5.4 \times 10^5}{(60)(0.73)(59.6)} = 207 \text{ ft}^2 \qquad (19.23 \text{ m}^2)$$

### Example 10-3

Water at the rate of 30,000 $lb_m$/hr is heated from 100 to 130°F in a shell-and-tube heat exchanger. On the shell side one pass is used with water as the heating fluid, 15,000 $lb_m$/hr entering the exchanger at 200°F. The overall heat-transfer coefficient is 250 Btu/hr-ft²-°F, and the average water velocity in the $\frac{3}{4}$-in.-ID tubes is 1.2 ft/sec. Because of space limitations, the exchanger must not be longer than 8 ft. Consistent with this restriction, calculate the number of tube passes, the number of tubes per pass, and the length of the tubes.

**Solution**   We first assume one tube pass and check to see if it satisfies the conditions of the problem. The exit temperature of the hot water is calculated from

$$q = \dot{m}_c c_c \, \Delta T_c = \dot{m}_h c_h \, \Delta T_h$$
$$\Delta T_h = \frac{(30,000)(1)(30)}{(15,000)(1)} = 60$$
$$T_{h_{exit}} = 200 - 60 = 140°F$$

For counterflow

$$\text{LMTD} = \Delta T_m = \frac{(200 - 130) - (140 - 100)}{\ln \left[(200 - 130)/(140 - 100)\right]} = 53.6°F$$
$$q = UA \, \Delta T_r = \dot{m}_c c_c \, \Delta T_c = (30,000)(1)(30) = 9 \times 10^5 \text{ Btu/hr}$$
$$A = \frac{9 \times 10^5}{(250)(53.6)} = 67.3 \text{ ft}^2 \qquad (6.25 \text{ m}^2)$$

Using the average water velocity in the tubes and the flow rate, we calculate the total flow area with

$$\dot{m}_c = \rho \, A u$$
$$A = \frac{30,000}{(62.4)(1.2)(3600)} = 0.111 \text{ ft}^2$$

This area is the product of the number of tubes and the flow area per tube.

$$0.111 = n \frac{\pi d^2}{4}$$
$$n = \frac{(0.111)(4)(144)}{\pi (0.75)^2} = 36.1$$

or $n = 36$ tubes. The surface area per tube per foot of length is

$$\pi d = \frac{\pi(0.75)}{12} = 0.196 \text{ ft}^2/\text{tube-ft}$$

We recall that the total surface area required for a one-tube-pass exchanger was 67.3 ft². We may thus compute the length of tube for this type of exchanger as

$$n\pi \, dL = 67.3$$
$$L = \frac{67.3}{(36)(0.196)} = 9.55 \text{ ft} \qquad (2.91 \text{ m})$$

This length is greater than the allowable 8 ft, so that we must use more than one tube pass. When we increase the number of passes, we correspondingly increase the total surface area required because of the reduction in LMTD caused by the correction factor $F$. We next try two tube passes. From Fig. 10-8

$$F = 0.88$$

Thus $A_{\text{total}} = \dfrac{q}{UF \, \Delta T_m} = \dfrac{9 \times 10^5}{(250)(0.88)(53.6)} = 76.3 \text{ ft}^2 \qquad (7.09 \text{ m}^2)$

The number of tubes per pass is still 36 because of the velocity requirement. For the two-tube-pass system the total surface area is now related to the length by

$$A_{\text{total}} = 2n\pi \, dL$$

so that $\qquad L = \dfrac{76.3}{(2)(36)(0.196)} = 5.41 \text{ ft} \qquad (1.65 \text{ m})$

This length is within the 8-ft requirement, so that the final design choice is

$$\text{Number of tubes per pass} = 36$$
$$\text{Number of passes} = 2$$
$$\text{Length of tubes per pass} = 5.4 \text{ ft}$$

## 10-6□EFFECTIVENESS – NTU METHOD

The log-mean-temperature-difference approach to heat-exchanger analysis is useful when the inlet and outlet temperatures are known or are easily determined. The LMTD is then easily calculated, and the heat flow, surface area, or overall heat-transfer coefficient may be determined. When the inlet or exit temperatures are to be evaluated for a given heat exchanger, the analysis frequently involves an iterative procedure because of the logarithmic function in the LMTD. In these cases the analysis is performed more easily by utilizing a method based on the effectiveness of the heat exchanger in transferring a given amount of heat. The effectiveness method also offers many advantages for analysis of problems in which a comparison between various types of heat exchangers must be made for purposes of selecting the type best suited to accomplish a particular heat-transfer objective.

We define the heat-exchanger effectiveness as

$$\text{Effectiveness} = \epsilon = \frac{\text{actual heat transfer}}{\text{maximum possible heat transfer}}$$

The actual heat transfer may be computed by calculating either the energy lost by the hot fluid or the energy gained by the cold fluid. Consider the parallel- and counterflow heat exchangers shown in Fig. 10-7. For the parallel-flow exchanger

$$q = \dot{m}_h c_h (T_{h_1} - T_{h_2}) = \dot{m}_c c_c (T_{c_2} - T_{c_1}) \tag{10-14}$$

and for the counterflow exchanger

$$q = \dot{m}_h c_h (T_{h_1} - T_{h_2}) = \dot{m}_c c_c (T_{c_1} - T_{c_2}) \tag{10-15}$$

To determine the maximum possible heat transfer for the exchanger, we first recognize that this maximum value could be attained if one of the fluids were to undergo a temperature change equal to the maximum temperature difference present in the exchanger, which is the difference in the entering temperatures for the hot and cold fluids. The fluid which might undergo this maximum temperature difference is the one having the *minimum* value of $\dot{m}c$ since the energy balance requires that the energy received by one fluid be equal to that given up by the other fluid; if we let the fluid with the larger value of $\dot{m}c$ go through the maximum temperature difference, this would require that the other fluid undergo a temperature difference greater than the maximum, and this is impossible. So, maximum possible heat transfer is expressed as

$$q_{\max} = (\dot{m}c)_{\min}(T_{h_{\text{inlet}}} - T_{c_{\text{inlet}}}) \tag{10-16}$$

The minimum fluid may be either the hot or cold fluid, depending on the mass-flow rates and specific heats. For the parallel-flow exchanger

$$\epsilon_h = \frac{\dot{m}_h c_h (T_{h_1} - T_{h_2})}{\dot{m}_h c_h (T_{h_1} - T_{c_1})} = \frac{T_{h_1} - T_{h_2}}{T_{h_1} - T_{c_1}} \tag{10-17}$$

$$\epsilon_c = \frac{\dot{m}_c c_c (T_{c_2} - T_{c_1})}{\dot{m}_c c_c (T_{h_1} - T_{c_1})} = \frac{T_{c_2} - T_{c_1}}{T_{h_1} - T_{c_1}} \tag{10-18}$$

The subscripts on the effectiveness symbols designate the fluid which has the minimum value of $\dot{m}c$.

For the counterflow exchanger

$$\epsilon_h = \frac{\dot{m}_h c_h (T_{h_1} - T_{h_2})}{\dot{m}_h c_h (T_{h_1} - T_{c_2})} = \frac{T_{h_1} - T_{h_2}}{T_{h_1} - T_{c_2}} \tag{10-19}$$

$$\epsilon_c = \frac{\dot{m}_c c_c (T_{c_1} - T_{c_2})}{\dot{m}_c c_c (T_{h_1} - T_{c_2})} = \frac{T_{c_1} - T_{c_2}}{T_{h_1} - T_{c_2}} \tag{10-20}$$

We may derive an expression for the effectiveness in parallel flow as follows: Rewriting Eq. (10-10),

$$\ln \frac{T_{h_2} - T_{c_2}}{T_{h_1} - T_{c_1}} = -UA \left( \frac{1}{\dot{m}_h c_h} + \frac{1}{\dot{m}_c c_c} \right) = \frac{-UA}{\dot{m}_c c_c} \left( 1 + \frac{\dot{m}_c c_c}{\dot{m}_h c_h} \right) \tag{10-21}$$

or

$$\frac{T_{h_2} - T_{c_2}}{T_{h_1} - T_{c_1}} = \exp \left[ \frac{-UA}{\dot{m}_c c_c} \left( 1 + \frac{\dot{m}_c c_c}{\dot{m}_h c_h} \right) \right] \tag{10-22}$$

If the cold fluid is the minimum fluid,

$$\epsilon = \frac{T_{c_2} - T_{c_1}}{T_{h_1} - T_{c_1}}$$

Rewriting the temperature ratio in Eq. (10-22),

$$\frac{T_{h_2} - T_{c_2}}{T_{h_1} - T_{c_1}} = \frac{T_{h_1} + (\dot{m}_c c_c / \dot{m}_h c_h)(T_{c_1} - T_{c_2}) - T_{c_2}}{T_{h_1} - T_{c_1}} \tag{10-23}$$

when the substitution

$$T_{h_2} = T_{h_1} + \frac{\dot{m}_c c_c}{\dot{m}_h c_h} (T_{c_1} - T_{c_2})$$

is made from Eq. (10-6). Equation (10-23) may now be rewritten

$$\frac{(T_{h_1} - T_{c_1}) + (\dot{m}_c c_c / \dot{m}_h c_h)(T_{c_1} - T_{c_2}) + (T_{c_1} - T_{c_2})}{T_{h_1} - T_{c_1}} = 1 - \left( 1 + \frac{\dot{m}_c c_c}{\dot{m}_h c_h} \right) \epsilon$$

Inserting this relation back in Eq. (10-22) gives for the effectiveness

$$\epsilon = \frac{1 - \exp \left[ (-UA/\dot{m}_c c_c)(1 + \dot{m}_c c_c / \dot{m}_h c_h) \right]}{1 + \dot{m}_c c_c / \dot{m}_h c_h} \tag{10-24}$$

COLD FLUID - PARALLEL FLOW

It may be shown that the same expression results for the effectiveness when the hot fluid is the minimum fluid, except that $\dot{m}_c c_c$ and $\dot{m}_h c_h$ are interchanged. As a consequence, the effectiveness is usually written

$$\epsilon = \frac{1 - \exp \left[ (-UA/C_{\min})(1 + C_{\min}/C_{\max}) \right]}{1 + C_{\min}/C_{\max}} \tag{10-25}$$

where $C = \dot{m}c$ is defined as the capacity rate.

A similar analysis may be applied to the counterflow case, and the following relation for effectiveness results:

$$\epsilon = \frac{1 - \exp \left[ (-UA/C_{\min})(1 - C_{\min}/C_{\max}) \right]}{1 - (C_{\min}/C_{\max}) \exp \left[ (-UA/C_{\min})(1 - C_{\min}/C_{\max}) \right]} \tag{10-26}$$

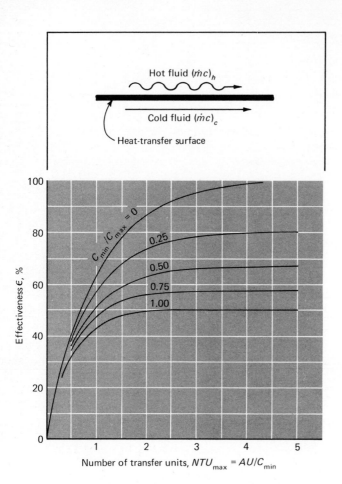

**Fig. 10-12 Effectiveness for parallel-flow exchanger performance.**

The grouping of terms $UA/C_{min}$ is called the *number of transfer units* (NTU) since it is indicative of the size of the heat exchanger.

Kays and London [3] have presented effectiveness ratios for various heat-exchanger arrangements, and some of the results of their analyses are available in chart form in Figs. 10-12 to 10-17. Examples 10-4 to 10-6 illustrate the use of the effectiveness–NTU method in heat-exchanger analysis.

### Example 10-4

The heat exchanger of Example 10-1 is used for heating water, as described in the example. Using the same entering fluid temperatures, calculate the exit water temperature when only 100 lb$_m$/min of water is

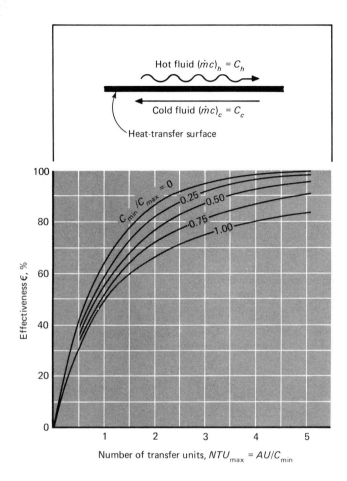

**Fig. 10-13 Effectiveness for counterflow exchanger performance.**

Hot fluid $(\dot{m}c)_h = C_h$

Cold fluid $(\dot{m}c)_c = C_c$

Heat-transfer surface

Effectiveness $\epsilon$, %

Number of transfer units, $NTU_{max} = AU/C_{min}$

$C_{min}/C_{max} = 0$

0.25

0.50

0.75

1.00

heated but the same quantity of oil is used. Also calculate the total heat transfer.

**Solution**   The flow rate of oil is calculated from the data for the original problem,

$$\dot{m}_h c_h \, \Delta T_h = \dot{m}_c c_c \, \Delta T_c$$
$$\dot{m}_h = \frac{(150)(1)(60)}{(0.45)(80)} = 250 \text{ lb}_m/\text{min}$$

The capacity rates for the new conditions are now calculated as

$$\dot{m}_h c_h = (250)(0.45)(60) = 6750 \text{ Btu/hr-°F}$$
$$\dot{m}_c c_c = (100)(1)(60) = 6000 \text{ Btu/hr-°F}$$

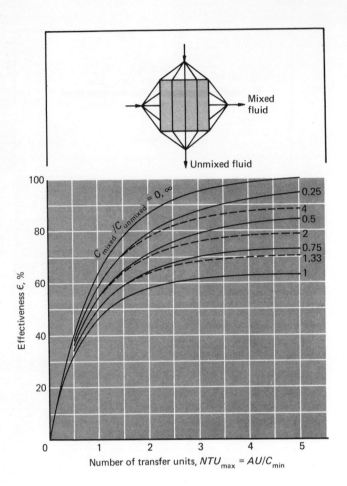

so that the water is the minimum fluid:

$$\frac{C_{\min}}{C_{\max}} = \frac{6000}{6750} = 0.89$$

$$\mathrm{NTU}_{\max} = \frac{UA}{C_{\min}} = \frac{(60)(151)}{6000} = 1.51$$

From Fig. 10-13, $\epsilon = 0.62$, so that

$$\epsilon = \frac{\Delta T_{\mathrm{cold}}}{230 - 100} = 0.62$$

$$\Delta T_{\mathrm{cold}} = (0.62)(130) = 80.6°\mathrm{F}$$

**Fig. 10-15 Effectiveness for cross-flow exchanger with fluids unmixed.**

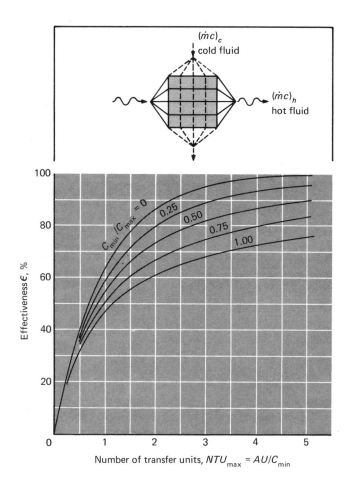

and the exit temperature of the water is

$$T_{w_{exit}} = 100 + 80.6 = 180.6°F$$

The total heat transfer is calculated as

$$q = \dot{m}_c c_c \, \Delta T_c = (6000)(80.6) = 483{,}000 \text{ Btu/hr} \qquad (141{,}500 \text{ W})$$

Notice that although the flow rate is reduced by 33 percent, the heat transfer is reduced by only 11 percent since the exchanger is more effective at the lower flow rate.

## Example 10-5

A finned-tube heat exchanger like that shown in Fig. 10-5 is used to heat 5000 cfm of air at 14.7 psia from 60 to 85°F. Hot water enters the

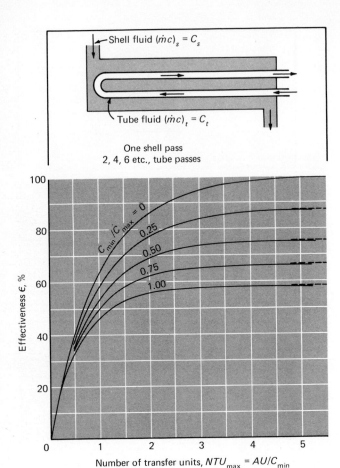

Fig. 10-16 Effectiveness for 1-2 parallel-counterflow exchanger performance.

tubes at 180°F, and the air flows across the tubes, producing an average overall heat-transfer coefficient of 40 Btu/hr-ft²-°F. The total surface area of the heat exchanger is 100 ft². Calculate the exit water temperature.

**Solution** From the statement of the problem we do not know whether the air or water is the minimum fluid. If the air is the minimum fluid, we may immediately calculate NTU and use Fig. 10-15 to determine the water-flow rate and hence the exit water temperature. If the water is the minimum fluid, a trial-and-error procedure must be used with Fig. 10-15. We assume that the air is the minimum fluid and then check our assumption. The air-flow rate is

$$\dot{m}_c = (5000)(0.0764)(60) = 22{,}900 \text{ lb}_m/\text{hr} \qquad (10{,}400 \text{ kg/h})$$
$$\dot{m}_c c_c = (22{,}900)(0.24) = 5500 \text{ Btu/hr-°F}$$

Fig. 10-17 **Effectiveness for 2-4 multipass counterflow exchanger performance.**

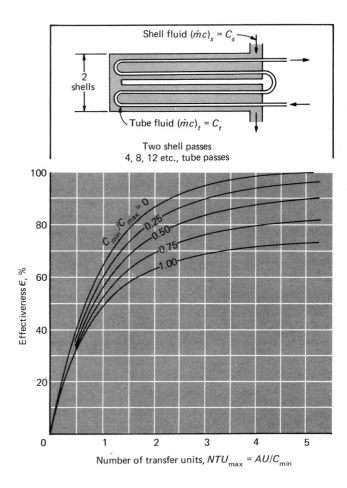

Two shell passes
4, 8, 12 etc., tube passes

The heat transfer is

$$q = \dot{m}_c c_c \, \Delta T_c = (5500)(25) = 137{,}500 \text{ Btu/hr}$$

If the air is the minimum fluid,

$$\text{NTU}_{\max} = \frac{(40)(100)}{5500} = 0.727$$

$$\epsilon = \frac{25}{180 - 60} = 0.208$$

Entering Fig. 10-15, we are unable to match these quantities with the curves. This requires that the hot fluid be the minimum. We must therefore assume values for the flow rate of the water until it is possible to match the performance as given by Fig. 10-15. The assumptions and calculations are listed in the following table:

| $\dot{m}_h c_h$ | $\dfrac{C_{min}}{C_{max}}$ | $NTU_{max}$ | $\Delta T_h$ | $\epsilon$ Calculated | From Fig. 10-15 |
|---|---|---|---|---|---|
| 2750 | 0.5 | 1.45 | 50 | 0.416 | |
| 1375 | 0.25 | 2.90 | 100 | 0.832 | 0.89 |
| 1210 | 0.22 | 3.31 | 113 | 0.94 | 0.93 |

We thus estimate the water-flow rate as 1210 $lb_m$/hr and the water-temperature drop as 113°F. The exit water temperature is, accordingly,

$$T_{w\,exit} = 180 - 113 = 67°F$$

## Example 10-6

A counterflow double-pipe heat exchanger is used to heat 10,000 $lb_m$/hr of water from 100 to 190°F by cooling an oil ($c_p = 0.5$) from 300 to 200°F. The overall heat-transfer coefficient is 150 Btu/hr-ft²-°F. A similar arrangement is to be built at another plant location, but it is desired to compare the performance of the single counterflow heat exchanger with two smaller counterflow heat exchangers connected in series on the water side and in parallel on the oil side, as shown in the sketch. The oil flow is split equally between the two exchangers, and it may be assumed that the overall heat-transfer coefficient for the smaller exchangers is the same as for the large exchanger. If the smaller exchangers cost 20 percent more per unit surface area, which would be the most economical arrangement—the one large exchanger or two equal-sized small exchangers?

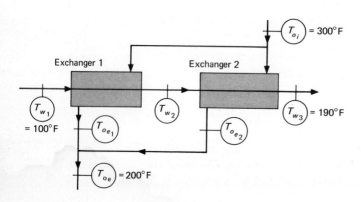

**Solution**   We calculate surface area required for both alternatives and then compare the costs. For the one large exchanger

$$q = \dot{m}_c c_c \, \Delta T_c = \dot{m}_h c_h \, \Delta T_h$$
$$= (10{,}000)(1)(190 - 100) = \dot{m}_h c_h (300 - 200)$$
$$= 9 \times 10^5 \text{ Btu/hr}$$
$$\dot{m}_c c_c = 10{,}000 \text{ Btu/hr-°F}$$
$$\dot{m}_h c_h = 9000 \text{ Btu/hr-°F}$$

so that the oil is the minimum fluid:

$$\epsilon_h = \frac{\Delta T_h}{300 - 100} = \frac{100}{200} = 0.5$$
$$\frac{C_{\min}}{C_{\max}} = \frac{9000}{10{,}000} = 0.9$$

From Fig. 10-13, $\text{NTU}_{\max} = 0.95$, so that

$$A = \text{NTU}_{\max} \frac{C_{\min}}{U}$$
$$= 0.95 \times \frac{9000}{150} = 57 \text{ ft}^2 \qquad (5.30 \text{ m}^2)$$

We now wish to calculate the surface-area requirement for the two exchangers shown in the sketch. We have

$$\dot{m}_h c_h = \frac{9000}{2} = 4500 \text{ Btu/hr-°F}$$
$$\dot{m}_c c_c = 10{,}000 \text{ Btu/hr-°F}$$
$$\frac{C_{\min}}{C_{\max}} = \frac{4500}{10{,}000} = 0.45$$

The number of transfer units is the same for each exchanger since $UA$ and $C_{\min}$ are the same for each exchanger. This requires that the effectiveness be the same for each exchanger. Thus

$$\epsilon_1 = \frac{T_{o_i} - T_{o_{e1}}}{T_{o_i} - T_{w_1}} = \epsilon_2 = \frac{T_{o_i} - T_{o_{e2}}}{T_{o_i} - T_{w_2}}$$
$$\epsilon_1 = \frac{300 - T_{o_{e1}}}{300 - 100} = \epsilon_2 = \frac{300 - T_{o_{e2}}}{300 - T_{w_2}} \qquad (a)$$

where the nomenclature for the temperature is indicated in the sketch. Since the average exit oil temperature must be 200°F, we may write

$$\frac{T_{o_{e1}} + T_{o_{e2}}}{2} = 200 \qquad (b)$$

An energy balance on the second exchanger gives

$$10{,}000(T_{w_3} - T_{w_2}) = 4500(T_{o_i} - T_{o_{e2}})$$
or
$$(10{,}000)(190 - T_{w_2}) = 4500(300 - T_{o_{e2}}) \qquad (c)$$

We now have the three equations $(a)$, $(b)$, and $(c)$ which may be solved for the three unknowns, $T_{o_{e1}}$, $T_{o_{e2}}$, and $T_{w_2}$. The solutions are

$$T_{o_{e1}} = 185°F$$
$$T_{o_{e2}} = 215°F$$
$$T_{w_2} = 152°F$$

The effectiveness is calculated as

$$\epsilon_1 = \frac{300 - 185}{300 - 100} = 0.575$$

From Fig. 10-13 we have

$$\text{NTU}_{\text{max}} = 1.1$$

so that $\quad A = \text{NTU}_{\text{max}} \dfrac{C_{\text{min}}}{U} = \dfrac{(1.1)(4500)}{150} = 33 \text{ ft}^2 \quad (3.07 \text{ m}^2)$

We thus find that 33 ft² of area is required in each of the small exchangers, or a total of 66 ft². This is greater than the 57 ft² required in the one large exchanger, and in addition, the cost per unit area is greater, so that the most economical choice would be the single large exchanger. It may be noted, however, that the pumping costs for the oil would probably be less with the two smaller exchangers, so that this could precipitate a decision in favor of the smaller exchangers if the pumping costs represented a sizable economic factor.

## 10-7□ANALYSIS FOR VARIABLE PROPERTIES

The convection heat-transfer coefficient is dependent on the fluid being considered. Correspondingly, the overall heat-transfer coefficient for a heat exchanger may vary substantially through the exchanger if the fluids are such that their properties are strongly temperature-dependent. In this circumstance the analysis is best performed on a numerical or finite-difference basis. To illustrate the technique let us consider the simple parallel-flow double-pipe heat exchanger of Sec. 10-5. The heat exchanger is divided into increments of surface area $\Delta A_j$. For this incremental surface area the hot and cold temperatures are $T_{h_j}$ and $T_{c_j}$, respectively, and we shall assume that the overall heat-transfer coefficient can be expressed as a function of these temperatures. Thus

$$U_j = U_j(T_{h_j}, T_{c_j})$$

The incremental heat transfer in $\Delta A_j$ is, according to Eq. (10-6),

$$\Delta q_j = -(\dot{m}_h c_h)_j (T_{h_{j+1}} - T_{h_j}) = (\dot{m}_c c_c)_j (T_{c_{j+1}} - T_{c_j}) \qquad (10\text{-}27)$$

Also, $$\Delta q_j = U_j \,\Delta A_j \,(T_h - T_c)_j \qquad (10\text{-}28)$$

The finite-difference equation analogous to Eq. (10-9) is

$$\frac{(T_h - T_c)_{j+1} - (T_h - T_c)_j}{(T_h - T_c)_j} = -U_j \left[ \frac{1}{(\dot{m}_h c_h)_j} + \frac{1}{(\dot{m}_c c_c)_j} \right] \Delta A_j$$
$$= -K_j(T_h, T_c) \Delta A_j \qquad (10\text{-}29)$$

where we have introduced the indicated definition for $K_j$. Reducing Eq. (10-29), we obtain

$$\frac{(T_h - T_c)_{j+1}}{(T_h - T_c)_j} = 1 - K_j \Delta A_j \qquad (10\text{-}30)$$

The numerical-analysis procedure is now clear when the inlet temperatures and flows are given:

1. Choose a convenient value of $\Delta A_j$ for the analysis.
2. Calculate the value of $U$ for the inlet conditions and through the initial $\Delta A$ increment.
3. Calculate the value of $q$ for this increment from Eq. (10-28).
4. Calculate the values of $T_h$, $T_c$, and $T_h - T_c$ for the *next* increment, using Eqs. (10-27) and (10-30).
5. Repeat the foregoing steps until all the increments in $\Delta A$ are employed.

The total heat-transfer rate is then calculated from

$$q_{total} = \sum_{j=1}^{n} \Delta q_j$$

where $n$ is the number of increments in $\Delta A$.

A numerical analysis such as the one discussed above is usually best performed with a digital computer. Heat-transfer rates calculated from a variable-properties analysis can frequently differ by substantial amounts from a constant-properties analysis. The most difficult part of the analysis is, of course, a determination of the values of $h$. The interested reader is referred to the heat-transfer literature for additional information on this complicated but important subject.

## 10-8□HEAT - EXCHANGER DESIGN CONSIDERATIONS

In the process and power industries, or related activities, many heat exchangers are purchased as off-the-shelf items, and a selection is made on the basis of cost and specifications furnished by the various manufacturers. In more specialized applications, such as the aerospace and electronics industries, a particular design is frequently called for. Where a heat exchanger forms a part of an overall machine or device to be manufactured, a standard item may be purchased; or if cost considerations and manufacturing quantities warrant, the heat exchanger may be specially designed for the application. Whether the heat exchanger is selected as an off-the-shelf item or designed especially for the application, the following considerations are almost always made:

1. Heat-transfer requirements
2. Cost
3. Physical size
4. Pressure-drop characteristics

The heat-transfer requirements must be met in the selection or design of any heat exchanger. The way that the requirements are met depends on the relative weights placed on items 2 to 4 above. By forcing the fluids through the heat exchanger at higher velocities the overall heat-transfer coefficient may be increased, but this higher velocity results in a larger pressure drop through the exchanger, and correspondingly larger pumping costs. If the surface area of the exchanger is increased, the overall heat-transfer coefficient, and hence the pressure drop, need not be so large; however, there may be limitations on the physical size which can be accommodated, and a larger physical size results in a higher cost for the heat exchanger. Prudent judgment and a consideration of all these factors will result in the proper design.

For purposes of making initial design estimates it is convenient to have a tabulation of values of the overall heat-transfer coefficient for various practical situations. Very comprehensive tabulations of this sort are available in Refs. 5 and 6. An abbreviated list is given in Table 10-2 to indicate the approximate values of $U$ encountered in practice.

## Table 10-2 Approximate Values of Overall Heat-transfer Coefficients

| Physical situation | $U$, Btu/hr-ft$^2$-°F |
| --- | --- |
| Brick exterior wall, plaster interior, uninsulated | 0.45 |
| Frame exterior wall, plaster interior: | |
|   Uninsulated | 0.25 |
|   With rock-wool insulation | 0.07 |
| Plate-glass window | 1.10 |
| Double plate-glass window | 0.40 |
| Steam condenser | 200–1000 |
| Feedwater heater | 200–1500 |
| Freon 12 condenser with water coolant | 50–150 |
| Water-to-water heat exchanger | 150–300 |
| Finned-tube heat exchanger, water in tubes, air across tubes | 5–10 |
| Water-to-oil heat exchanger | 20–60 |

## REVIEW QUESTIONS

1. Define the overall heat-transfer coefficient.
2. What is a fouling factor?
3. Why does a "mixed" or "unmixed" fluid arrangement influence heat exchanger performance?
4. When is the LMTD method most applicable to heat exchanger calculations?
5. Define effectiveness.
6. What advantage does the effectiveness-NTU method have over the LMTD method?
7. What is meant by the "minimum" fluid?
8. Why is a counterflow exchanger more effective than a parallel flow exchanger?

## PROBLEMS

10-1. Hot exhaust gases are used in a finned-tube cross-flow heat exchanger to heat 40 gpm of water from 100 to 180°F. The gases ($c_p = 0.26$) enter at 400°F and leave at 200°F. The overall heat-transfer coefficient is 32 Btu/hr-ft²-°F. Calculate the area of the heat exchanger using (a) the LMTD approach and (b) the effectiveness–NTU method.

10-2. For the exchanger in Prob. 10-1 the water-flow rate is reduced by half, while the gas-flow rate is maintained constant along with the fluid inlet temperatures. Calculate the percentage reduction in heat transfer as a result of this reduced flow rate. Assume that the overall heat-transfer coefficient remains the same.

10-3. Repeat Prob. 10-1 for a shell-and-tube exchanger with two tube passes. The gas is the shell fluid.

10-4. Repeat Prob. 10-2 using the shell-and-tube exchanger of Prob. 10-3.

10-5. It is desired to heat 500 lb$_m$/hr of water from 100 to 200°F with oil ($c_p = 0.5$) having an initial temperature of 350°F. The mass flow of oil is also 500 lb$_m$/hr. Two double-pipe heat exchangers are available:

Exchanger 1: $U = 100$ Btu/hr-ft²-°F
  $A = 5$ ft²
Exchanger 2: $U = 64.6$ Btu/hr-ft²-°F
  $A = 10$ ft²

Which exchanger should be used?

10-6. Water at the rate of 500 lb$_m$/hr at 100°F is available for use as a coolant in a double-pipe heat exchanger whose total surface area is 15 ft². The water is to be used to cool oil ($c_p = 0.5$) from an initial temperature of 250°F. Because of other circumstances, an exit water temperature greater than 210°F cannot be allowed. The exit temperature of the oil must not be below 140°F. The overall heat-transfer coefficient is 50 Btu/hr-ft²-°F.

Estimate the maximum flow rate of oil which may be cooled, assuming the flow rate of water is fixed at 500 $lb_m$/hr.     *8000 lbm/hr* , *1.6×10⁶ BTU/hr*

10-7.   Hot water enters a counterflow heat exchanger at 210°F. It is used to heat a cool stream of water from 40 to 90°F. The flow rate of the cool stream is 10,000 $lb_m$/hr, and the flow rate of the hot stream is 20,000 $lb_m$/hr. The overall heat-transfer coefficient is 146.2. What is the area of the heat exchanger? Calculate the effectiveness of the heat exchanger.

10-8.   Starting with a basic energy balance, derive an expression for the effectiveness of a heat exchanger in which a condensing vapor is used to heat a cooler fluid. Assume that the hot fluid (condensing vapor) remains at a constant temperature throughout the process.

10-9.   Water at 180°F enters a counterflow heat exchanger. It leaves at 100°F. The water is used to heat an oil from 80 to 120°F. What is the effectiveness of the heat exchanger?

10-10.   Derive Eq. (10-12), assuming that the heat exchanger is a counterflow double-pipe arrangement.

10-11.   Derive Eq. (10-26).

10-12.   Saturated steam at 100 psia is to be used to heat carbon dioxide in a cross-flow heat exchanger consisting of four hundred $\frac{1}{4}$-in.-OD brass tubes in a square in-line array. The distance between tube centers is $\frac{3}{8}$ in. in both the normal- and parallel-flow direction. The carbon dioxide flows across the tube bank, while the steam is condensed on the inside of the tubes. A flow rate of 1 $lb_m$/sec of $CO_2$ at 15 psia and 70°F is to be heated to 200°F. Estimate the length of the tubes to accomplish this heating. Assume that the steam-side heat-transfer coefficient is 1000 Btu/hr-ft²-°F, and neglect the thermal resistance of the tube wall.

10-13.   Repeat Prob. 10-12 with the $CO_2$ flowing on the inside of the tubes and the steam condensing on the outside of the tubes. Compare these two designs on the basis of $CO_2$ pressure drop through the exchanger.

10-14.   A small shell-and-tube exchanger with one tube pass ($A = 50$ ft² and $U = 50$ Btu/hr-ft²-°F) is to be used to heat high-pressure water initially at 70°F with hot air initially at 500°F. If the exit water temperature is not to exceed 200°F and the air-flow rate is 1 $lb_m$/sec, calculate the water-flow rate.   *≈ 500 lbm/hr*

10-15.   Replot Figs. 10-12 and 10-13 as $\epsilon$ versus log $NTU_{max}$ over the range $0.1 < NTU_{max} < 100$.

10-16.   A counterflow double-pipe heat exchanger is currently used to heat 20,000 $lb_m$/hr of water from 80 to 150°F by cooling an oil ($c_p = 0.5$) from 280 to 200°F. It is desired to "bleed off" 5000 $lb_m$/hr of water at 120°F so that the single exchanger will be replaced by a two-exchanger arrangement which will permit this. The overall heat-transfer coefficient is 80 Btu/hr-ft²-°F for the single exchanger, and may be taken as this same value for each of the smaller exchangers. The same total oil flow is used for the two-exchanger arrangement, except that the flow is split between the two exchangers. Determine the areas of each of the smaller exchangers and the oil-flow rates through each. Assume that the water flows in series through the two exchangers, with the "bleed off" taking place between them. Assume the smaller exchangers have the same areas.

10-17.    Repeat Prob. 10-16, assuming that condensing steam at 280°F is used instead of the hot oil and that the exchangers are of the shell-and-tube type, with the water making two passes on the tube side. The overall heat-transfer coefficient may be taken as 300 Btu/hr-ft²-°F for this application.

10-18.    The condenser on a certain automobile air conditioner is designed to remove 60,000 Btu/hr from Freon 12 when the automobile is moving at 40 mph and the ambient temperature is 95°F. The Freon 12 temperature is 150°F under these conditions, and it may be assumed that the air-temperature rise across the exchanger is 10°F. The overall heat-transfer coefficient for the finned-tube heat exchanger under these conditions is 35 Btu/hr-ft²-°F. If the overall heat-transfer coefficient varies as the 0.7 power of velocity and air-mass flow varies directly as the velocity, plot the percentage reduction in performance of the condenser as a function of velocity between 10 and 40 mph. Assume that the Freon temperature remains constant at 150°F.

10-19.    A shell-and-tube heat exchanger with four tube passes is used to heat 20,000 $lb_m$/hr of water from 80 to 160°F. Hot water at 200°F is available for the heating process, and a flow rate of 40,000 $lb_m$/hr may be used. The cooler fluid is used on the tube side of the exchanger. The overall heat-transfer coefficient is 140 Btu/hr-ft²-°F. Assuming that the flow rate of the hot fluid and the overall heat-transfer coefficient remain constant, plot the percentage reduction in heat transfer as a function of mass-flow rate of the cooler fluid.

10-20.    Two identical double-pipe heat exchangers are constructed of a 2-in. standard pipe placed inside a 3-in. standard pipe. The length of the exchangers is 10 ft. Forty gpm of water initially at 80°F is to be heated by passing through the inner pipes of the exchangers in a series arrangement. Thirty gpm of water at 120°F and 30 gpm of water at 200°F are available to accomplish the heating. The two heating streams may be mixed in any way desired before and after they enter the heat exchangers. Determine the flow arrangement for optimum performance (maximum heat transfer) and the total heat transfer under these conditions.

10-21.    A long steel pipe with 2 in. ID and $\frac{1}{8}$ in. wall thickness passes through a large room maintained at 100°F and atmospheric pressure. Ten gpm of hot water enters one end of the pipe at 180°F. If the pipe is 50 ft long, calculate the exit water temperature, considering both free convection and radiation heat loss from the outside of the pipe.

10-22.    Some of the brine from a large refrigeration system is to be used to furnish chilled water for air-conditioning part of an office building. The brine is available at 0°F, and 30 tons of cooling are required (1 ton = 12,000

Btu-hr). The chilled water from the conditioned air coolers enters a shell-and-tube heat exchanger at 50°F, and the exchanger is to be so designed that the exit chilled-water temperature is not below 40°F. The overall heat-transfer coefficient for the heat exchanger is 150 Btu/hr-ft²-°F. If the chilled water is used on the tube side and two tube passes are used, plot the heat-exchanger area required as a function of the brine exit temperature.

10-23. A gas-turbine regenerator is a heat exchanger which uses the hot exhaust gases from the turbine to preheat the air delivered to the conbustion chamber. In an air standard analysis of gas-turbine cycles it is assumed that the mass of fuel is small in comparison with the mass of air, and consequently the hot-gas flow through the turbine is essentially the same as the air flow into the combustion chamber. Using this assumption, and also assuming that the specific heat of the hot exhaust gases is the same as that of the incoming air, derive an expression for the effectiveness of a regenerator under both counterflow and parallel-flow conditions.

10-24. Water at 190°F enters a double-pipe heat exchanger and leaves at 130°F. It is used to heat a certain oil from 80 to 120°F. Calculate the effectiveness of the heat exchanger.

10-25. A counterflow double-pipe heat exchanger is to be used to heat 5000 $lb_m$/hr of water from 100 to 200°F with an oil flow of 7500 $lb_m$/hr. The oil has a specific heat of 0.5 and enters the heat exchanger at a temperature of 350°F. The overall heat-transfer coefficient is 75 Btu/hr-ft²-°F. Calculate the area of the heat exchanger and the effectiveness.

10-26. A shell-and-tube heat exchanger is to be designed to heat 60,000 $lb_m$/hr of water from 185 to 210°F. The heating process is accomplished by condensing steam at 50 psia. One shell pass is used along with two tube passes, each consisting of 30 1-in.-OD tubes. Assuming a value of $U$ as 500 Btu/hr-ft²-°F, calculate the length of tubes required in the heat exchanger.

10-27. Suppose the heat exchanger in Prob. 10-26 has been in operation an extended period of time, so that the fouling factors in Table 10-1 apply. Calculate the exit water temperature for "fouled" conditions, assuming the same total flow rate.

10-28. High-temperature flue gases at 800°F ($c_p = 0.28$) are employed in a crossflow heat exchanger to heat an engine oil from 100 to 180°F. Using the information given in this chapter, obtain an approximate design for the heat exchanger for an oil-flow rate of 10 gpm.

10-29. A cross-flow finned-tube heat exchanger uses hot water to heat an appropriate quantity of air from 60 to 80°F. The water enters the heat exchanger at 160°F and leaves at 105°F, and the total heat-transfer rate is to be 100,000 Btu/hr. The overall heat-transfer coefficient is 8 Btu/hr-ft²-°F. Calculate the area of the heat exchanger.

10-30. Calculate the heat-transfer rate for the exchanger in Prob. 10-29 when the water-flow rate is reduced to half that of the design value.

10-31.   A small steam condenser is designed to condense 100 $lb_m$/hr of steam at 12 psia with cooling water at 50°F. The exit water temperature is not to exceed 135°F. The overall heat-transfer coefficient is 600 Btu/hr-ft²-°F. Calculate the area required for a double-pipe heat exchanger.

10-32.   Suppose the inlet water temperature in the exchanger of Prob. 10-31 is raised to 85°F. What percentage increase in flow rate would be necessary to maintain the same rate of condensation?

10-33.   A counterflow double-pipe heat exchanger is used to heat water from 80°F to 120°F by cooling an oil from 210°F to 150°F. The exchanger is designed for a total heat transfer of 100,000 Btu/hr with an overall heat-transfer coefficient of 60 Btu/hr-ft²-°F. Calculate the surface area of the exchanger.

10-34.   Suppose that the oil in Prob. 10-33 is sufficiently dirty so that a fouling factor of 0.004 must be used in the analysis. What is the surface area under these conditions. How much would the heat transfer be reduced if the exchanger in Prob. 10-33 were used with this fouling factor and the same inlet fluid temperatures?

10-35.   A shell and tube exchanger with one shell pass and two tube passes is used as a water-to-water heat-transfer system with the hot fluid in the shell side. The hot water is cooled from 180°F to 140°F and the cool fluid is heated from 50°F to 150°F. Calculate the surface area for a heat transfer of $2 \times 10^5$ Btu/hr and a heat-transfer coefficient of 200 Btu/hr-ft²-°F.

10-36.   What is the heat transfer for the exchanger in Prob. 10-35 if the flow rate of the hot fluid is reduced in half while the inlet conditions and heat-transfer coefficient remain the same?

10-37.   Rework Example 6-3 using the LMTD concept. Repeat for an inlet air temperature of 100°F.

10-38.   Air at 30 psia and 400°F enters a 1.0-in.-ID tube at 20 ft/sec. The tube is constructed of copper with a thickness of 0.032 in. and a length of 40 ft. Atmospheric air at 14.7 psia and 70°F flows normal to the outside of the tube with a free-stream velocity of 40 ft/sec. Calculate the air temperature at exit from the tube. What would be the effect of reducing the hot-air flow in half?

## REFERENCES

1.   Kreith, F.: "Radiation Heat Transfer for Spacecraft and Solar Power Plant Design," International Textbook Company, Scranton, Pa., 1962.
2.   "Standards of Tubular Exchanger Manufacturers Association," 4th ed., 1959.
3.   Kays, W. M., and A. L. London: "Compact Heat Exchangers," 2d ed., McGraw-Hill Book Company, New York, 1964.
4.   Bowman, R. A., A. E. Mueller, and W. M. Nagle: Mean Temperature Difference in Design, *Trans. ASME*, vol. 62, p. 283, 1940.
5.   Perry, J. H. (ed.): "Chemical Engineers' Handbook," 4th ed., McGraw-Hill Book Company, New York, 1963.
6.   American Society of Heating and Air Conditioning Engineers Guide, annually.

7.  Sparrow, E. M., and R. D. Cess: "Radiation Heat Transfer," Wadsworth Publishing Co., Inc., New York, 1966.

# CHAPTER
# 11
# MASS TRANSFER

## 11-1□INTRODUCTION

Mass transfer can result from several different phenomena. There is a mass transfer associated with convection in that mass is transported from one place to another in the flow system. This type of mass transfer occurs on a macroscopic level, and is usually treated in the subject of fluid mechanics. When a mixture of gases or liquids is contained such that there exists a concentration gradient of one or more of the constituents across the system, there will be a mass transfer on a microscopic level as the result of diffusion from regions of high concentration to regions of low concentration. In this chapter we are primarily concerned with some of the simple relations which may be used to calculate mass diffusion and their relation to heat transfer. Nevertheless, one must remember that the general subject of mass transfer encompasses both mass diffusion on a molecular scale and the bulk mass transport, which may result from a convection process.

Not only may mass diffusion occur on a molecular basis, but also in turbulent-flow systems accelerated diffusion rates will occur as a result of the rapid-eddy mixing processes, just as these mixing processes created increased heat transfer and viscous action in turbulent flow.

Although beyond the scope of our discussion, it is well to mention that mass diffusion may also result from a temperature gradient in a system; this is called thermal diffusion. Similarly, a concentration gradient can give rise to a temperature gradient and a consequent heat transfer. These two effects are termed *coupled phenomena* and may be treated by the methods of irreversible thermodynamics. The reader is referred to the monographs by Prigogine [1] and de Groot [2] for a discussion of irreversible thermodynamics and coupled phenomena and their application to diffusion processes.

## 11-2□FICK'S LAW OF DIFFUSION

Consider the system shown in Fig. 11-1. A thin partition separates the two gases $A$ and $B$. When the partition is removed, the two gases diffuse through one another until equilibrium is established and the concentration of the gases is uniform throughout the box. The diffusion rate is given by Fick's

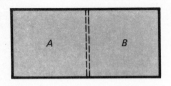

**Fig. 11-1  Diffusion of component $A$ into component $B$,**

law of diffusion, which states that the mass flux of a constituent per unit area is proportional to the concentration gradient. Thus

$$\frac{\dot{m}_A}{A} = -D\frac{\partial C_A}{\partial x} \tag{11-1}$$

where proportionality constant $D$ = diffusion coefficient

$\dot{m}_A$ = mass flux per unit time

$C_A$ = mass concentration of component $A$ per unit volume

The units of $D$ are in square feet per hour when $\dot{m}_A$ is in pounds mass per hour and $C_A$ is in pounds mass per cubic foot. An expression similar to Eq. (11-1) could also be written for the diffusion of constituent $A$ in either the $y$ or $z$ direction.

Notice the similarity between Eq. (11-1) and the Fourier law of heat conduction,

$$\left(\frac{q}{A}\right)_x = -k\frac{\partial T}{\partial x}$$

and the equation for shear stress between fluid layers,

$$\tau = \mu\frac{\partial u}{\partial y}$$

The heat-conduction equation describes the transport of energy, the viscous-shear equation describes the transport of momentum across fluid layers, and the diffusion law describes the transport of mass.

To understand the physical mechanism of diffusion, consider the imaginary plane shown by the dashed line in Fig. 11-2. The concentration of component $A$ is greater on the left side of this plane than on the right side. A higher concentration means that there are more molecules per unit volume. If the system is a gas or a liquid, the molecules move about in a random fashion, and the higher the concentration, the more molecules will cross a given plane per unit time. Thus, on the average, there are more molecules moving from left to right across the plane than in the opposite direction. This results in a net mass transfer from the region of high concentration to the region of low concentration. The fact that the molecules collide with each other influences the diffusion process strongly. In a mixture

**Fig. 11-2   Sketch illustrating diffusion dependence on concentration profile.**

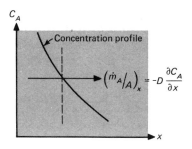

of gases there is a decided difference between a collision of like molecules and a collision of unlike molecules. The collision between like molecules does not appreciably alter the basic molecular movement because the two molecules are identical and it does not make any difference whether one or the other of the two molecules crosses a certain plane. The collision of two unlike molecules, say, molecules $A$ and $B$, might result in molecule $B$ crossing some particular plane instead of molecule $A$. The molecules would, in general, have different masses; thus the mass transfer would be influenced by the collision. By using the kinetic theory of gases it is possible to predict analytically the diffusion rates for some systems by taking into account the collision mechanism and molecular weights of the constituent gases. The reader is referred to the discussion by Present [3] for an outline of the problems involved.

In gases the diffusion rates are clearly dependent on the molecular speed, and consequently we should expect a dependence of the diffusion coefficient on temperature since the temperature indicates the average molecular speed.

## 11-3□DIFFUSION IN GASES

Gilliland [4] has proposed a semiempirical equation for the diffusion coefficient in gases:

$$D = 0.0069 \frac{T^{\frac{3}{2}}}{p(V_A^{\frac{1}{3}} + V_B^{\frac{1}{3}})^2} \sqrt{\frac{1}{M_A} + \frac{1}{M_B}} \qquad (11\text{-}2)$$

where $D$ is in square feet per hour, $T$ is in degrees Rankine, $p$ is the total system pressure in atmospheres, and $V_A$ and $V_B$ are the molecular volumes of constituents $A$ and $B$ as calculated from the atomic volumes in Table 11-1. $M_A$ and $M_B$ are the molecular weights of constituents $A$ and $B$. Example 11-1 illustrates the use of Eq. (11-2) for calculation of diffusion coefficients.

Equation (11-2) offers a convenient expression to calculate the diffusion coefficient for various compounds and mixtures, but it should not be used as a substitute for experimental values of the diffusion coefficient when they are available for a particular system. References 3 and 5 to 9 present more information on calculation of diffusion coefficients. An abbreviated table of diffusion coefficients is given in Table A-8.

## Table 11-1    Atomic Volumes

| | | | |
|---|---|---|---|
| Air | 29.9 | $N_2$ in secondary amines | 12.0 |
| Bromine | 27.0 | Oxygen, molecule ($O_2$) | 7.4 |
| Carbon | 14.8 | Oxygen coupled to two other | |
| Carbon dioxide | 34.0 | elements: | |
| Chlorine: | | In aldehydes and ketones | 7.4 |
| Terminal as in R—Cl | 21.6 | In methyl esters | 9.1 |
| Medial as in R—CHCl—R | 24.6 | In ethyl esters | 9.9 |
| Fluorine | 8.7 | In higher esters and ethers | 11.0 |
| Hydrogen, molecule ($H_2$) | 14.3 | In acids | 12.0 |
| In compounds | 3.7 | In union with S, P, N | 8.3 |
| Iodine | 37.0 | Phosphorus | 27.0 |
| Nitrogen, molecule ($N_2$) | 15.6 | Sulfur | 25.6 |
| $N_2$ in primary amines | 10.5 | Water | 18.8 |

For three-membered ring, deduct 6.0. For four-membered ring, deduct 8.5. For five-membered ring, deduct 11.5. For six-membered ring, deduct 15.0. For naphthalene ring, deduct 30.0.

## Example 11-1

Calculate the diffusion coefficient for $CO_2$ in air at atmospheric pressure and 77°F using Eq. (11-2), and compare this value with that in Table A-8.
**Solution**    From Table 11-1

$$V_{CO_2} = 34.0 \qquad M_{CO_2} = 44$$
$$V_{air} = 29.9 \qquad M_{air} = 28.9$$
$$D = \frac{(0.0069)(537)^{\frac{3}{2}}}{(1)[(34.0)^{\frac{1}{3}} + (29.9)^{\frac{1}{3}}]^2} \sqrt{\frac{1}{44} + \frac{1}{28.9}}$$
$$= 0.51 \text{ ft}^2/\text{hr}$$

From Table A-8

$$D = 0.164 \text{ cm}^2/\text{sec} = 0.62 \text{ ft}^2/\text{hr}$$

so that the two values are in fair agreement.

We realize from the discussion pertaining to Fig. 11-1 that the diffusion process is occurring two ways at the same time; i.e., gas $A$ is diffusing into gas $B$ at the same time that gas $B$ is diffusing into gas $A$. We thus could refer to the diffusion coefficient for either of these processes.

In working with Fick's law one may use mass flux per unit area and mass concentration as in Eq. (11-1), or the equation may be expressed in terms of molal concentrations and fluxes. There is no general rule to say which type of expression will be most convenient, and the specific problem under consideration will determine the one to be used. For gases, Fick's law

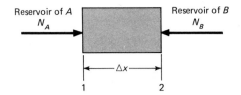

**Fig. 11-3   Sketch illustrating equimolal diffusion.**

may be expressed conveniently in terms of partial pressures by making use of the perfect gas equation of state. (This transformation holds only for gases at low pressures or at a state where the perfect gas equation of state applies.)

$$p = \rho R T \qquad (11\text{-}3)$$

The density $\rho$ represents the mass concentration to be used in Fick's law. The gas constant $R$ for a particular gas may be expressed in terms of the universal gas constant $R_0$ and the molecular weight of the gas.

$$R_A = \frac{R_0}{M_A} \qquad (11\text{-}4)$$

where $\qquad R_0 = 1545 \ \text{lb}_f\text{-ft/lb}_m\text{-mole-}°\text{R}$

Then $\qquad C_A = \rho_A = \dfrac{p_A M_A}{R_0 T}$

Consequently, Fick's law of diffusion for component $A$ into component $B$ could be written

$$\frac{\dot{m}_A}{A} = -D_{AB} \frac{M_A}{R_0 T} \frac{dp_A}{dx} \qquad (11\text{-}5)$$

if isothermal diffusion is considered. For the system in Fig. 11-1 we could also write for the diffusion of component $B$ into component $A$

$$\frac{\dot{m}_B}{A} = -D_{BA} \frac{M_B}{R_0 T} \frac{dp_B}{dx} \qquad (11\text{-}6)$$

still considering isothermal conditions. Notice the different subscripts on the diffusion coefficient. Now consider a physical situation called equimolal counterdiffusion, as indicated in Fig. 11-3. $N_A$ and $N_B$ represent the steady-state molal diffusion rates of components $A$ and $B$, respectively. In this steady-state situation each molecule of $A$ is replaced by a molecule of $B$, and vice versa. The molal diffusion rates are given by

$$N_A = \frac{\dot{m}_A}{M_A} = -D_{AB} \frac{A}{R_0 T} \frac{dp_A}{dx}$$

$$N_B = \frac{\dot{m}_B}{M_B} = -D_{BA} \frac{A}{R_0 T} \frac{dp_B}{dx}$$

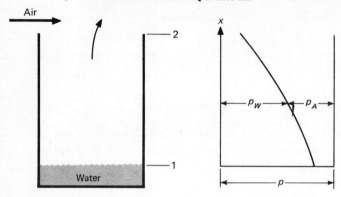

**Fig. 11-4**  Diffusion of water vapor into air.

The total pressure of the system remains constant at steady state, so that

$$p = p_A + p_B$$

and

$$\frac{dp_A}{dx} + \frac{dp_B}{dx} = 0$$

or

$$\frac{dp_A}{dx} = \frac{-dp_B}{dx} \tag{11-7}$$

Since each molecule of $A$ is replacing a molecule of $B$, we may set the molal diffusion rates equal.

$$-N_A = N_B$$

or

$$-D_{AB} \frac{A}{R_0 T} \frac{dp_A}{dx} = -D_{BA} \frac{A}{R_0 T} \frac{dp_A}{dx}$$

where Eq. (11-7) has been used to express the pressure gradient of component $B$. We thus find

$$D_{AB} = D_{BA} = D \tag{11-8}$$

The calculation of $D$ may be made with Eq. (11-2).

We may integrate Eq. (11-5) to obtain the mass flux of component $A$ as

$$\frac{\dot{m}_A}{A} = \frac{-D M_A}{R_0 T} \frac{p_{A_2} - p_{A_1}}{\Delta x} \tag{11-9}$$

corresponding to the nomenclature of Fig. 11-3.

Now consider the isothermal evaporation of water from a surface and the subsequent diffusion through a stagnant air layer as shown in Fig. 11-4. The free surface of the water is exposed to air in the tank, as shown. We assume that the system is isothermal and that the total pressure remains constant. We further assume that the system is in steady state. This requires that there be a slight air movement over the top of the tank to remove the water vapor which diffuses to that point. Whatever air movement may be necessary to accomplish this, it is assumed that it does not create turbulence or otherwise alter the concentration profiles in the air in the tank. We further assume that both the air and water vapor behave as ideal gases.

As the water evaporates, it will diffuse upward through the air, and at steady state this upward movement must be balanced by a downward diffusion of air so that the concentration at any $x$ position will remain constant. But at the surface of the water there can be no net mass movement of air downward. Consequently there must be a bulk mass movement upward with a velocity just large enough to balance the diffusion of air downward. This bulk mass movement then produces an *additional* mass flux of water vapor upward.

The diffusion of air downward is given by

$$\dot{m}_A = \frac{-DAM_A}{R_0 T} \frac{dp_A}{dx} \qquad (11\text{-}10)$$

where $A$ denotes the cross-sectional area of the tank. This must be balanced by the bulk mass transfer upward so that

$$-\rho_A A v = -\frac{p_A M_A}{R_0 T} A v \qquad (11\text{-}11)$$

where $v$ is the bulk mass velocity upward.

Combining Eqs. (11-10) and (11-11), we find

$$v = \frac{D}{p_A} \frac{dp_A}{dx} \qquad (11\text{-}12)$$

The mass diffusion of water vapor upward is

$$\dot{m}_w = -DA \frac{M_w}{R_0 T} \frac{dp_w}{dx} \qquad (11\text{-}13)$$

and the bulk transport of water vapor is

$$\rho_w A v = \frac{p_w M_w}{R_0 T} A v \qquad (11\text{-}14)$$

The total mass transport is the sum of those given in Eqs. (11-13) and (11-14). Adding these quantities and making use of Eq. (11-12) gives

$$\dot{m}_{w\text{total}} = -\frac{DAM_w}{R_0 T} \frac{dp_w}{dx} + \frac{p_w M_w}{R_0 T} A \frac{D}{p_A} \frac{dp_A}{dx}$$

The partial pressure of the water vapor may be related to the partial pressure of the air by making use of Dalton's law,

$$p_A + p_w = p$$

or

$$\frac{dp_A}{dx} = -\frac{dp_w}{dx}$$

since the total pressure is constant. The total mass flow of water vapor then becomes

$$\dot{m}_{w\text{total}} = -\frac{DM_wA}{R_0T}\frac{p}{p-p_w}\frac{dp_w}{dx} \tag{11-15}$$

This relation is called Stefan's law. It may be integrated to give

$$\dot{m}_{w\text{total}} = \frac{DpM_wA}{R_0T(x_2-x_1)}\ln\frac{p-p_{w_2}}{p-p_{w_1}} = \frac{DpM_wA}{R_0T(x_2-x_1)}\ln\frac{p_{A_2}}{p_{A_1}} \tag{11-16}$$

## Example 11-2

Estimate the diffusion rate of water from the bottom of a test tube $\frac{1}{2}$ in. in diameter and 6 in. long into dry atmospheric air at 77°F.

**Solution**  We use Eq. (11-16) to calculate the mass flux. The partial pressure of the water vapor at the bottom of the test tube is the saturation pressure corresponding to 77°F, and the water-vapor pressure may be taken as zero at the top of the test tube since it is diffusing into dry air. Accordingly,

$$p_{A_1} = p - p_{w_1} = 14.696 - 0.4593 = 14.237 \text{ psia}$$
$$p_{A_2} = p - p_{w_2} = 14.696 - 0 = 14.696$$

From Table A-8

$$D = 0.256 \text{ cm}^2/\text{sec} = 0.99 \text{ ft}^2/\text{hr}$$

$$
\begin{aligned}
\dot{m}_w &= \frac{DpM_wA}{R_0T(x_2-x_1)}\ln\frac{p_{A_2}}{p_{A_1}} \\
&= \frac{(0.99)(14.7)(18)\pi(0.5)^2}{(4)(1545)(537)(0.5)}\ln\frac{14.696}{14.237} \\
&= 3.95 \times 10^{-6} \text{ lb}_m/\text{hr} \quad (1.79 \times 10^{-6} \text{ kg/h})
\end{aligned}
$$

## 11-4□DIFFUSION IN LIQUIDS AND SOLIDS

Fick's law of diffusion is also used for problems involving liquid and solid diffusion, and the main difficulty is one of determining the value of the diffusion coefficient for the particular liquid or solid. Unfortunately, only approximate theories are available for predicting diffusion coefficients in these systems. Furthermore, there is a scarcity of experimental data, and considerably more work needs to be performed in both a theoretical and experimental sense. Bird, Stewart, and Lightfoot [9] discuss the calculation of diffusion in liquids, and Jost [6] gives a detailed discussion of the various theories which have been employed to predict values of the diffusion coefficient. The reader is referred to these books for more information on diffusion in liquids and solids.

Diffusion in solids is complex because of the strong influence of the molecular force fields on the process. For these systems Fick's law [Eq. (11-1)] is often used, along with an experimentally determined diffusion coefficient, although there is some indication that this relation may not adequately describe the physical processes.

The numerical value of the diffusion coefficient for liquids and solids is much smaller than for gases, primarily because of the larger molecular force fields, the increased number of collisions, and the consequent reduction in the freedom of movement of the molecules.

## 11-5□THE MASS - TRANSFER COEFFICIENT

We may define a mass-transfer coefficient in a manner similar to that used for defining the heat-transfer coefficient. Thus

$$\dot{m}_A = h_{D_A} A (C_{A_1} - C_{A_2}) \tag{11-17}$$

where $\dot{m}_A$ = diffusive mass flux of component $A$

$h_{D_A}$ = mass-transfer coefficient

$C_{A_1}, C_{A_2}$ = concentrations through which diffusion occurs

If one considers steady-state diffusion across a layer of thickness $\Delta x$,

$$\dot{m}_A = - \frac{DA(C_{A_2} - C_{A_1})}{\Delta x} = h_{D_A} A (C_{A_1} - C_{A_2})$$

and

$$h_{D_A} = \frac{D}{\Delta x} \tag{11-18}$$

For the water-vaporization example discussed above,

$$C_{w_1} - C_{w_2} = \frac{M_w}{R_0 T} (p_{w_1} - p_{w_2})$$

so that the mass-transfer coefficient for this situation could be written

$$h_{D_w} = \frac{Dp}{(x_2 - x_1)(p_{w_1} - p_{w_2})} \ln \frac{p - p_{w_2}}{p - p_{w_1}} \tag{11-19}$$

Note that the units of the mass-transfer coefficients are in feet per hour in the engineering system of units.

We have already seen that the phenomenological laws governing heat, mass, and momentum transfer are similar. In Chap. 5 it was shown that the energy and momentum equations of a laminar boundary layer are similar, viz.,

$$u \frac{\partial u}{\partial x} + v \frac{\partial u}{\partial y} = \nu \frac{\partial^2 u}{\partial y^2} \tag{11-20}$$

$$u \frac{\partial T}{\partial x} + v \frac{\partial T}{\partial y} = \alpha \frac{\partial^2 T}{\partial y^2} \tag{11-21}$$

We further observed that the ratio $\nu/\alpha$, the Prandtl number, was the connecting link between the velocity and temperature field and was thus an important parameter in all convection heat-transfer problems. If we considered a laminar boundary layer on a flat plate in which diffusion was occurring as a result of some mass-transfer condition at the surface, we could derive an equation for the concentration of a particular component in the boundary layer. This equation would be

$$u \frac{\partial C_A}{\partial x} + v \frac{\partial C_A}{\partial y} = D \frac{\partial^2 C_A}{\partial y^2} \tag{11-22}$$

where $C_A$ is the concentration of the component which is diffusing through the boundary layer. Note the similarity between Eq. (11-22) and Eqs. (11-20) and (11-21). The concentration and velocity profiles will have the same shape when $\nu = D$ or $\nu/D = 1$. The dimensionless ratio $\nu/D$ is called the *Schmidt number*

$$\text{Sc} = \frac{\nu}{D} = \frac{\mu}{\rho D} \tag{11-23}$$

and is important in problems where both convection and mass transfer are important. Thus the Schmidt number plays a role similar to that of the Prandtl number in convection heat-transfer problems. Whereas in convection heat-transfer problems we write for the functional dependence of the heat-transfer coefficient

$$\frac{hx}{k} = f(\text{Re, Pr})$$

in convection mass-transfer problems we should write the functional relation

$$\frac{h_D x}{D} = f(\text{Re, Sc})$$

The temperature and concentration profiles will be similar when $\alpha = D$ or $\alpha/D = 1$ and the ratio $\alpha/D$ is called the *Lewis number*

$$\text{Le} = \frac{\alpha}{D} \tag{11-24}$$

The similarities between the governing equations for heat, mass, and momentum transfer suggest that empirical correlations for the mass-transfer coefficient would be similar to those for the heat-transfer coefficient. This turns out to be the case, and some of the empirical relations for mass-transfer coefficients are presented below. Gilliland [4] presented the equation

$$\frac{h_D d}{D} = 0.023 \left(\frac{\rho u_m d}{\mu}\right)^{0.83} \left(\frac{\nu}{D}\right)^{0.44} \tag{11-25}$$

for the vaporization of liquids into air inside circular columns where the liquid wets the surface and the air is forced through the column. The grouping of terms $h_D x/D$ or $h_D d/D$ is called the *Sherwood number*

$$\text{Sh} = \frac{h_D x}{D} \tag{11-26}$$

Note the similarity between Eq. (11-25) and the Dittus-Boelter equation (6-1). Equation (11-25) is valid for

$$2000 < \text{Re}_d < 35{,}000 \qquad \text{and} \qquad 0.6 < \text{Sc} < 2.5$$

and is applicable to flow in smooth tubes.

The Reynolds analogy for pipe flow may be extended to mass-transfer problems to express the mass-transfer coefficient in terms of the friction factor. The analogy is written

$$\frac{h_D}{u_m} \text{Sc}^{\frac{2}{3}} = \frac{f}{8} \tag{11-27}$$

which may be compared with the analogy for heat transfer [Eq. (6-7)].

$$\frac{h}{u_m c_p \rho} \text{Pr}^{\frac{2}{3}} = \frac{f}{8} \tag{11-28}$$

For flow over smooth flat plates the Reynolds analogy for mass transfer becomes

Laminar:
$$\frac{h_D}{u_\infty} \text{Sc}^{\frac{2}{3}} = \frac{C_f}{2} = 0.332 \, \text{Re}_x^{-\frac{1}{2}} \tag{11-29}$$

Turbulent:
$$\frac{h_D}{u_\infty} \text{Sc}^{\frac{2}{3}} = \frac{C_f}{2} = 0.0288 \, \text{Re}_x^{-\frac{1}{5}} \tag{11-30}$$

Equations (11-29) and (11-30) are analogous to Eqs. (5-51) and (5-50).

When both heat and mass transfer are occurring simultaneously, the mass- and heat-transfer coefficients may be related by dividing Eq. (11-28) by Eq. (11-27).

$$\frac{h}{h_D} = \rho c_p \left(\frac{\text{Sc}}{\text{Pr}}\right)^{\frac{2}{3}} = \rho c_p \left(\frac{\alpha}{D}\right)^{\frac{2}{3}} = \rho c_p \, \text{Le}^{\frac{2}{3}} \tag{11-31}$$

## Example 11-3

Dry air at atmospheric pressure blows across a thermometer which is enclosed in a dampened cover. This is the classical "wet-bulb" thermometer. The thermometer reads a temperature of 65°F. What is the temperature of the dry air?

**Solution**   We solve this problem by first noting that the thermometer exchanges no net heat at steady-state conditions and that the heat which must be used to evaporate the water from the cover must come from the air. We therefore make the energy balance

$$hA(T_\infty - T_w) = \dot{m}_w h_{fg}$$

where $h$ is the heat-transfer coefficient and $\dot{m}_w$ is the mass of water evaporated. Now

$$\dot{m}_w = h_D A (C_w - C_\infty)$$

so that

$$hA(T_\infty - T_w) = h_D A (C_w - C_\infty) h_{fg}$$

Using Eq. (11-31),

$$\rho c_p \left(\frac{\alpha}{D}\right)^{\frac{2}{3}} (T_\infty - T_w) = (C_w - C_\infty) h_{fg}$$

The concentration at the surface $C_w$ is that corresponding to saturation conditions at the temperature measured by the thermometer. Thus

$$C_w = \frac{p_w}{R_w T_w} = \frac{(0.3056)(144)(18)}{(1545)(525)} = 0.000976 \text{ lb}_m/\text{ft}^3$$

The other properties are

$$C_\infty = 0 \text{ (since the free stream is dry air)}$$
$$\rho = 0.076 \text{ lb}_m/\text{ft}^3$$
$$c_p = 0.24 \text{ Btu/lb}_m\text{-}°\text{F}$$
$$\alpha/D = \text{Sc/Pr} = 0.845$$
$$h_{fg} = 1057 \text{ Btu/lb}_m$$

Then

$$T_\infty - T_w = \frac{(0.000976 - 0)(1057)}{(0.076)(0.24)(0.845)^{\frac{2}{3}}} = 63$$
$$T_\infty = 65 + 63 = 128°\text{F}$$

This calculation should now be corrected by recalculating the density at the arithmetic-average temperature between wall and free-stream conditions. Making this calculation,

$$\rho = 0.072$$

and

$$T_\infty = 131°\text{F}$$

It is not necessary to correct the ratio Sc/Pr because this parameter does not change appreciably over this temperature range.

### Example 11-4

If the airstream in Example 11-3 is at 90°F while the wet-bulb temperature remains at 65°F, calculate the relative humidity of the airstream. **Solution** From thermodynamics we recall that the relative humidity is defined as the ratio of concentration of vapor to the concentration at saturation conditions for the airstream. We therefore calculate the actual water-vapor concentration in the airstream from

$$\rho c_p \left(\frac{\alpha}{D}\right)^{\frac{2}{3}} (T_\infty - T_w) = (C_w - C_\infty) h_{fg}$$

and then compare this with the saturation concentration to determine the relative humidity. The properties are

$$\rho = 0.076 \ \text{lb}_m/\text{ft}^3$$
$$c_p = 0.24 \ \text{Btu}/\text{lb}_m\text{-}°\text{F}$$
$$\alpha/D = \text{Sc}/\text{Pr} = 0.845$$
$$T_\infty = 90°\text{F}$$
$$T_w = 65°\text{F}$$
$$C_w = 0.000976 \ \text{lb}_m/\text{ft}^3$$
$$h_{fg} = 1057 \ \text{Btu}/\text{lb}_m$$

We thus obtain for $C_\infty$ from the relation of Example 11-3:

$$C_\infty = 0.000589 \ \text{lb}_m/\text{ft}^3$$

The saturation concentration at the free-stream temperature is 0.00214 $\text{lb}_m/\text{ft}^3$, so that the relative humidity is

$$\text{RH} = \frac{0.000589}{0.00214} = 27.5\%$$

## REVIEW QUESTIONS
1.  How is the diffusion coefficient defined?
2.  Define the mass-transfer coefficient.
3.  Define the Schmidt and Lewis numbers. What is the physical significance of each?

## PROBLEMS
11-1.  Using physical reasoning, justify the $T^{\frac{3}{2}}$ dependence of the diffusion coefficient as shown by Eq. (11-2). HINT: Recall that mean molecular velocity is proportional to $T^{\frac{1}{2}}$ and that the density of a perfect gas is inversely proportional to temperature.

11-2.  Calculate the diffusion coefficient for benzene in atmospheric air at 77°F using Eq. (11-2).

11-3.  Dry air at atmospheric pressure and 77°F blows across a flat plate at a velocity of 5 ft/sec. The plate is 1 ft square and is covered with a film of water which may evaporate into the air. Plot the heat flow from the plate as a function of the plate temperature between $T_w = 60°\text{F}$ and $T_w = 150°\text{F}$.

11-4.  The cover on a wet-bulb thermometer is soaked in benzene, and the thermometer is exposed to a stream of dry air. The thermometer indicates a temperature of 79°F. Calculate the free-stream air temperature. The vapor pressure of benzene is 1.93 psia, and the enthalpy of vaporization is 90 cal/g at 79°F.

11-5.  Dry air at 77°F and atmospheric pressure flows inside a 2-in.-diameter pipe at a velocity of 10 ft/sec. The wall is coated with a thin film of water, and the wall temperatures is 77°F. Calculate the water-vapor concentration in the air at exit of a 10-ft length of the pipe.

**11-6.** An open pan 6 in. in diameter and 3 in. deep contains water at 77°F and is exposed to atmospheric air at 77°F and 50 percent relative humidity. Calculate the evaporation rate of water in pounds mass per hour.

**11-7.** A test tube $\frac{1}{2}$ in. in diameter and 6 in. deep contains benzene at 79°F and is exposed to dry atmospheric air at 79°F. Using the properties given in Prob. 11-4, calculate the evaporation rate of benzene in pounds mass per hour.

**11-8.** Dry air at 77°F and atmospheric pressure blows over a 1-ft-square surface of ice at a velocity of 5 ft/sec. Estimate the amount of moisture evaporated per hour, assuming that the block of ice is perfectly insulated except for the surface exposed to the airstream.

**11-9.** The temperature of an airstream is to be measured, but the thermometer available does not have a sufficiently high range. Accordingly, a dampened cover is placed around the thermometer before it is placed in the airstream. The thermometer reads 90°F. Estimate the true air temperature, assuming that it is dry at atmospheric pressure.

**11-10.** Assume that a human forearm may be approximated by a cylinder 4 in. in diameter and 1 ft long. The arm is exposed to a dry-air environmental temperature of 115°F in a 10-mph breeze on the desert and receives a radiant heat flux from the sun of 350 Btu/hr-ft² of view area (view area for the cylinder $= Ld$). If the arm is perspiring so that it is covered with a thin layer of water, estimate the arm surface temperature. Neglect internal heat generation of the arm. Assume an emissivity of unity for the water film.

**11-11.** A 1-ft-square plate is placed in a low-speed wind tunnel; the surface is covered with a thin layer of water. The dry air is at atmospheric pressure and 110°F and blows over the plate at a velocity of 40 ft/sec. The enclosure walls of the wind tunnel are at 50°F. Calculate the equilibrium temperature of the plate, assuming an emissivity of unity for the water film.

**11-12.** Calculate the rate of evaporation for the system in Prob. 11-11.

**11-13.** Refine the analysis of Prob. 11-10 by assuming a body heat generation of 180 Btu/hr-ft³.

**11-14.** A small tube $\frac{1}{4}$ in. in diameter and 5 in. deep contains water with the top open to atmospheric air at 70°F, 14.696 psia, and 50 percent relative humidity. Heat is added to the bottom of the tube. Plot the diffusion rate of water as a function of water temperature over the range of 70°F to 180°F.

**11-15.** Dry air at 70°F enters a 0.5-in.-ID tube where the interior surface is coated with liquid water. The mean flow velocity is 10 ft/sec and the tube wall is maintained at 70°F. Calculate the diffusion rate of water vapor at the entrance conditions. How much moisture is picked up by the air for a 3-ft-long tube?

**11-16.** Dry air at 150°F blows over a 1-ft-square plate at a velocity of 20 ft/sec. The plate is covered with a smooth porous material and water is supplied to the material at 80°F. Assuming that the underside of the plate is insulated, estimate the amount of water that must be supplied to maintain the plate temperature at 100°F. Assume that the radiation temperature of the surroundings is 150°F and that the porous surface radiates as a blackbody.

## REFERENCES

1. Prigogine, I.: "Introduction to Thermodynamics of Irreversible Processes," Charles C Thomas, Publisher, Springfield, Ill., 1955.
2. de Groot, S. R.: "Thermodynamics of Irreversible Processes," North Holland Publishing Company, Amsterdam, 1952.
3. Present, R. D.: "Kinetic Theory of Gases," McGraw-Hill Book Company, New York, 1958.
4. Gilliland, E. R.: Diffusion Coefficients in Gaseous Systems, *Ind. Eng. Chem.*, vol. 26, p. 681, 1934.
5. Perry, J. H. (ed.): "Chemical Engineers' Handbook," 4th ed., McGraw-Hill Book Company, New York, 1963.
6. Jost, W.: "Diffusion in Solids, Liquids and Gases," Academic Press Inc., New York, 1952.
7. Reid, R. C., and T. K. Sherwood: "The Properties of Gases and Liquids," McGraw-Hill Book Company, New York, 1958.
8. "Handbook of Chemistry and Physics," Chemical Rubber Publishing Company, Cleveland, Ohio, 1960.
9. Bird, R., W. E. Stewart, and E. N. Lightfoot: "Transport Phenomena," John Wiley & Sons, Inc., New York, 1960.

# CHAPTER 12

## SPECIAL TOPICS IN HEAT TRANSFER

### 12-1□INTRODUCTION

A number of specialized subjects in heat transfer are important in modern technology. In this chapter we shall present a brief introduction to four such topics. The treatment here is not intended to be comprehensive, but rather to indicate the basic physical mechanism of the processes involved. It would be possible, of course, to include these topics in other chapters of the book, but we choose to group them together in order to focus more specialized attention on the subject matter. Since the discussions which follow represent only cursory treatments of the subjects, the serious reader will wish to consult the appropriate references for additional information.

### 12-2□HEAT TRANSFER IN MAGNETOFLUIDYNAMIC (MFD) SYSTEMS

It is known that an electrical conductor moving in a magnetic field generates an emf which is proportional to the speed of motion and the magnetic field strength. Correspondingly, the magnetic field exerts a restraining force on the conductor which tends to impede its motion. The current flow in the conductor generates joulean heat in accordance with the familiar $I^2R$ relation. Similar effects are experienced when a conductive fluid moves through a magnetic field. An ionized gas, whether it occurs as the result of an elevation of temperature or by a suitable seeding process, is electrically conductive, and may be influenced by a magnetic field.

Our primary concern is with the heat transfer to a conducting fluid under the influence of a magnetic field. This problem area is important in systems involving high-temperature plasmas, liquid-metal systems, and magnetofluidynamic power-generation systems. Our discussion will be limited to a very simple case which serves to illustrate the influence a magnetic field may have on the flow and heat transfer in a conductive fluid.

First, let us examine some basic electromagnetic concepts. For a neutrally charged system the current density $\mathbf{J}$ is given by

$$\mathbf{J} = \sigma\mathbf{E} \tag{12-1}$$

where $\sigma$ is the electrical conductivity and $\mathbf{E}$ is the electric field vector. The magnetic field strength $\mathbf{B}$ is expressed by

$$\mathbf{B} = \mu_0\mathbf{H} \tag{12-2}$$

where $\mu_0$ is called the magnetic permeability and $\mathbf{H}$ is the magnetic field intensity. The force exerted on a system of charged particles by an electric field is given by

$$\mathbf{F}_e = \rho_e\mathbf{E} \tag{12-3}$$

where $\rho_e$ is the charge density (charge per unit volume). The magnetic force exerted on a current-carrying conductor is

$$\mathbf{F}_m = \mathbf{J} \times \mathbf{B} \tag{12-4}$$

The total electromagnetic force is given by the sum of Eqs. (12-3) and (12-4).

$$\mathbf{F}_{em} = \rho_e\mathbf{E} + \mathbf{J} \times \mathbf{B} \tag{12-5}$$

The work done on the system per unit time by the electromagnetic force is

$$W_{em} = \mathbf{F}_{em} \cdot \mathbf{V} \tag{12-6}$$

where $\mathbf{V}$ is the velocity of the conductor. The magnetic field induces a voltage in the conductor of magnitude $\mathbf{V} \times \mathbf{B}$, and subsequently induces a current of

$$\mathbf{J}_{ind} = \sigma(\mathbf{V} \times \mathbf{B}) \tag{12-7}$$

There is a further transport of charge resulting from the macroscopic velocity $V$, given by

$$\mathbf{J}_{trans} = \rho_e\mathbf{V} \tag{12-8}$$

The *conduction current* is now defined as

$$\mathbf{J}_c = \sigma(\mathbf{E} + \mathbf{V} \times \mathbf{B}) \tag{12-9}$$

and the total current flow is

$$\mathbf{J} = \mathbf{J}_c + \rho_e\mathbf{V} \tag{12-10}$$

A careful manipulation of the foregoing equations yields the following relation for the electromagnetic work:

$$W_{em} = \mathbf{E} \cdot \mathbf{J} - \frac{\mathbf{J}_c \cdot \mathbf{J}_c}{\sigma} \tag{12-11}$$

The second term in Eq. (12-11) is called the ohmic heating and is given as the product of the resistivity and conduction current squared.

We now wish to apply these relations to a very simple boundary-layer

**Fig. 12-1  Hydromagnetic boundary layer.**

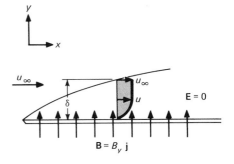

flow system. Consider the flow over the flat plate shown in Fig. 12-1. Impressed on the flow is a constant magnetic field **B** directed in the $y$ direction. We assume that the magnetic field is uniform throughout the boundary layer and that the impressed electric field is zero. In writing a momentum equation for this flow system we need then only consider the magnetic force as given by Eq. (12-4). If all properties are assumed constant, including electrical conductivity, there results

$$\rho\left(u\frac{\partial u}{\partial x} + v\frac{\partial u}{\partial y}\right) = -\frac{\partial p}{\partial x} + \mu\frac{\partial^2 \mu}{\partial y^2} + (\mathbf{J}_c \times \mathbf{B})_x \qquad (12\text{-}12)$$

This is the momentum equation, which is equivalent to Eq. (5-11) for conventional flow. Note that only the $x$ component of the magnetic force is considered since the equation represents a force summation in the $x$ direction. The fluid velocity is written

$$\mathbf{V} = u\mathbf{i} + v\mathbf{j} \qquad (12\text{-}13)$$

so that

$$\mathbf{J}_c = \sigma(\mathbf{E} + \mathbf{V} \times \mathbf{B}) = \sigma(\mathbf{V} \times \mathbf{B})$$
$$= \sigma(u\mathbf{i} + v\mathbf{j}) \times B_y\mathbf{j} = \sigma u B_y\mathbf{k} \qquad (12\text{-}14)$$

Thus
$$\mathbf{J}_c \times \mathbf{B} = -\sigma u B_y{}^2\mathbf{i} \qquad (12\text{-}15)$$

and
$$(\mathbf{J}_c \times \mathbf{B})_x = -\sigma u B_y{}^2 \qquad (12\text{-}16)$$

The momentum equation becomes

$$u\frac{\partial u}{\partial x} + v\frac{\partial u}{\partial y} = -\frac{1}{\rho}\frac{\partial p}{\partial x} + v\frac{\partial^2 u}{\partial y^2} - \frac{\sigma B_y{}^2}{\rho}u \qquad (12\text{-}17)$$

The equivalent integral momentum equation for zero pressure gradient is

$$\mu\frac{\partial u}{\partial y}\bigg)_{y=0} + \int_0^\delta \sigma B_y{}^2 u\, dy = \frac{d}{dx}\left[\int_0^\delta \rho u(u_\infty - u)\, dy\right] \qquad (12\text{-}18)$$

We are able to write the same conditions on the velocity profile as in Sec. 5-2, so that the cubic-parabola profile is employed to effect the integration,

$$\frac{u}{u_\infty} = \frac{3}{2}\frac{y}{\delta} - \frac{1}{2}\left(\frac{y}{\delta}\right)^3 \tag{12-19}$$

Inserting (12-19) in (12-18) and performing the integration gives

$$\frac{3\nu u_\infty}{2\delta} + \frac{5\sigma u_\infty B_y^2}{8\rho} = \frac{39}{280}u_\infty^2\frac{d\delta}{dx} \tag{12-20}$$

This differential equation is linear in $\delta^2$, and may be solved to give

$$\left(\frac{\delta}{x}\right)^2 = \frac{2.40}{\text{Re}_x\,N}\left(e^{8.97N} - 1\right) \tag{12-21}$$

where $N$ is called the *magnetic influence number*,

$$N = \frac{\sigma B_y^2 x}{\rho u_\infty} \tag{12-22}$$

The Reynolds number is defined in the conventional way, i.e.,

$$\text{Re}_x = \frac{\rho u_\infty x}{\mu}$$

The relation for laminar-boundary-layer thickness from Sec. 5-4 was

$$\frac{\delta_0}{x} = \frac{4.64}{\text{Re}_x^{\frac{1}{2}}} \tag{12-23}$$

We may thus form the comparative ratio

$$\frac{\delta_m}{\delta_0} = \frac{0.334}{N^{\frac{1}{2}}}\left(e^{8.97N} - 1\right) \tag{12-24}$$

where now $\delta_m$ is the boundary-layer thickness in the presence of the magnetic field and $\delta_0$ is the thickness for zero magnetic field. The functional relationship given in Eq. (12-24) is plotted in Fig. 12-2. The effect of an increased magnetic field is to increase the boundary-layer thickness as a result of the increased retarding force. The effect is analogous to flow against a positive pressure gradient.

We next wish to examine the effect of the magnetic field on the heat transfer from the plate. Again, constant properties are assumed, as well as a zero impressed electric field. The electromagnetic work term then becomes

$$W_{em} = -\frac{\mathbf{J}_c \cdot \mathbf{J}_c}{\sigma} \tag{12-25}$$

and the boundary-layer energy equation is written

$$\rho c_p\left(u\frac{\partial T}{\partial x} + v\frac{\partial T}{\partial y}\right) = k\frac{\partial^2 T}{\partial y^2} + \mu\left(\frac{\partial u}{\partial y}\right)^2 + \frac{\mathbf{J}_c \cdot \mathbf{J}_c}{\sigma} \tag{12-26}$$

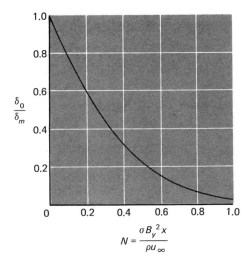

**Fig. 12-2   Influence of magnetic field on boundary-layer thickness.**

$$\frac{\delta_0}{\delta_m}$$

$$N = \frac{\sigma B_y^{\,2} x}{\rho u_\infty}$$

Now consider the flat plate shown in Fig. 12-3. The plate surface is maintained at the constant temperature $T_w$, the free-stream temperature is $T_\infty$, and the thermal-boundary-layer thickness is designated by the conventional symbol $\delta_t$. To simplify the analysis we consider low-speed incompressible flow so that the viscous-heating effects are negligible. The integral energy equation then becomes

$$\frac{d}{dx}\left[\int_0^{\delta_t} (T_\infty - T)u\,dy\right] = \alpha \left.\frac{dT}{dy}\right)_{y=0} - \frac{\sigma}{\rho c_p}\int_0^{\delta_t} B_y^{\,2}u^2\,dy \quad (12\text{-}27)$$

since

$$\frac{\mathbf{J}_c \cdot \mathbf{J}_c}{\sigma} = \frac{[\sigma(\mathbf{V} \times \mathbf{B})] \cdot [\sigma(\mathbf{V} \times \mathbf{B})]}{\sigma} = \sigma B_y^{\,2}u^2 \quad (12\text{-}28)$$

In Eq. (12-27) it is assumed that the magnetic heating effects are confined

**Fig. 12-3   Thermal boundary layer in magneto-fluidynamic flow.**

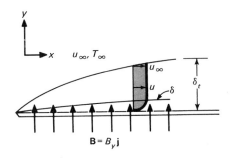

to the boundary-layer region. Again, as in Sec. 5-6, we are able to write a cubic-parabola type of function for the temperature distribution, so that

$$\frac{\theta}{\theta_\infty} = \frac{T - T_w}{T_\infty - T_w} = \frac{3}{2}\frac{y}{\delta_t} - \frac{1}{2}\left(\frac{y}{\delta_t}\right)^3 \tag{12-29}$$

If the same procedure were followed at this point as in Sec. 5-6, the temperature and velocity functions given by Eqs. (12-19) and (12-29) would be inserted in (12-27) in order to arrive at a differential equation to be solved for $\delta_t$, the thermal-boundary-layer thickness in the presence of the magnetic field. The problem with this approach is that a nonlinear equation results which must be solved by numerical methods.

Suppose the fluid is highly conducting, such as a liquid metal. In this case, the thermal-boundary-layer thickness will be much greater than the hydrodynamic thickness. This is evidenced by the fact that the Prandtl numbers for liquid metals are very low, of the order of 0.01. For such a fluid, then, we might approximate the actual fluid behavior with a slug flow model for energy transport in the thermal boundary layer, as outlined in Sec. 6-5. We assume a constant velocity profile

$$u = u_\infty \tag{12-30}$$

Now, inserting Eqs. (12-29) and (12-30) in the integral energy equation gives

$$\frac{3u_\infty\theta_\infty}{8}\frac{d\delta_t}{dx} = \frac{3\alpha\theta_\infty}{2\delta_t} - \frac{\sigma B_y{}^2 u_\infty{}^2}{\rho c_p}\delta_t \tag{12-31}$$

This is a linear differential equation in $\delta_t{}^2$, and has the solution

$$\delta_t{}^2 = \frac{4\alpha}{Ku_\infty}\left(1 - e^{-2Kx}\right) \tag{12-32}$$

where

$$K = \frac{8\sigma B_y{}^2 u_\infty}{3\theta_\infty\rho c_p} \tag{12-33}$$

The heat-transfer coefficient in the presence of the magnetic field may now be calculated from

$$h_m = \frac{-k(\partial T/\partial y)_{y=0}}{T_w - T_\infty} = \frac{3k}{2\delta_t} \tag{12-34}$$

Expressed in dimensionless form,

$$\mathrm{Nu}_m = \sqrt{\tfrac{3}{2}}\,\sqrt{N\,\mathrm{Ec}}\,\mathrm{Re}_x\,\mathrm{Pr}\,(1 - e^{-5.33N\,\mathrm{Ec}})^{-\frac{1}{2}} \tag{12-35}$$

where

$$\mathrm{Nu}_m = \text{Nusselt number} = \frac{h_m x}{k} \tag{12-36}$$

$$\mathrm{Ec} = \text{Eckert number} = \frac{u_\infty{}^2}{c_p\theta_\infty} \tag{12-37}$$

and Pr is the Prandtl number of the fluid. If Eq. (12-27) is solved for the

**Fig. 12-4   Influence of magnetic field on heat transfer from flat plate.**

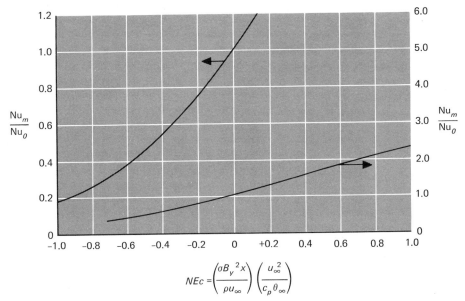

$$NEc = \left( \frac{\sigma B_y^{\,2} x}{\rho u_\infty} \right) \left( \frac{u_\infty^2}{c_p \theta_\infty} \right)$$

case of zero magnetic field but with the same slug flow model as in Sec. 6-5, there results

$$\mathrm{Nu}_0 = \frac{h_0 x}{k} = \frac{3\sqrt{2}}{8} \sqrt{\mathrm{Re}_x \, \mathrm{Pr}} \qquad (12\text{-}38)$$

The comparative ratio of interest in the heat-transfer case is thus

$$\frac{\mathrm{Nu}_m}{\mathrm{Nu}_0} = \left( \frac{5.33N \, \mathrm{Ec}}{1 - e^{-5.33N \, \mathrm{Ec}}} \right)^{\!\frac{1}{2}} \qquad (12\text{-}39)$$

This equation is plotted in Fig. 12-4. It may be noted that the magnetic field will increase the heat-transfer rate for positive values of the Eckert number $(T_\infty > T_w)$ and decrease the heat transfer for negative Eckert numbers $(T_w > T_\infty)$. This behavior results from the fact that the magnetic field tends to heat the fluid, thereby reducing or increasing the temperature gradient between it and the plate.

It is necessary to caution that the foregoing analysis is a highly idealized one, which has been used primarily for the purpose of illustrating the effects of magnetic fields on heat transfer. A more realistic analysis would consider the variation of electrical conductivity of the fluid and take into account the exact velocity profile rather than the slug flow model. A survey of more

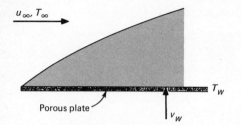

**Fig. 12-5  Porous flat plate with fluid injection.**

exact relations for heat transfer in magnetofluidynamic systems is given by Romig [1], and a complete survey of the entire subject of magnetofluidynamics is available in the text by Sutton and Sherman [2].

## 12-3□TRANSPIRATION COOLING

When high-velocity heat-transfer situations are encountered as described in Sec. 5-11 the adiabatic wall temperature of the surface exposed to the flow stream can become quite large, and significant amounts of cooling may be required in order to reduce the surface temperature to a reasonable value. One technique for cooling the surface is called transpiration, or sweat, cooling. It operates on the principle shown in Fig. 12-5. A porous flat plate is exposed to the high-velocity flow stream while a fluid is forced through the plate into the boundary layer. This fluid could be the same as the free stream or some different fluid. The injection process carries additional energy away from the region close to the plate surface, over and above that which would normally be conducted into the boundary layer. There is also an effect on the velocity profile of the boundary layer and, correspondingly, on the frictional-drag characteristics.

For incompressible flow without viscous heating and for zero pressure gradient the boundary-layer equations to be solved are the familiar ones presented in Chap. 5 when the injected fluid is the same as the free-stream fluid.

$$u\,\frac{\partial u}{\partial x} + v\,\frac{\partial u}{\partial y} = \nu\,\frac{\partial^2 u}{\partial y^2}$$
$$u\,\frac{\partial T}{\partial x} + v\,\frac{\partial T}{\partial y} = \alpha\,\frac{\partial^2 T}{\partial y^2} \tag{12-40}$$

The boundary conditions, however, are different from those used in Chap. 5. Now we must take

$$v = v_w \qquad \text{at } y = 0 \tag{12-41}$$

instead of $v = 0$ at $y = 0$, since a finite injection velocity is involved. We

**Fig. 12-6   Laminar-boundary-layer velocity profiles for injection on a flat plate, according to Eckert and Hartnett [3].**

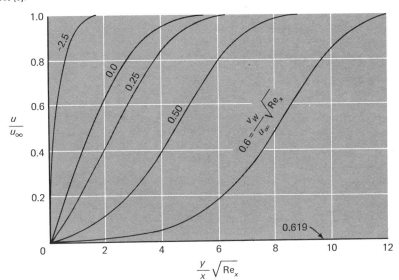

still have

$$u = 0 \qquad \text{at } y = 0$$
$$u = u_\infty \qquad \text{at } y \to \infty$$
$$\frac{\partial u}{\partial y} = 0 \qquad \text{at } y \to \infty$$

The boundary-layer equations may be solved by the technique outlined in App. B or by the integral method of Chap. 5. Eckert and Hartnett [3] have developed a comprehensive set of solutions for the transpiration-cooling problem, and we shall present the results of their analysis without exploring the techniques employed for solution of the equations.

Figure 12-6 shows the boundary-layer velocity profiles which result from various injection rates in a laminar boundary layer. The injection parameter

$$\frac{v_w}{u_\infty} \sqrt{\text{Re}_x}$$

uses the conventional definition of Reynolds number as $\text{Re}_x = \rho u_\infty x / \mu$. Negative values of the injection parameter indicate *suction* on the plate and produce much blunter velocity profiles. For values of the injection parameter greater than 0.619 the boundary layer is completely blown away from the

**Fig. 12-7   Effect of fluid injection on flat-plate heat transfer, according to Eckert and Hartnett [3].**

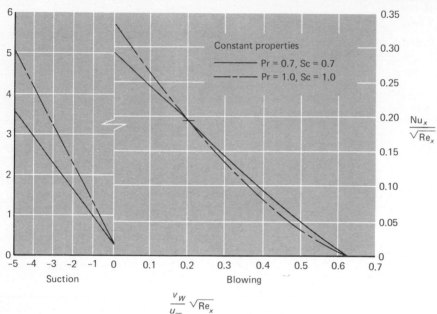

plate. Temperature profiles have a similar shape for the various injection rates. The overall effect of the injection (or suction) process on heat transfer is indicated in Fig. 12-7. As expected, blowing causes a reduction in heat transfer, while suction causes an increase.

An important application of transpiration cooling is that of plane stagnation flow, as illustrated in Fig. 12-8. Solutions for the influence of transpiration on heat transfer in the neighborhood of such a stagnation line have also been worked out in Ref. 3, and the results are shown in Fig.

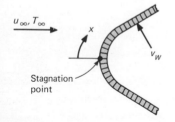

**12-8   Plane stagnation flow with fluid injection.**

**Fig. 12-9  Effects of fluid injection on plane-stagnation heat transfer, according to Eckert and Hartnett [3].**

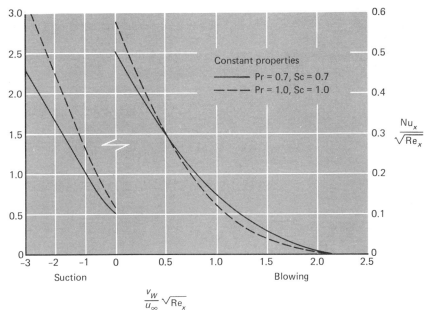

12-9. As would be expected, gas injection or suction can exert a significant effect on the temperature recovery factor for flow over a flat plate. These effects are indicated in Fig. 12-10, where the recovery factor $r$ is defined in the conventional way as

$$r = \frac{T_{aw} - T_\infty}{T_0 - T_\infty} = \frac{T_{aw} - T_\infty}{u_\infty^2 / 2c_p} \qquad (12\text{-}42)$$

It is well to remind the reader at this point that the curves presented above all involve the injection of a gas which is identical with the free-stream gas (usually air). When different injection gases are used, the mass diffusion in the boundary layer must be taken into proper account. Several papers have analyzed this situation, and the interested reader is referred to Refs. 4, 5, and 7 for a discussion of the problems. Experimental results of Ref. 6 indicate the approximate behavior which results from the use of different coolants, as shown in Fig. 12-11. In this figure the product $\rho_w v_w$ refers to the density and velocity of the injection gas at the wall. The ratio $h/h_0$ represents the ratio of the heat-transfer coefficient with injection to that without injection. In general, lightweight gases exhibit a stronger cooling effect than air injection because of their larger specific heats.

Fig. 12-10    Effects of fluid injection on recovery factor for flow over a flat plate, according to Eckert and Hartnett [3].

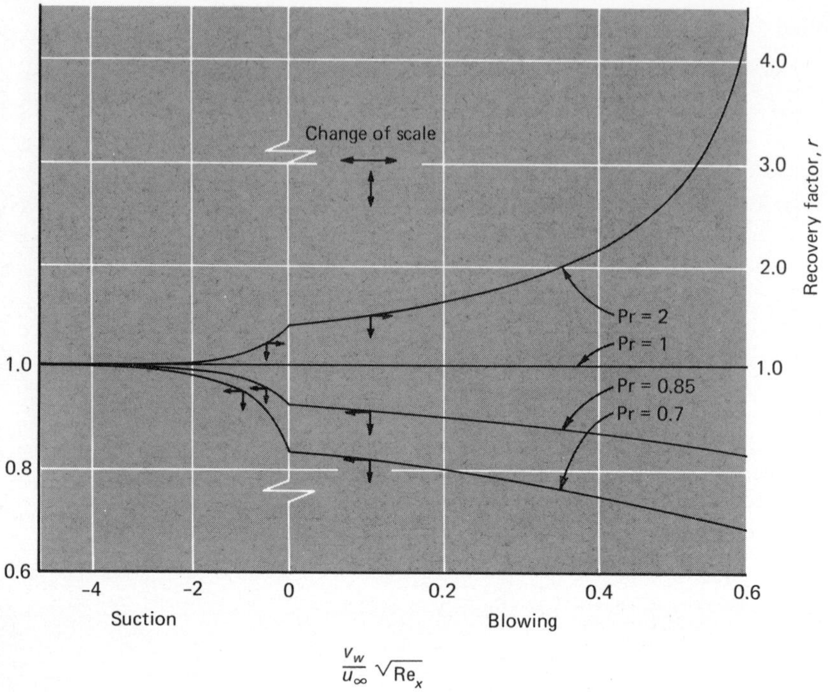

## Example 12-1

For the flow conditions of Example 5-7 calculate the percent reduction in heat transfer at a position where $Re_x = 5 \times 10^4$ and the injection parameter is 0.2. For this calculation assume constant properties evaluated at the average temperature between the recovery and free-stream temperatures. Also calculate the mass flow of coolant air required per unit area at this location.

**Solution**  We first calculate the recovery (adiabatic wall) temperature. From Fig. 12-10,

$$r = 0.79 \qquad \text{for Pr} = 0.7$$

From Example 5-7

$$T_0 = 1175°R$$
$$T_\infty = 420°R$$
$$u_\infty = 3018 \text{ ft/sec}$$

Thus

$$T_{aw} - T_\infty = 0.79(1175 - 420) = 596$$
$$T_{aw} = 596 + 420 = 1016°R$$

**Fig. 12-11   Reduction in surface heat transfer for various injection rates, according to Leadon and Scott [6].** $M = 3.0$, $\mathrm{Re}_x = 4 \times 10^6$, $T_w = 507°\mathrm{R}$, $T_\infty = 474°\mathrm{R}$.

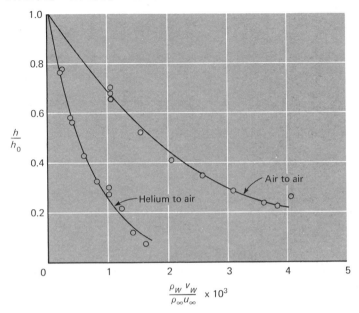

We shall evaluate properties at

$$T_r = \frac{1016 + 420}{2} = 718°\mathrm{R} = 258°\mathrm{F}$$

According to Fig. 12-7, with

$$\frac{v_w}{u_\infty} \sqrt{\mathrm{Re}_x} = 0.2 \qquad \text{and} \qquad \mathrm{Pr} = 0.7$$

we obtain

$$\frac{\mathrm{Nu}_x}{\sqrt{\mathrm{Re}_x}} = 0.19$$

For zero injection we have

$$\frac{\mathrm{Nu}_x}{\sqrt{\mathrm{Re}_x}} = 0.29$$

The percent reduction in heat transfer is obtained by comparing these last two numbers.

$$\text{Reduction in heat transfer} = \frac{0.29 - 0.19}{0.29} \times 100 = 34.5\%$$

The mass flow of coolant air at the wall is calculated from

$$\frac{\dot{m}_w}{A} = \rho_w v_w$$

We are assuming a constant-properties analysis; so these properties must be evaluated at the selected reference temperature of 258°F. Thus

$$\rho_w = \frac{(14.7)(144)}{(20)(718)(53.35)} = 0.00277 \text{ lb}_m/\text{ft}^3$$

$$v_w = \left(\frac{v_w}{u_\infty} \sqrt{\text{Re}_x}\right) \frac{u_\infty}{\sqrt{\text{Re}_x}}$$

$$= (0.2)(3018)(5 \times 10^4)^{-\frac{1}{2}}$$

$$= 2.96 \text{ ft/sec}$$

Thus

$$\frac{\dot{m}_w}{A} = (0.00277)(296) = 8.19 \times 10^{-3} \text{ lb}_m/\text{sec-ft}^2$$

## 12-4□LOW - DENSITY HEAT TRANSFER

A number of practical situations involve heat transfer between a solid surface and a low-density gas. In employing the term "low density" we shall mean those circumstances where the mean-free path of the gas molecules is no longer small in comparison with a characteristic dimension of the heat-transfer surface. The mean-free path is the distance a molecule travels, on the average, between collisions. The larger this distance becomes, the greater the distance required to communicate the temperature of a hot surface to a gas in contact with it. This means that we shall not necessarily be able to assume that a gas in the immediate neighborhood of the surface will have the same temperature as the heated surface, as was done in the boundary-layer analyses of Chap. 5. Because the mean-free path is also related to momentum transport between molecules, we shall also be forced to abandon our assumption of zero fluid velocity near a stationary surface for those cases where the mean-free path is not negligible in comparison with the surface dimensions.

Three general flow regimes may be anticipated for the flow over a flat plate shown in Fig. 12-12. First, the continuum flow region is encountered when the mean-free path $\lambda$ is very small in comparison with a characteristic body dimension. This is the convection heat-transfer situation analyzed in preceding chapters. At lower gas pressures, when $\lambda \sim L$, the flow seems to "slip" along the surface and $u \neq 0$ at $y = 0$. This situation is appropriately called slip flow. At still lower densities all momentum and energy exchange is the result of strictly molecular bombardment of the surface. This regime is called free-molecule flow, and must be analyzed apart from conventional continuum fluid mechanics.

Evidently, the parameter which is of principal interest is a ratio of the

**Fig. 12-12   Three types of flow regimes for a flat plate. (a) Continuum flow; (b) slip flow; (c) free-molecule flow.**

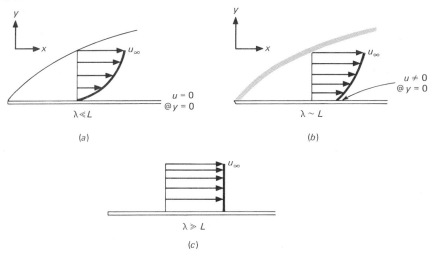

mean-free path to a characteristic body dimension. This grouping is called the *Knudsen number*,

$$\text{Kn} = \frac{\lambda}{L} \tag{12-43}$$

According to the kinetic theory of gases, the mean-free path may be calculated from

$$\lambda = \frac{0.707}{4\pi r^2 n} \tag{12-44}$$

where $r$ is the effective molecular radius for collisions and $n$ is the molecular density. An approximate relation for the mean-free path of air molecules is given by

$$\lambda = 8.64 \times 10^{-7}\,\frac{T}{p} \qquad \text{ft} \tag{12-45}$$

where $T$ is in degrees Rankine and $p$ is in $lb_f/ft^2$.

The speed of sound in a gas is related to mean molecular speed $\bar{v}$ through the relation

$$a = \bar{v}\,\sqrt{\frac{\pi\gamma}{8}} \tag{12-46}$$

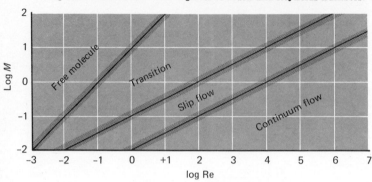

**Fig. 12-13    Relation of flow regimes to Mach and Reynolds numbers.**

The mean molecular speed may be expressed by

$$\bar{v} = \sqrt{\frac{8RT}{\pi}} \qquad (12\text{-}47)$$

where $R$ is the gas constant for the particular gas. It can be shown that the transport properties (viscosity, thermal conductivity, diffusion coefficient) of a gas are directly related to the mean molecular speed. Based on this relation, the kinetic-theory representation of the Reynolds number may be derived as

$$\text{Re} = \frac{2u_\infty L}{\bar{v}\lambda} \qquad (12\text{-}48)$$

Using Eq. (12-46), the Mach number is expressed as

$$M = \frac{u}{a} = \frac{u}{\bar{v}} \sqrt{\frac{8}{\pi\gamma}} \qquad (12\text{-}49)$$

Now, combining Eqs. (12-43), (12-48), and (12-49), the Knudsen number may be expressed as

$$\text{Kn} = \frac{\lambda}{L} = \sqrt{\frac{\pi\gamma}{2}} \frac{M}{\text{Re}} \qquad (12\text{-}50)$$

This relation enables us to interpret a low-density flow regime in terms of the conventional flow parameters, Mach and Reynolds numbers. Figure 12-13 gives an approximate representation of these effects.

As a first example of low-density heat transfer let us consider the two parallel infinite plates shown in Fig. 12-14. The plates are maintained at different temperatures and separated by a gaseous medium. Let us neglect natural convection effects. If the gas density is sufficiently high so that

**Fig. 12-14  Effect of mean-free path on conduction heat transfer between parallel plates. (a) Physical model; (b) anticipated temperature profiles.**

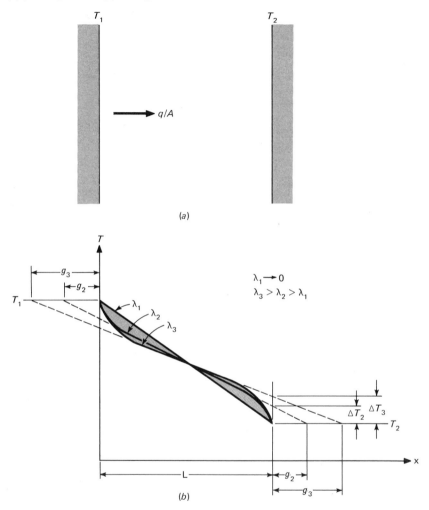

(a)

(b)

$\lambda \rightarrow 0$, a linear temperature profile through the gas will be experienced as shown for the case of $\lambda_1$. As the gas density is lowered, the larger mean-free paths require a greater distance from the heat-transfer surfaces in order for the gas to accommodate to the surface temperatures. The anticipated temperature profiles are shown in Fig. 12-14b. Extrapolating the straight portion of the low-density curves to the wall produces a temperature "jump" $\Delta T$, which may be calculated by making the following energy balance:

$$\frac{q}{A} = k\frac{T_1 - T_2}{g + L + g} = k\frac{\Delta T}{g} \tag{12-51}$$

In this equation we are assuming that the extrapolation distance $g$ is the same for both plate surfaces. In general, the temperature jump will depend on the type of surface, and these extrapolation distances will not be equal unless the materials are identical. For different types of materials we should have

$$\frac{q}{A} = k\frac{T_1 - T_2}{g_1 + L + g_2} = k\frac{\Delta T_1}{g_1} = k\frac{\Delta T_2}{g_2} \tag{12-52}$$

where now $\Delta T_1$ and $\Delta T_2$ are the temperature jumps at the two heat-transfer surfaces and $g_1$ and $g_2$ are the corresponding extrapolation distances. For identical surfaces the temperature jump would then be expressed as

$$\Delta T = \frac{g}{2g + L}(T_1 - T_2) \tag{12-53}$$

Similar expressions may be developed for low-density conduction between concentric cylinders. In order to predict the heat-transfer rate it is necessary to establish relations for the temperature jump for various gas-to-solid interfaces.

We have already mentioned that the temperature-jump effect arises as a result of the failure of the molecules to "accommodate" to the surface temperature when the mean-free path becomes of the order of a characteristic body dimension. The parameter which describes this behavior is called the *accommodation coefficient* $\alpha$, defined by

$$\alpha = \frac{E_i - E_r}{E_i - E_w} \tag{12-54}$$

where $E_i$ = energy of incident molecules on a surface
  $E_r$ = energy of molecules reflected from the surface
  $E_w$ = energy molecules would have if they acquired energy of wall at temperature $T_w$

Values of the accommodation coefficient must be determined from experiment, and a brief summary of such measurements is given in Table 12-1.

It is possible to employ the kinetic theory of gases along with values of $\alpha$ to determine the temperature jump at a surface. The result of such an analysis is

$$T_{y=0} - T_w = \frac{2 - \alpha}{\alpha}\frac{2\gamma}{\gamma + 1}\frac{\lambda}{\text{Pr}}\frac{\partial T}{\partial y}\bigg)_{y=0} \tag{12-55}$$

**Fig. 12-15  Nomenclature for use with Eq. (12-55).**

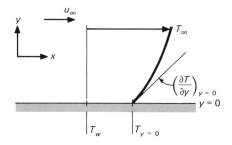

The nomenclature for Eq. (12-55) is noted in Fig. 12-15. This temperature jump is denoted by $\Delta T$ in Fig. 12-14, and the temperature gradient for use with Fig. 12-14 would be

$$\frac{T_1 - T_2 - 2\dot{\Delta}T}{L}$$

When free-molecule flow is encountered, Oppenheim [8] has given convenient charts for calculating recovery factors and heat-transfer coefficients for flow over standard geometric shapes. Figures 12-16 and 12-17 give samples of these charts. The molecular speed ratio $S$ used in these charts is defined by

$$S = \frac{u_\infty}{v_m} = \frac{u_\infty}{\sqrt{2RT}} = M\sqrt{\frac{\gamma}{2}} \qquad (12\text{-}56)$$

## Table 12-1  Thermal Accommodation Coefficients for Air According to Weidmann and Trumpler [10]†

| Surface | Accommodation coefficient |
|---|---|
| Flat black lacquer on bronze | 0.88–0.89 |
| Bronze, polished | 0.91–0.94 |
|    Machined | 0.89–0.93 |
|    Etched | 0.93–0.95 |
| Cast iron, polished | 0.87–0.93 |
|    Machined | 0.87–0.88 |
|    Etched | 0.89–0.96 |
| Aluminum, polished | 0.87–0.95 |
|    Machined | 0.95–0.97 |
|    Etched | 0.89–0.97 |

† See also Refs. 9 and 11.

**Fig. 12-16   Recovery factors for free-molecule flow, according to Oppenheim [8].**

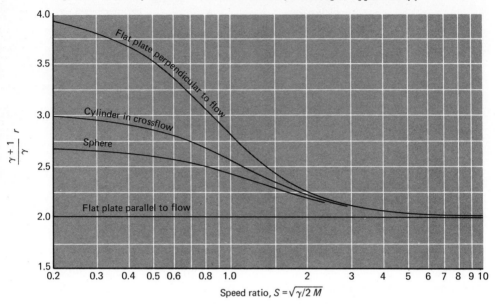

**Fig. 12-17   Stanton numbers for free-molecule flow, according to Oppenheim [8].**

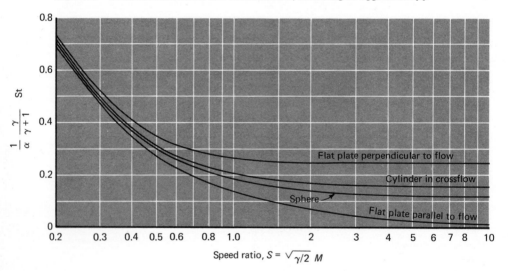

where $v_m$ is called the most probable molecular speed. The recovery factor $r$ is defined in the usual way as

$$r = \frac{T_{aw} - T_\infty}{T_0 - T_\infty} \tag{12-57}$$

where $T_{aw}$ is the adiabatic wall temperature and $T_0$ is the free-stream stagnation temperature. The Stanton number is also defined in the conventional way,

$$St = \frac{h}{\rho c_p u_\infty} \tag{12-58}$$

with the convection heat-transfer coefficient defined in terms of the adiabatic wall temperature,

$$q = hA(T_w - T_{aw}) \tag{12-59}$$

It is well to mention that the equilibrium temperature of a surface in free-molecule flow is usually influenced strongly by radiation heat transfer because of the high values of $T_{aw}$ encountered for high flow velocities. The radiation equilibrium temperature for the surface $T_{rw}$ is obtained by equating the convection gain to the radiation loss. Thus

$$hA(T_{aw} - T_{rw}) = \sigma \epsilon A(T_{rw}^4 - T_s^4) \tag{12-60}$$

where $\epsilon$ is the surface emissivity and $T_s$ is an effective radiation temperature of the surroundings, usually taken as $T_\infty$.

An excellent summary of low density heat transfer is given by Springer [13].

## Example 12-2

Two polished aluminum plates are separated by a distance of 1 in. in air at $10^{-6}$ atm pressure. The plates are maintained at 200 and 100°F, respectively. Calculate the conduction heat transfer through the air gap. Compare this with the radiation heat transfer and the conduction for air at normal atmospheric pressure.

**Solution**   We first calculate the mean-free path to determine if low-density effects are important. From Eq. (12-45)

$$\lambda = \frac{(8.64 \times 10^{-7})(610)}{(14.7)(144)(10^{-6})} = 0.249 \text{ ft}$$

The plate spacing is $L = 1$ in. $= 0.0833$ ft; so we should expect low-density effects to be important. Evaluating properties at the mean air temperature of 150°F, we have

$k = 0.0167$ Btu/hr-ft-°F
$\gamma = 1.40$
$Pr = 0.699$
$\alpha \approx 0.9$ from Table 12-1

Combining Eq. (12-55) with the central-temperature-gradient relation gives

$$\Delta T = \frac{2 - \alpha}{\alpha} \frac{2\gamma}{\gamma + 1} \frac{\lambda}{Pr} \frac{T_1 - T_2 - 2\Delta T}{L}$$

Inserting the appropriate properties gives

$$\Delta T = \left(\frac{2 - 0.9}{0.9}\right)\left(\frac{2.8}{2.4}\right)\left(\frac{0.249}{0.699}\right)\left(\frac{100 - 2\Delta T}{0.0833}\right)$$

Solving this relation for $\Delta T$ gives

$$\Delta T = 46.2°F$$

The conduction heat transfer is thus

$$\frac{q}{A} = k \frac{T_1 - T_2 - 2\Delta T}{L} = \frac{(0.0167)(100 - 92.4)}{0.0833}$$
$$= 1.52 \text{ Btu/hr-ft}^2 \quad (4.79 \text{ W/m}^2)$$

At normal atmospheric pressure the conduction would be

$$\frac{q}{A} = k \frac{T_1 - T_2}{L} = 20 \text{ Btu/hr-ft}^2 \quad (63.1 \text{ W/m}^2)$$

The radiation heat transfer is calculated with Eq. (8-40), using $\epsilon_1 = \epsilon_2 = 0.06$.

$$\left(\frac{q}{A}\right)_{rad} = \frac{\sigma(T_1{}^4 - T_2{}^4)}{2/\epsilon - 1} = 4.87 \text{ Btu/hr-ft}^2 \quad (15.36 \text{ W/m}^2)$$

## Example 12-3

A thermocouple is constructed by welding the two wires together so that a spherical bead is formed with a diameter of 0.03 in. The bead is exposed to a high-velocity stream at $M = 6$ and $p = 10^{-6}$ atm, $T = -100°F$. Estimate the temperature of the thermocouple bead assuming a surface emissivity of 0.7. Assume that the bead has an accommodation coefficient equal to that of cast iron.

**Solution**   We first establish the flow regime. Evaluating properties at free-stream conditions, we have

$$\mu = 0.0319 \text{ lb}_m/\text{hr-ft}$$
$$k = 0.0104 \text{ Btu/hr-ft-°F}$$
$$c_p = 0.239 \text{ Btu/lb}_m\text{-°F}$$

The gas density is

$$\rho = \frac{p}{RT} = \frac{(14.7)(144)(10^{-6})}{(53.35)(360)} = 1.105 \times 10^{-7} \text{ lb}_m/\text{ft}^3$$

The acoustic velocity is

$$a = \sqrt{\gamma g_c R T} = 930 \text{ ft/sec}$$

so that
$$u_\infty = (6)(930) = 5580 \text{ ft/sec}$$

The Reynolds number is now calculated as

$$\text{Re} = \frac{\rho u_\infty d}{\mu} = \frac{(1.105 \times 10^{-7})(5580)(3600)(0.03/12)}{0.0319}$$

$$= 0.174$$

An inspection of Fig. 12-13 shows that the free-molecule range is encountered. Accordingly, we can make use of Figs. 12-16 and 12-17 to compute the heat-transfer parameters.

The molecular speed ratio is calculated as

$$S = \sqrt{\frac{\gamma}{2}} M = (0.7)^{\frac{1}{2}}(6) = 5.02$$

From the figures

$$\frac{\gamma + 1}{\gamma} r = 2.04 \qquad \frac{1}{\alpha} \frac{\gamma}{\gamma + 1} \text{St} = 0.127$$

Then, using $\alpha = 0.9$,

$$r = \frac{(2.04)(1.4)}{2.4} = 1.18$$

$$\text{St} = \frac{(0.127)(2.4)(0.9)}{1.4} = 0.196$$

The stagnation temperature is calculated from Eq. (5-76),

$$T_0 = (360)[1 + (0.2)(6)^2] = 2950°\text{R}$$

and the adiabatic wall temperature is

$$T_{aw} = T_\infty + r(T_0 - T_\infty)$$
$$= 360 + (1.18)(2950 - 360)$$
$$= 3420°\text{R}$$

The Stanton number is now used to calculate the heat-transfer coefficient from Eq. (12-58).

$$h = \rho c_p u_\infty \text{St} = (1.105 \times 10^{-7})(0.239)(5580)(3600)(0.196)$$
$$= 0.104 \text{ Btu/hr-ft}^2\text{-°F} \qquad (0.59 \text{ W/m}^2\text{-°C})$$

We now employ the energy balance indicated in Eq. (12-60) to calculate the radiation equilibrium temperature $T_{rw}$, taking the radiation surroundings at $-100°\text{F}$.

$$(0.104)(3420 - T_{rw}) = (0.1714 \times 10^{-8})(0.7)[T_{rw}^4 - (360)^4]$$

**Fig. 12-18  Schematic of ablating solid surface.**

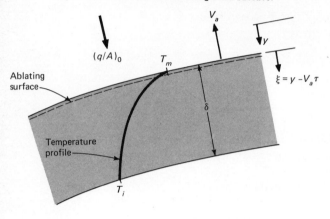

A solution of this equation gives

$$T_{rw} = 709°R = 249°F$$

It is easy to see from this example that radiation heat transfer is very important for low-density work, even when high velocities are encountered.

## 12-5□ABLATION

The very high speeds encountered in missile reentry situations present many interesting and unusual heat-transfer problems. The major concern is usually with a total energy which must be absorbed in the reentry body rather than with a heat-transfer *rate*, since the reentry times are very short. In these instances a common cooling technique is that of *ablation* whereby part of the solid body exposed to the hot, high-speed flow is allowed to melt and blow away. Thus, part of the heat is expended to melt the material rather than being conducted into the interior of the vehicle. A number of glasses and plastic materials normally find application as ablators.

We shall employ a simplified analysis of the ablation problem utilizing the coordinate system and nomenclature shown in Fig. 12-18. The solid wall is exposed to a constant heat flux of $(q/A)_0$ at the surface. This heat flux may result from combined convection- and radiation-energy transfer from the high-speed boundary layer. As a result of the high-heat flux the solid body melts and a portion of the surface is removed at the ablation velocity $V_a$. We shall assume that a steady-state situation is attained so that the surface ablates at a constant rate. It is further assumed that the thickness of the solid is very large compared with the depth of material removed by the melting process. If the process is essentially one-dimensional the appropriate differential equation for constant properties is

$$\frac{\partial^2 T}{\partial y^2} = \frac{1}{\alpha} \frac{\partial T}{\partial \tau} \qquad (12\text{-}61)$$

This problem is most easily solved with a simple change of variable into a moving coordinate system. If time is measured from the start of ablation then a convenient variable change is

$$\xi = y - V_a \tau \qquad (12\text{-}62)$$

The effect of this transformation is to fix the origin at the surface in the moving coordinate system. The derivatives in Eq. (12-61) may now be expressed as

$$\frac{\partial^2 T}{\partial y^2} = \frac{\partial^2 T}{\partial \xi^2} \qquad \frac{\partial T}{\partial \tau} = V_a \frac{\partial T}{\partial \xi}$$

so that an ordinary differential equation results:

$$\frac{d^2 T}{d\xi^2} + \frac{V_a}{\alpha} \frac{dT}{d\xi} = 0$$

This equation has the solution

$$T = C_1 + C_2 \exp \frac{-V_a \xi}{\alpha} \qquad (12\text{-}63)$$

At the surface the temperature is the melting temperature $T_m$, while the temperature in the interior of the body is $T_i$. The appropriate boundary conditions are thus

$$T = T_m \qquad \text{at } \xi = 0$$
$$T = T_i \qquad \text{as } \xi \to \infty$$

Evaluating the constants in (12-63) the final solution is

$$\frac{T - T_i}{T_m - T_i} = \exp \frac{-V_a \xi}{\alpha} \qquad (12\text{-}64)$$

The total energy incident on the body is either conducted into the slab or used to melt the material. Thus,

$$\left(\frac{q}{A}\right)_0 = \left(\frac{q}{A}\right)_{cond} + \left(\frac{q}{A}\right)_{abl}$$
$$= -k \frac{\partial T}{\partial \xi}\bigg)_{\xi=0} + \rho V_a H_{ab} \qquad (12\text{-}65)$$

where $H_{ab}$ is the heat of ablation. Evaluating the surface temperature gradient from Eq. (12-64) gives

$$\left(\frac{q}{A}\right)_0 = \rho V_a c(T_m - T_i) + \rho V_a H_{ab} \qquad (12\text{-}66)$$

which may be solved for the ablation velocity $V_a$ to give

$$V_a = \frac{(q/A)_0}{\rho H_{ab}[1 + c(T_m - T_i)/H_{ab}]} \tag{12-67}$$

As mentioned previously the main purpose of ablation is to reduce the quantity of energy conducted into the solid. It is therefore of interest to compare the total energy conducted in time $\tau$ with the total incident energy in this same time. The total conduction energy is

$$\left(\frac{Q}{A}\right)_{\text{cond}} = \int_0^\tau - k\left(\frac{\partial T}{\partial \xi}\right)_{\xi=0} d\tau$$
$$= \rho V_a c(T_m - T_i) \tag{12-68}$$

In this time the total energy is $(q/A)_0 \tau$. We therefore form the following ratio of interest:

$$\frac{(Q/A)_{\text{cond}}}{(Q/A)_{\text{total}}} = \frac{\rho V_a c(T_m - T_i)}{(q/A)_0} \tag{12-69}$$

Making use of Eq. (12-67) this becomes

$$\frac{(Q/A)_{\text{cond}}}{(Q/A)_{\text{total}}} = \frac{c(T_m - T_i)}{H_{ab}[1 + c(T_m - T_i)/H_{ab}]} \tag{12-70}$$

A few quick observations are in order at this point. The conducted energy is clearly reduced for large values of the heat of ablation. Similarly, the rate of material removal $\rho V_a$ is dependent on the heat of ablation and decreases for increased values of $H_{ab}$. In order for this very simplified solution to apply, the overall slab thickness $\delta$ must be large compared with the depth of penetration by melting. In terms of the above parameters this means that

$$\delta \gg V_a \tau$$

The reader should realize that the above analysis has several serious faults. It does not take into account the transient nature of the incident heat flux or the many complications involved in the melting-mass transfer process which occurs at the surface. Some of these problems are discussed by Dorrance [12].

## 12-6□THE HEAT PIPE

We have seen that one of the objectives of heat-transfer analysis can be the design of heat exchangers to transfer energy from one location to another. The smaller and more compact the heat exchanger, the better the design. The heat pipe is a novel device that allows the transfer of very substantial quantities of heat through small surface areas. The basic configuration of the device is shown in Fig. 12-19. A circular pipe has a layer of wicking material covering the inside surface as shown with a hollow core in the center. A condensible fluid is also contained in the pipe, and the liquid

Fig. 12-19  Basic heat pipe configuration.

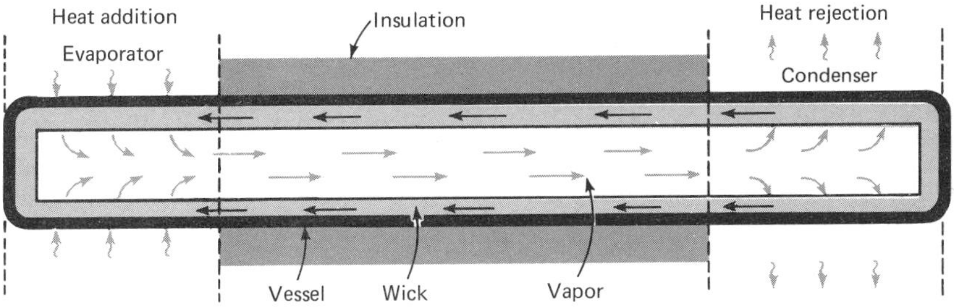

permeates the wicking material by capillary action. When heat is added to one end of the pipe (the evaporator) liquid is vaporized in the wick and the vapor moves to the central core. At the other end of the pipe, heat is removed (the condenser) and the vapor condenses back into the wick. Liquid is replenished in the evaporator section by capillary action; thus, the action of the heat pipe is essentially independent of spatial orientation and gravity forces.

A variety of fluid and pipe materials have been used for heat-pipe construction, and some typical operating characteristics are summarized in Table 12-2. Very high heat fluxes are obtained; and for this reason intensive research efforts are being devoted to optimum wick designs, novel configurations for specialized applications, etc. Two of the earliest theoretical analyses of heat pipes are presented by Cotter et al. [16, 17], but the device is in such a rapid state of development that the current research literature should be consulted for the latest information. For our purposes we shall be concerned primarily with applications for the device.

The basic configuration of Fig. 12-19 can be modified by the use of a flexible connecting tube as illustrated in Fig. 12-20 from Ref. 14. A proposed application for this kind of arrangement is a quick-starting heating system for an automobile. The heat-input reservoir could be connected to the exhaust manifold; the heat-output reservoir is connected to the interior-air heat exchanger. Thus, heating for the interior would be available almost immediately without the long warmup period required in conventional heaters that operate off the engine coolant system.

Cooling problems in microelectric circuits are particularly critical because the heat generation must be dissipated from such small surface areas and the performance of the electronic devices is strongly temperature-dependent. The heat-pipe concept offers a convenient way to transfer the heat from the small area to a larger area where it can be dissipated more easily. One way of doing this is shown in Fig. 12-21 according to Ref. 15. Of course, the finned heat dissipation surface could be water-cooled if the need required. The advantage of the heat pipe in electronic cooling applications is its nearly isothermal operation regardless of heat flux, within the

**Table 12-2   Some Typical Operating Characteristics of Heat Pipes, According to Ref. 15**

| Temperature range, °C | Working fluid | Vessel material | Measured axial† heat flux, kw/sq in. | Measured surface† heat flux, watts/sq in. |
|---|---|---|---|---|
| −200 to −80 | Liquid nitrogen | Stainless steel | 0.431 @ −163°C | 6.5 @ −163°C |
| −70 to +60 | Liquid ammonia | Nickel, aluminum, stainless steel | 1.9 | 19 |
| −45 to +120 | Methanol | Copper, nickel, stainless steel | 2.9 @ 100°C‡ | 487 @ 100°C |
| +5 to +230 | Water | Copper, nickel | 4.3 @ 200°C | 942 @ 170°C |
| +190 to +550 | Mercury§ +0.02% magnesium +0.001% tantalum | Stainless steel | 162 @ 360°C¶ | 1,170 @ 360°C |
| +400 to +800 | Potassium§ | Nickel, stainless steel | 36 @ 750°C | 1,170 @ 750°C |
| +500 to +900 | Sodium§ | Nickel, stainless steel | 60 @ 850°C | 1,443 @ 760°C |
| +900 to +1,500 | Lithium§ | Niobium +1% zirconium | 13 @ 1,250°C | 1,334 @ 1,250°C |
| 1,500 to +2,000 | Silver§ | Tantalum +5% tungsten | 26 | 2,665 |

† Varies with temperature.
‡ Using threaded artery wick.
§ Tested at Los Alamos Scientific Laboratory.
¶ Measured value based on reaching the sonic limit of mercury in the heat pipe.

operating range of the unit.

The basic design of a heat pipe may be modified to operate as a temperature-control device as shown in Fig. 12-22. A reservoir containing a noncondensible gas is connected to the heat-removal end of the heat pipe. This gas may then form an interface with the vapor and "choke off" part of the condensation to the wick. With increased heat addition, more vapor is generated with an increase in vapor pressure and the noncondensible gas is forced back into the reservoir, thereby opening up additional condenser area to remove the additional heat. For a reduction in heat addition, just the reverse operation is observed. If the heat-source temperature drops below a certain minimum value, depending on the specific fluid and gas combinations in the heat pipe, a complete shutoff can occur. So the control feature can be particularly useful for fast warmup applications in addition to its value as a temperature leveler for variable load conditions.

**Fig. 12-20  Modification of basic heat pipe configuration to permit flexible location of heat input and output reservoirs. From Ref. 14.**

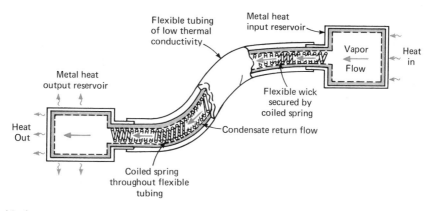

## Example 12-4

Compare the axial heat flux in a heat pipe using water as the working fluid at 200°C with that of a solid copper rod 3 in. long with a temperature differential of 100°C.

**Solution**    For the copper rod the heat flux is

$$q'' = \frac{q}{A} = -k\frac{\Delta T}{\Delta x}$$

The thermal conductivity of copper is 216 Btu/hr-ft-°F, so that

$$q'' = \frac{-(216)(-100)(9/5)}{(3/12)} = 1.554 \times 10^5 \text{ Btu/hr-ft}^2$$

From Table 12-2, the typical axial heat flux for a water heat pipe is

$$q''_{\text{axial}} = 4.3 \text{ kw/in.}^2 = 2.11 \times 10^6 \text{ Btu/hr-ft}^2$$

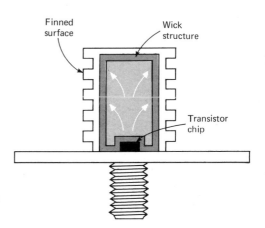

**Fig. 12-21  Application of heat pipe principle to cooling of power transistor, according to Ref. 15.**

**Fig. 12-22  Heat pipe combined with noncondensable gas reservoir to provide a temperature control device. From Ref. 14.**

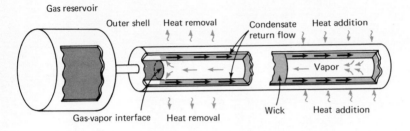

Thus, this heat pipe transfers more than ten times the heat of a pure copper rod with a substantial temperature gradient. This example illustrates why the heat pipe enjoys wide application possibilities.

### REVIEW QUESTIONS

1.  Under what conditions will a magnetic field cause an increase in heat transfer; a decrease in heat transfer, when impressed on a conducting fluid?

2.  How do the boundary conditions for transpiration cooling differ from those for ordinary flow over a flat plate?

3.  How is a distinction made between continuum, slip, and free-molecule flows in a physical sense?

4.  What is the Knudsen number?

5.  What is the accommodation coefficient?

6.  What is meant by a radiation equilibrium temperature?

7.  How is ablation used for high-speed cooling?

8.  What is a heat pipe? How does it work?

9.  Why is the heat pipe so useful?

10.  Describe how a heat pipe may be used as a temperature-control device.

### PROBLEMS

12-1.  Air flows over a flat plate at $M = 4.0$ and $-40°F$, 0.1 psia. Calculate the adiabatic wall temperature for an injection parameter of 0.1 and at a position where $Re_x = 10^5$.

12-2.  Calculate the heat-transfer rate $q/A$ for the conditions of Prob. 12-1. Compare this with what would be experienced with no injection.

12-3.  An air flow at 1000°F and 15 psia impinges on the front side of a porous horizontal cylinder 3 in. in diameter at a velocity of 2000 ft/sec. Air is injected through the porous material to maintain the surface temperature at 300°F. Calculate the heat-transfer rate at a distance of 0.3 in. from the stagnation point and with an injection parameter of 0.5.

12-4.   Derive an expression for the ratio of the heat conducted through a gas layer at low density to that conducted for $\lambda = 0$. Plot this ratio versus $\lambda/L$ for $\alpha = 0.9$ and air properties evaluated at 100°F.

12-5.   Develop an expression similar to Eq. (12-25) for low-density conduction between concentric cylindrical surfaces.

12-6.   A fine iron wire 0.001 in. in diameter is exposed to a high-velocity airstream at $10^{-6}$ atm, $-60°F$, and $M = 5$. Estimate the wire surface temperature, assuming $\epsilon = 0.4$ and $\alpha = 0.9$. Assume the radiation surroundings temperature at the free-stream temperature.

12-7.   Suppose the wire of Prob. 12-6 is employed as a resistance thermometer to sense the free-stream temperature for the flow conditions stated. Plot the indicated temperature of the wire versus the actual free-stream temperature for the range $-120$ to $0°F$.

12-8.   A superinsulating material is to be constructed of polished aluminum sheets separated by a distance of $\frac{1}{32}$ in. The space between the sheets is sealed and evacuated to a pressure of $10^{-5}$ atm pressure. Four sheets are used. The two outer sheets are maintained at 100°F and 200°F and have a thickness of 0.03 in., whereas the inner sheets have a thickness of 0.007 in. Calculate the conduction and radiation transfer across the layered section per unit area. For this calculation, allow the inner sheets to "float" in the determination of the radiation heat transfer. Evaluate properties at 150°F.

12-9.   Air flows over a flat plate at $p = 10^{-10}$ atm and $T_\infty = -80°F$. The velocity corresponds to $M = 5.0$ and the plate is 1 ft square. Calculate the radiation equilibrium temperature for an accommodation coefficient of 0.88 and surface emissivities of 0.3 and 0.8. Assume the radiation temperature of the surroundings as $-100°F$.

12-10.   What cooling rate would be necessary to maintain the plate of Prob. 12-9 at a surface temperature of 150°F?

# REFERENCES

1.   Romig, M.: The Influence of Electric and Magnetic Fields on Heat Transfer to Electrically Conducting Fluids, *Advan. Heat Transfer*, vol. 1, pp. 268–352, 1964.

2.   Sutton, G. W., and A. Sherman: "Engineering Magnetohydrodynamics," McGraw-Hill Book Company, New York, 1965.

3.   Eckert, E. R. G., and J. P. Hartnett: Mass Transfer Cooling in a Laminar Boundary Layer with Constant Fluid Properties, *Trans. ASME*, vol. 79, pp. 247–254, 1957.

4.   Sziklas, E. A.: An Analysis of the Compressible Laminar Boundary Layer with Foreign Gas Injection, *United Aircraft Corp. Res. Dept. Rept.* SR-0539-8, 1956.

5.   Leontev, A. I.: Heat and Mass Transfer in Turbulent Boundary Layers, *Advan. Heat Transfer*, vol. 3, 1966.

6.   Leadon, B. M., and C. J. Scott: Transpiration Cooling Experiments in a Turbulent Boundary Layer at $M = 3.0$, *J. Aeron. Sci.*, vol. 23, pp. 798–799, 1956.

7.   Eckert, E. R. G., et al.: Mass Transfer Cooling of a Laminar Boundary Layer

by Injection of a Light Weight Foreign Gas, *Jet Propulsion*, vol. 28, pp. 34–39, 1958.

8.  Oppenheim, A. K.: Generalized Theory of Convective Heat Transfer in a Free-molecule Flow, *J. Aeron. Sci.*, vol. 20, p. 49, 1953.

9.  Wachman, H. Y.: The Thermal Accommodation Coefficient: A Critical Survey, *ARS J.*, vol. 32, p. 2, 1962.

10. Weidmann, M. L., and P. R. Trumpler: Thermal Accommodation Coefficients, *Trans. ASME*, vol. 68, p. 57, 1946.

11. Devienne, F. M.: Low Density Heat Transfer, *Advan. Heat Transfer*, vol. 2, p. 271, 1965.

12. Dorrance, W. H.: "Viscous Hypersonic Flow," McGraw-Hill Book Company, New York, 1962.

13. Springer, G. S.: Heat Transfer in Rarefied Gases: A Survey, *Fluid Dyn. Lab. Publ.* 69-1, Univ. of Mich., May, 1969.

14. Feldman, K. T., and G. H. Whiting: Applications of the Heat Pipe, *Mech. Eng.*, vol. 90, p. 48, November, 1968.

15. Dutcher, C. H., and M. R. Burke: Heat Pipes—A Cool Way to Cool Circuits, *Electronics*, pp. 93–100, Feb. 16, 1970.

16. Grover, G. M., and T. P. Cotter, and G. F. Erikson: Structures of Very High Thermal Conductance, *J. Appl. Phys.*, vol. 35, p. 1990, 1964.

17. Cotter, T. P.: Theory of Heat Pipes, *Los Alamos Sci. Lab. Rept.* LA-3246-MS, February, 1965.

# CHAPTER
# 13
# HEAT TRANSFER
# IN THE ENVIRONMENT

## 13-1 □ INTRODUCTION

Because the behavior of the environment and proper control, modification, and alleviation of human indiscretions to the atmosphere are important to all elements of society, it is relevant to see how the discipline of heat transfer is applicable to a better understanding of the problems that arise. This chapter will present analyses of several atmospheric and environmental processes that fall in the general province of heat-transfer and transport phenomena study. The reader should recognize that this subject matter is treated in a variety of overlapping scientific disciplines such as meteorology, hydrology, geophysics, and climatology. Some appropriate references for further study are given at the end of the chapter. It is also worthwhile to point out that the subject matter in this chapter, like that in Chap. 12, could be dispersed through the specific chapters dealing with convection and radiation, but we choose to group the environmental topics together in one location in order to give added emphasis to the discussion.

## 13-2 □ RADIATION PROPERTIES OF THE ENVIRONMENT

We have already described the radiation spectrum of the sun in Chap. 8 and noted that the major portion of solar energy is concentrated in the short wavelength region. It was also noted that as a consequence of this spectrum, real surfaces may exhibit substantially different absorption properties for solar radiation than for long wavelength "earthbound" radiation.

Meteorologists and hydrologists use the term *insolation* to describe the intensity of direct solar radiation incident upon a horizontal surface per unit area and per unit time, designated with the symbol $I$. Although we shall emphasize other units, it will be helpful to mention a unit that appears in the meteorological literature:

$$1 \; \textit{langley} \; (\text{ly}) = 1 \; \text{cal/cm}^2 \quad (3.687 \; \text{Btu/ft}^2)$$

Insolation and radiation intensity are frequently expressed in ly per unit time, e.g., the Stefan-Boltzmann constant would be

$$\sigma = 0.826 \times 10^{-10} \text{ ly/min-}°K^4$$

Radiation heat transfer in the environment is governed by the absorption, scattering, and reflection properties of the atmosphere and natural surfaces. Two types of scattering phenomena occur in the atmosphere. *Molecular scattering* is observed because of the interaction of radiation with individual molecules. The blue color of the sky results from the scattering of the violet (short) wavelengths by the air molecules. *Particulate scattering* in the atmosphere results from the interaction of radiation with the many types of particles that may be suspended in the air. Dust, smog, and water droplets are all important types of particulate scattering centers. The scattering process is governed mainly by the size of the particle in comparison with the wavelength of radiation. Maximum scattering occurs when wavelength and particle size are equal and decreases progressively for longer wavelengths. For wavelengths smaller than the particle size, the radiation tends to be reflected.

Reflection phenomena in the atmosphere occur for wavelengths less than the particle size and are fairly independent of wavelength in this region. The term *albedo* is used to describe the reflective properties of surfaces and is defined by

$$A = \text{albedo} = \frac{\text{reflected energy}}{\text{incident energy}} \tag{13-1}$$

## Table 13-1   Albedo of Natural Surfaces, According to Ref. 3

| Surface | Albedo A | Surface | Albedo A |
|---|---|---|---|
| Water | 0.03–0.40 | Spring wheat | 0.10–0.25 |
| Black, dry soil | 0.14 | Winter wheat | 0.16–0.23 |
| Black, moist soil | 0.08 | Winter rye | 0.18–0.23 |
| Gray, dry soil | 0.25–0.30 | High, dense grass | 0.18–0.20 |
| Gray, moist soil | 0.10–0.12 | Green grass | 0.26 |
| Blue, dry loam | 0.23 | Grass dried in sun | 0.19 |
| Blue, moist loam | 0.16 | Tops of oak | 0.18 |
| Desert loam | 0.29–0.31 | Tops of pine | 0.14 |
| Yellow sand | 0.35 | Tops of fir | 0.10 |
| White sand | 0.34–0.40 | Cotton | 0.20–0.22 |
| River sand | 0.43 | Rice field | 0.12 |
| Bright, fine sand | 0.37 | Lettuce | 0.22 |
| Rock | 0.12–0.15 | Beets | 0.18 |
| Densely urbanized areas | 0.15–0.25 | Potatoes | 0.19 |
| Snow | 0.40–0.85 | Heather | 0.10 |
| Sea ice | 0.36–0.50 | | |

Fig. 13-1  Effect of solar altitude angle and cloud cover on albedo for a horizontal water surface according to Ref. 4.

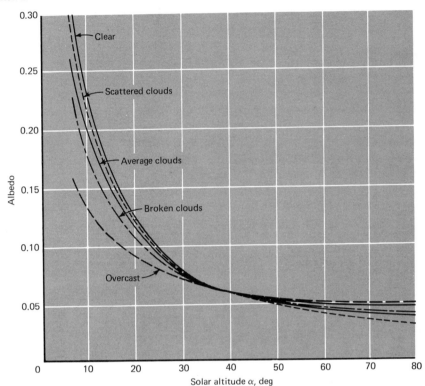

The albedos of some natural surfaces are given in Table 13-1. The effect of solar incident angle on the albedo of water is shown in Fig. 13-1, where $\alpha$ is the angle the incoming rays make with the horizontal.

The atmosphere absorbs radiation quite selectively in narrow wavelength bands. The absorption for solar radiation occurs in entirely different bands from the absorption of the radiation from the earth because of the different spectrums for the two types of radiation. In Fig. 13-2 we see the approximate spectrums for solar and earth radiation with some important absorption bands superimposed on the diagram. Note the scale differential on the two curves. A quick inspection of these curves will show that the atmosphere transmits most of the short wavelength radiation while absorbing most of the back radiation from the earth. Therefore, the atmosphere acts very much like the greenhouse discussed in Chap. 8, trapping the incoming solar radiation to provide energy and warmth for man on the planet earth. Some concern is voiced that man may upset the energy budget of the earth through excessive contamination of the atmosphere with pollutants. Such a possibility does exist but is beyond the scope of our discussion.

**Fig. 13-2** Thermal radiation spectrums for the sun and earth with primary absorptions bands indicated by shaded areas. Note different scales.

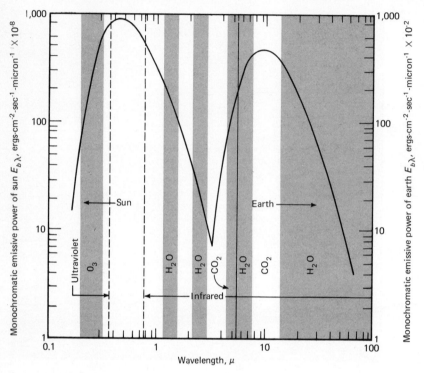

The absorption and scattering of radiation may be described with Beer's law [Eq. (8-46)], which we repeat here for convenience

$$\frac{I_{\lambda_x}}{I_{\lambda_o}} = \exp\left(-a_\lambda x\right) \tag{13-2}$$

where $a_\lambda$ is the monochromatic absorption coefficient and $x$ is the thickness of the layer absorbing the radiation. For a scattering process, we would replace $a_\lambda$ by a *scattering coefficient* $k_\lambda$.

The conventional approach in atmospheric problems is to assume that the absorption and scattering processes are superimposed on one another and may be expressed in the form of Eq. (13-2) over all wavelengths. The appropriate coefficients are defined as

$a_{ms}$ = average molecular scattering coefficient over all wavelengths
$a_{ps}$ = average particulate scattering coefficient over all wavelengths
$a$ = average absorption coefficient over all wavelengths
$a_t = a_{ms} + a_{ps} + a$ = total attenuation coefficient over all wavelengths

Using these coefficients, the radiation insolation at the earth's surface is expressed as

$$\frac{I_c}{I_o} = \exp(-a_t m) = \exp(-n a_{ms} m) \tag{13-3}$$

where $m$ is the relative thickness of the air mass and $n$ is defined as the *turbidity factor* of the air:

$$n = \frac{a_t}{a_{ms}} \tag{13-4}$$

The insolations are defined as

$I_c$ = direct, cloudless sky insolation at earth's surface
$I_o$ = insolation at outer limits of earths atmosphere.

The molecular scattering coefficient for air at atmospheric pressure is given as [3]

$$a_{ms} = 0.128 - 0.054 \log m \tag{13-4a}$$

The relative thickness of the air mass is calculated as the cosecant of the solar altitude $\alpha$. The turbidity factor is thus a convenient means of specifying atmospheric purity and clarity; its value ranges from about 2.0 for very clear air to 4 or 5 for very smoggy industrial environments.

The insolation at the outer edge of the atmosphere is expressed in terms of the solar constant $E_{b_o}$ by

$$I_o = E_{b_o} \sin \alpha \tag{13-4b}$$

where $\alpha$ is again the angle the rays make with the horizontal. In Chap. 8 we gave the solar constant as

$$\begin{aligned} E_{b_o} &= 442.4 \text{ Btu/hr-ft}^2 \\ &= 2.00 \text{ cal/cm}^2\text{-min} \\ &= 1,395 \text{ watts/m}^2 \end{aligned}$$

An average variation of incident solar radiation for cloudy and cloudless situations as a function of solar altitude angle is given in Table 13-2.

## Example 13-1

A certain smoggy atmosphere has a turbidity of 4.0. Calculate the direct, cloudless sky insolation for a solar altitude angle of 75°. How much is this reduced from that of a clear sky?

**Solution** We shall be using Eq. (13-3) for this calculation, so that we need to make the following intermediate calculations.

$$\begin{aligned} I_o &= E_{b_o} \sin \alpha = (442.4) \sin 75° = 427.3 \text{ Btu/hr-ft}^2 \\ m &= \csc 75° = 1.035 \\ a_{ms} &= 0.128 - 0.054 \log m = 0.1272 \end{aligned}$$

**Table 13-2    Solar Irradiation (Insolation) on a Horizontal Surface under Average Atmospheric Conditions†**

| Solar altitude $\alpha$, deg | Average total insolation, ly hr$^{-1}$ |
|---|---|
| 5 | 3.6 |
| 10 | 9.7 |
| 15 | 17.2 |
| 20 | 25.0 |
| 25 | 32.8 |
| 30 | 40.6 |
| 35 | 47.7 |
| 40 | 54.7 |
| 45 | 61.1 |
| 50 | 67.2 |
| 60 | 77.5 |
| 70 | 85.3 |
| 80 | 89.7 |
| 90 | 91.4 |

† According to Ref. 4.

We have $n = 4.0$, so from Eq. (13-3),

$$I_c = (427.3) \exp\left[-(0.1272)(4.0)(1.035)\right]$$
$$= 252.4 \text{ Btu/hr-ft}^2$$

For a very clear day, $n = 2.0$ and we would have

$$I_c = (427.3) \exp\left[-(0.1272)(2.0)(1.035)\right]$$
$$= 328.4 \text{ Btu/hr-ft}^2$$

Thus the insolation has been reduced by 23 percent as a result of the smoggy environment.

The absorption of radiation in natural bodies of water is an important process because of its influence on evaporation rates and the eventual dispersion of water vapor in the atmosphere. Experimental measurements [16] show that solar radiation is absorbed very rapidly in the top layers of the water followed by an approximately exponential decay with depth in the water. The incident radiation follows a variation of

$$\frac{I_z}{I_s} = (1 - \beta)e^{-az} \tag{13-5}$$

where $I_s$ is the intensity at the surface and $I_z$ is the intensity at a depth $z$.

$\beta$ represents the fraction of energy absorbed at the surface and $a$ is an absorption or *extinction coefficient* for the material. $\beta$ may be interpreted as a measure of the long wavelength content of solar radiation, since the shorter wavelengths penetrate into the water more readily. This coefficient appears to have the value 0.4 for all lakes for which data are available, and it is assumed to be independent of time. Values of the extinction coefficient can vary considerably, e.g., it has the value 0.05 for the extremely clear water of Lake Tahoe [17] and a value of 0.27 for the more turbid water of Lake Castle, California [18].

## Example 13-2

Calculate the heat-generation rate resulting from solar radiation absorption in a lake with an extinction coefficient of 0.10 and a solar altitude of 90° on a clear day. Perform the calculation for a depth of 1 ft.

**Solution**    The heat-generation rate is obtained by differentiating Eq. (13-5):

$$\dot{q} = \frac{q_{absorbed}}{A\,dz} = \frac{dI_z}{dz} = I_s a(1 - \beta)e^{-az}$$

The surface insolation $I_s$ is calculated with Eq. (13-3).

$$I_o = E_{b_o} \sin \alpha = 442.4 \text{ Btu/hr-ft}^2$$
$$m = \csc 90° = 1.0$$
$$a_{ms} = 0.128 - 0.054 \log m = 0.128$$
$$n = 2.0$$
$$I_c = I_s = (442.4) \exp[-(0.128)(2.0)(1.0)]$$
$$= 342.5 \text{ Btu/hr-ft}^2$$

We also have $a = 0.10$ and $\beta = 0.4$, so that

$$\dot{q} = (342.5)(0.10)(1 - 0.4) \exp[-(0.10)(1.0)]$$
$$= 18.59 \text{ Btu/hr-ft}^3$$

Additional information on radiation in the atmosphere is given by Kondratyev [1] and Geiger [2].

## 13-3☐CONVECTIVE STABILITY IN THE ATMOSPHERE

Convection currents in the atmosphere clearly govern the transport of moisture and pollutants and provide the major climatological changes that man experiences as "weather." Our purpose in this brief section is to show how temperature gradients in the atmosphere influence the stability of the convection processes. A detailed treatment of the many facets of atmospheric convection is presented in the literature of meteorology, like that of Refs. 19 and 20.

Let us consider a small parcel of dry air that may be moved about in the atmosphere. If we assume that the parcel is adiabatic, i.e., that no heat is transferred to it by either conduction or radiation, it will experience a change in temperature as a result of the change in position in the atmospheric pressure field. For a reversible adiabatic process, the temperature and pressure are related by (Ref. 8)

$$Tp^{1-\gamma/\gamma} = \text{const} \tag{13-6}$$

where $\gamma = c_p/c_v$ is the ratio of specific heats. The variation of pressure with elevation in a gravity field was given by Eq. (7-2) as

$$\frac{dp}{dz} = \frac{-\rho g}{g_c} \tag{13-7}$$

where $g_c$ is inserted to adjust the units and $z$ is the elevation coordinate. Differentiating Eq. (13-6), we obtain

$$\frac{1-\gamma}{\gamma}\frac{1}{p}\frac{dp}{dz} + \frac{1}{T}\frac{dT}{dz} = 0 \tag{13-8}$$

The air density is expressed through the ideal gas law as

$$\rho = \frac{p}{RT} \tag{13-9}$$

Combining Eqs. (13-7), (13-8), and (13-9), we obtain for the temperature gradient $\Gamma$

$$\Gamma = \frac{dT}{dz} = -\frac{g(\gamma - 1)}{\gamma R g_c} \tag{13-10}$$

$\Gamma$ is called the *dry adiabatic lapse rate* (DALR) and has the approximate value for $\gamma = 1.4$,

$$\Gamma = -9.8°\text{C/km} \qquad (-5.355°\text{F}/1{,}000 \text{ ft}) \tag{13-11}$$

According to this concept, we should then expect a drop in temperature of about 5°F with each rise in elevation of 1,000 ft. In the real atmosphere, the actual temperature profile is seldom given by the DALR because of the many convection currents, heating from the earth or by radiation, and precipitation processes. The actual *ambient atmospheric lapse rate* (temperature gradient) is designated by the symbol $\alpha$. It may also be designated simply as the lapse rate (LR).

Figure 13-3 shows some of the different temperature gradients that may occur in the atmosphere. Imagine a parcel of air that is moved about in each of these cases. In (a) a parcel that is lifted adiabatically will have a higher temperature than its surroundings (and a lower density) and thus will continue to rise even after removal of an initial lifting force. This condition is unstable, and the process will terminate when the parcel cools to the temperature of its surroundings. Condition (b) is one of neutral stability

**Fig. 13-3** Typical temperature profiles in the atmosphere. (a) Superadiabatic or unstable condition; (b) adiabatic or neutral stability; (c) subadiabatic or stable condition; (d) temperature inversion or highly stable condition.

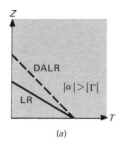

(a)

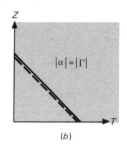

(b)

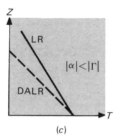

(c)

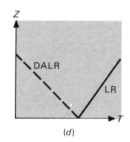

(d)

where the actual and adiabatic lapse rates are equal. In (c) a parcel of air that is lifted adiabatically will be cooler than its surroundings (and more dense) and thus will tend to fall back to its original position, producing a stable condition. The circumstance in (d) is a *temperature inversion* where the air temperature actually increases with elevation. This condition is highly stable and tends to trap the ground layers in place with very little vertical convection. These different conditions are pertinent to the discussion of free convection in Sec. 7-9 and a Benard cell type of phenomenon can sometimes be observed in the atmosphere.

## Influence of Moisture

When a moist parcel of air is raised in the gravitational field, we have an adiabatic cooling phenomenon until the air reaches the point of saturation, that is, 100 percent relative humidity. Upon a further rise in elevation, cooling will cause vapor to condense and precipitate out of the air parcel. The energy removed by this condensation process is then available to heat the surrounding air. The cooling that takes place during the precipitation is called a *pseudoadiabatic* process and may be analyzed as follows. For a rise in elevation $dz$, the amount of water removed is $dm_s$ and the heat liberated is

$$dq = -h_{fg}\, dm_s \qquad (13\text{-}12)$$

This heat then acts to raise the temperature of the remaining air. From thermodynamics

$$dq = c_{v_a} dT + p_a dv \tag{13-13}$$

where all the properties are those of the dry air, that is, $p_a$ is the partial pressure of the air. Using the ideal gas relations,

$$p_a v = R_a T \tag{13-14}$$
$$R_a = c_{p_a} - c_{v_a} \tag{13-15}$$

Eq. (13-13) may be rewritten as

$$dq = c_{p_a} dT - RT \frac{dp_a}{p_a} \tag{13-16}$$

But $p_a = p - p_s$ where $p$ is the total pressure and $p_s$ is the saturation pressure of the vapor at the temperature of the mixture. We may now combine Eqs. (13-12) and (13-16) to give

$$-h_{fg} dm_s = c_{p_a} dT - RT \frac{d(p - p_s)}{p - p_s} \tag{13-17}$$

The mass of water vapor per unit mass of dry air is

$$m_s = \frac{\rho_s}{\rho_a} = 0.622 \frac{p_s}{p - p_s} \tag{13-18}$$

The saturation pressure $p_s$ is a function of temperature and may be obtained from thermodynamic property tables. Thus, Eqs. (13-17) and (13-18) may be combined with such property data to obtain a relation between pressure and temperature for the pseudoadiabatic process. The important point is that the cooling of a moist parcel of air produces a different phenomenon from that for the dry air parcel and may strongly influence the atmospheric motion through the energy released during the condensation processes.

Let us proceed a few steps further with the analysis. We define the *potential temperature* $T_p$ as the temperature a dry parcel of air would attain if it were raised in pressure adiabatically to a reference pressure $p_o$. Then, from the adiabatic relation (13-6),

$$\frac{T}{T_p} = \left(\frac{p}{p_o}\right)^{\frac{\gamma - 1}{\gamma}} \tag{13-19}$$

Differentiating and clearing terms, we can obtain

$$\frac{dT}{T} = \frac{R}{c_p} \frac{dp}{p} + \frac{dT_p}{T_p} \tag{13-20}$$

Now if $p_s$ is very small in comparison with $p$ in Eq. (13-17), we can write

$$-h_{fg} \frac{dm_s}{T} = c_{p_a} \frac{dT}{T} - R_a \frac{dp}{p}$$

Substituting Eq. (13-20), we obtain

$$-h_{fg} \frac{dm_s}{T} = c_{p_a} \frac{dT_p}{T_p} \tag{13-21}$$

During the condensation process, the temperature does not change markedly and we may assume

$$\frac{dm_s}{T} \approx d\left(\frac{m_s}{T}\right)$$

so that Eq. (13-21) becomes

$$\frac{dT_p}{T_p} = \frac{-h_{fg}}{c_{p_a}} d\left(\frac{m_s}{T}\right) \tag{13-22}$$

The potential temperature when all the moisture has been condensed out is now used as a boundary condition

$$T_p = T_{p_o} \qquad \text{at } m_s = 0$$

Integrating Eq. (13-22) with this condition gives

$$\frac{T_p}{T_{p_o}} = \exp\left(\frac{-h_{fg} m_s}{c_p T}\right) \tag{13-23}$$

$T_{p_o}$ is sometimes called the *equivalent potential temperature*.

The importance of the above concepts is that a parcel of dry air will move along a line of constant potential temperature in an adiabatic process and a parcel of saturated air will move along a constant $T_{p_o}$ line in a pseudo-adiabatic process. These concepts can then form the basis for construction of charts that are very useful in meteorological calculations.

As a final discussion of stability, we may investigate a typical diurnal variation of the temperature profiles that might be experienced in the atmosphere as shown in Fig. 13-4. In the afternoon the sun will have heated the ground and we might experience the convectively unstable circumstance where the temperature drops more sharply than predicted by the DALR. As the sun sets, the ground begins to cool by radiation (assuming a clear sky) and a ground inversion layer may be formed as shown. Cooling continues throughout the night and into early morning so that the inversion layer thickens to its maximum value at about the time the sun rises. As morning arrives and progressively greater amounts of solar radiation heat up the ground, we may encounter the transition profile shown. With sufficient heating the unstable afternoon profile is established once again, with its vertical natural convection transport processes. The vertical temperature gradients obviously depend on local wind and weather conditions and can be strongly influenced by the moisture content of the air.

Vertical temperature gradients provide a strong influence on the dispersion of pollutants and contaminants in the atmosphere. The afternoon profile

**Fig. 13-4** Typical diurnal variation of atmospheric temperature profiles.

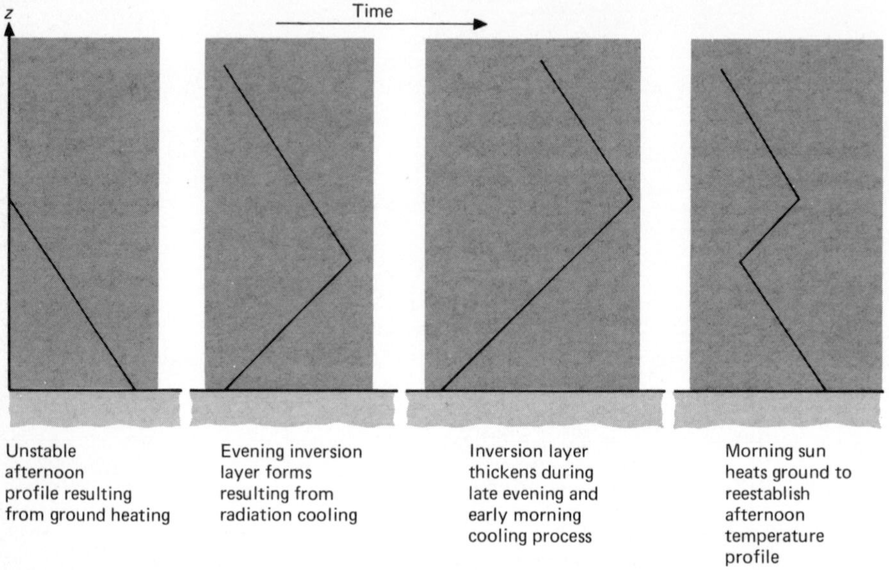

| Unstable afternoon profile resulting from ground heating | Evening inversion layer forms resulting from radiation cooling | Inversion layer thickens during late evening and early morning cooling process | Morning sun heats ground to reestablish afternoon temperature profile |

shown in Fig. 13-4 furnishes strong vertical transport that tends to carry pollutants away from the ground, while the thick early morning inversion layer tends to trap contaminants in the lower atmosphere. If the upper-level inversion layer of later morning occurs at a height where contaminants are discharged from smokestacks and the like, the pollutants will tend to fall to the ground and be mixed and churned about below the inversion layer.

The remarks above explain, of course, why a community may experience heavier concentrations of air pollution during rush-hour traffic on a clear, still morning before the inversion layer has been removed by solar heating. The same rush-hour traffic is less likely to produce high concentration in the afternoon because of the vertical mixing currents that disperse the pollutants. For more information on this interesting subject, the reader may consult more extensive works like Refs. 5, 6, 7, and 9.

### 13-4□EVAPORATION PROCESSES

We have already described some evaporation processes in Chap. 11 and indicated relations between heat and mass transfer. In the atmosphere, the continuous evaporation and condensation of water from the soil, oceans, and lakes influences every form of life and provides many of the day-to-day varieties of climate that govern man's environment. These processes are very complicated because in practice they are governed by substantial atmospheric convection currents that are difficult to describe analytically. Let us first consider the diffusion of water vapor from a horizontal

Fig. 13-5   Diffusion of water vapor from a horizontal surface.

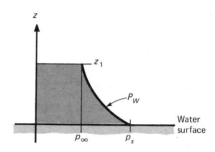

Fig. 13-5   Diffusion of water vapor from a horizontal surface.

surface into quiescent air as indicated in Fig. 13-5. At the surface, the partial pressure of the vapor is $p_s$. The vapor pressure steadily drops with a rise in elevation $z$ to the "free atmosphere" value of $p_\infty$. The molecular diffusion of the water vapor may be written in the form of Eq. (11-13) as

$$\frac{\dot{m}_w}{A} = -D_w \frac{M_w}{R_o T} \frac{dp_w}{dz} \qquad (13\text{-}24)$$

where $A$ is the surface area under consideration. In hydrologic applications, it is convenient to express this relation in terms of the local atmospheric density and pressure. The total pressure may be expressed as

$$p = \rho \frac{R_o}{M} T \qquad (13\text{-}25)$$

where $\rho$ and $M$ are the density and molecular weight of the moist air, respectively. Because the molar concentration of water vapor is so small in atmospheric applications, the molecular weight of the moist air is essentially that of dry air, and Eqs. (13-24) and (13-25) can be combined to give

$$\frac{\dot{m}_w}{A} = -D_w \frac{M_w}{M_a} \frac{\rho}{p} \frac{dp_w}{dz}$$

But, $M_w/M_a = 0.622$, so that

$$\frac{\dot{m}_w}{A} = -0.622 D_w \frac{\rho}{p} \frac{dp_w}{dz} \qquad (13\text{-}26)$$

Using the boundary conditions

$$p_w = p_s \qquad \text{at } z = 0$$
$$p_w = p_\infty \qquad \text{at } z = z_1$$

Eq. (13-26) may be integrated to give

$$\frac{\dot{m}_w}{A} = 0.622 D_w \frac{\rho}{p} \frac{p_s - p_\infty}{z_1} \qquad (13\text{-}27)$$

Evaporation processes in the atmosphere are much more complicated than

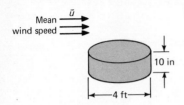

$P_S, P_W$     **Fig. 13-6   Class A standard pan for measurement of evaporation rates.**

indicated by the simple form of Eq. (13-27) for two reasons:

1.  The diffusion process involves substantial turbulent eddy motion so that the diffusion coefficient $D_w$ may vary significantly with the height $z$.
2.  The air is seldom quiescent and wind currents contribute substantially to the evaporation rate.

As in the many convection problems we have encountered previously, the solution to a complicated problem of this sort is frequently obtained by appealing to carefully controlled measurements in search of an empirical relationship to predict evaporation rates.

In this problem, a "standard pan" is used as shown in Fig. 13-6. The mean wind movement is measured 6 in. above the pan rim and water evaporation rates are measured with the pan placed on the ground (*land pan*) or in a body of water (*floating pan*). For the land pan and with a *convectively stable* atmosphere, the evaporation rate has been correlated experimentally [13] as

$$E_{lp} = (0.37 + 0.0041\bar{u})(p_s - p_w)^{0.88} \qquad (13\text{-}28)$$

where $E_{lp}$ = land-pan evaporation, in./day
  $\bar{u}$ = daily wind movement, miles/day measured 6 in. above the pan rim
  $p_s$ = saturation vapor pressure at dry-bulb air temperature 5 ft above ground surface, in. Hg
  $p_w$ = actual vapor pressure of air under temperature and humidity conditions 5 ft above the ground surface, in. Hg

Heat transfer to the pan influences the evaporation rate differently for the ground or water experiments. To convert the pan measurements to those for a natural surface, Eq. (13-28) is multiplied by a *pan coefficient* that is 0.7 for the land pan and 0.8 for the floating pan. If the atmosphere is not convectively stable, vertical density gradients can cause substantial deviations from Eq. (13-28). These problems are discussed in Refs. 10, 11, 12, and 13.

Over extended periods of time, the evaporation from lakes is strongly influenced by the temperature distribution in the lake and the problem must be attacked on the basis of an overall energy balance. Reference 3 may be consulted for an example of such an analysis.

## Example 13-3

A standard land pan is used to measure the evaporation rate in atmospheric air at $100°F$ and 30 percent relative humidity. The mean wind speed is 10 mph. What is the evaporation rate in $lb_m/hr\text{-}ft^2$ on the land under these conditions.

**Solution**   For this calculation we can make use of Eq. (13-28). From thermodynamic steam tables

$$p_s = p_g \qquad \text{at } 100°F = 0.9492 \text{ psia} = 1.933 \text{ in. Hg}$$

$$\text{Relative humidity} = \frac{p_w}{p_s}$$

$$p_w = (0.30)(1.933) = 0.580 \text{ in. Hg}$$

Also, $\bar{u} = 10$ mph $= 240$ miles/day
Equation (13-28) now yields, with the application of the 0.7 factor,

$$E_{lp} = (0.7)[0.37 + (0.0041)(240)](1.933 - 0.580)^{0.88}$$
$$= 1.237 \text{ in./day}$$

Noting that the standard pan has a diameter of 4 ft, the above figure may be used to calculate the mass evaporation rate per unit area as

$$\frac{\dot{m}_w}{A} = \frac{E_{lp}}{12} \rho_w$$

$$= \frac{(1.237)(62.4)}{12} = 6.432 \text{ lb}_m/\text{day-ft}^2$$

$$\frac{\dot{m}_w}{A} = 0.268 \text{ lb}_m/\text{hr-ft}^2$$

As a matter of interest, we might calculate the molecular diffusion rate for water vapor from Eq. (13-27), taking $z_1$ as the 5-ft dimension above the standard pan. Since

$$\rho = \frac{p}{RT}$$

Eq. (13-27) can be written as

$$\frac{\dot{m}_w}{A} = 0.622 \frac{D_w}{RT} \frac{p_s - p_w}{z_1}$$

From Table A-8,

$$D_w = 0.256 \text{ cm}^2/\text{sec} = 0.99 \text{ ft}^2/\text{hr}$$

so that,

$$\frac{\dot{m}_w}{A} = \frac{(0.622)(0.99)(0.9492)(1-0.3)}{(53.35)(560)(5.0)}$$
$$= 2.74 \times 10^{-6} \text{ lb}_m/\text{hr-ft}^2$$

This number is negligibly small in comparison with the previous calculation. This means that in the actual evaporation process, turbulent diffusion and convective transport play dominant roles in comparison with molecular diffusion.

## 13-5□TEMPERATURE DISTRIBUTIONS IN LAKES

The temperature distribution in lakes represents an interesting application of a number of heat-transfer principles. Let us first consider the energy balance in a qualitative way. Basically, the temperature distribution in a deep lake depends on the energy absorbed from solar insolation at the surface, vertical mixing currents, macroscopic convection currents, loss of energy by surface evaporation, and reemission of long wavelength radiation from the surface of the lake.

In general, the temperature distribution in the lake follows a seasonal variation. In the early spring the lake will be nearly isothermal. As the calender advances, the solar heat load on the surface increases and surface absorption causes the surface layers to warm most rapidly. Eddy and convective mixing near the surface create a water layer that is fairly isothermal. This layer is called the *epilimnion*. Beneath this layer, the temperature drops off sharply with depth with the deep water still maintained at its winter temperature. The deep water is called the *hypolimnion*. As the calender moves ahead into summer, the epilimnion becomes warmer and thicker until with the cooler weather of autumn, the surface temperature begins to drop. Then we have a circumstance of warmer water beneath cooler water and substantial natural convection currents act to promote vertical mixing. Eventually, nearly isothermal conditions are again established in winter. In this latter cooling and vertical mixing process we say that the lake "turns over." The region where the steep temperature gradients are observed during the spring-summer heating process is called the *metalimnion*. The point at which the temperature gradient is a maximum is called the *thermocline*. Figure 13-7 illustrates the regions that have been described above.

Thermal gradients and transport phenomena in lakes are extremely important because of their influence on the distribution of nutrients, sediments, pollutants, and biological elements both during the seasons and over extended periods of time. Additional descriptive information is given by Hutchinson [15].

Let us now turn our attention to a consideration of the details of the thermal energy distribution in the lake. The presentation here follows the development of Dake and Harleman [16]. Consider a horizontal fluid layer

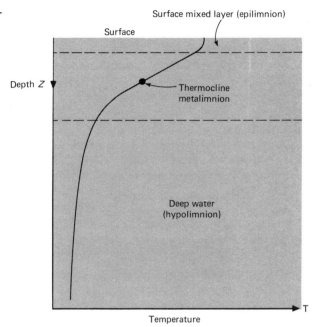

**Fig. 13-7  Classification of temperature regions in deep lakes.**

of thickness $dz$. For the transient heating process the energy balance is

| Heat conducted in (molecular and turbulent) + heat transported in by vertical macroscopic fluid movement = + radiant energy absorbed | Heat conducted out (molecular and turbulent) + heat transported out by macroscopic fluid movement + energy accumulation in the fluid layer with time |
|---|---|

The radiant energy absorbed per unit area may be obtained by differentiating Eq. (13-5). Thus

$$dI_z = -I_s(1 - \beta)ae^{-az}\, dz$$

This is the decrement of the intensity through the distance $dz$. The absorption per unit volume $\dot{q}$ is therefore

$$\dot{q} = I_s a(1 - \beta)e^{-az}$$

The overall energy balance now becomes, for unit horizontal area,

$$-(k + k_t)\frac{\partial T}{\partial z} + \rho c \bar{w} T + I_s a(1 - \beta)e^{-az}\, dz = -(k + k_t)\left[\frac{\partial T}{\partial z} + \frac{\partial^2 T}{\partial z^2}\, dz\right]$$
$$+ \rho c \bar{w}\left(T + \frac{\partial T}{\partial z}\, dz\right) + \rho c\, \frac{\partial T}{\partial \tau}$$

Canceling like terms and setting $\epsilon_h = k_t/\rho c$ as the eddy diffusivity for heat gives

$$\frac{\partial T}{\partial \tau} + \bar{w}\frac{\partial T}{\partial z} = (\epsilon_h + \alpha)\frac{\partial^2 T}{\partial z^2} + \frac{I_s a}{\rho c}(1 - \beta)e^{-az} \qquad (13\text{-}29)$$

In this equation $\bar{w}$ is the mean macroscopic fluid velocity in the vertical direction and $\alpha = k/\rho c$ is the thermal diffusivity. The surface boundary condition may be established by noting that

Surface absorption of    Energy conducted into    Energy lost by evapora-
radiant energy        = lake at surface      + tion and radiation from
                                                          surface

Thus,

$$\beta I_s = -\rho c(\epsilon_h + \alpha)\frac{\partial T}{\partial z}\bigg|_{z=0} + q_L(\tau) \qquad (13\text{-}30)$$

where $q_L(\tau)$ is the energy lost per unit area from the surface as a function of time. If we take an initial condition of uniform temperature, the additional boundary conditions are

$$T = T_0 \qquad \text{at } \tau = 0$$

$$T \to T_0 \qquad \text{and} \qquad \frac{\partial T}{\partial z} \to 0 \qquad \text{as } z \to h$$

where $h$ is the depth of the lake. Outside the epilimnion it has been determined [16, 18] that macroscopic convection and turbulent diffusion can be neglected, thereby reducing Eq. (13-29) to

$$\frac{\partial T}{\partial \tau} = \alpha \frac{\partial^2 T}{\partial z^2} + \frac{a I_s}{\rho c}(1 - \beta)e^{-az} \qquad (13\text{-}31)$$

Dake and Harleman [16] used a superposition technique to obtain a solution to Eq. (13-29) with the simplifying assumptions of:

1. Constant radiant flux at the surface, that is, $I_s = \text{const}$
2. Surface heat losses expressable as a constant fraction $k_L$ of the surface insolation absorbed, that is,

$$q_L = k_L \beta I_s \qquad (13\text{-}32)$$

The solution to Eq. (13-29) is then expressed as

$$T = T_o + T_s + T_i \qquad (13\text{-}33)$$

where $T_o$ = initial uniform temperature
     $T_s$ = temperature distribution due to surface absorbed radiation
     $T_i$ = temperature distribution due to internal absorbed radiation
$T_s$ is obtained by solving Eq. (13-29) without the last term on the right, and with the surface boundary condition given by Eq. (13-30). $T_i$ is obtained as

a solution to Eq. (13-20) with the following boundary and initial conditions.

$$T_i = 0 \quad \text{at } \tau = 0, \text{ for all } z$$
$$T_i = 0 \quad \text{at } z = h, \text{ for all } \tau$$
$$T_i = 0 \quad \text{at } z = 0$$

The solution is rather cumbersome, and we present only the results:

$$T_s = \frac{2\beta(1 - k_L)}{\rho c \alpha} T_s \left[ \left(\frac{\alpha\tau}{\pi}\right)^{\frac{1}{2}} \exp\left(\frac{-z^2}{4\alpha\tau}\right) - \frac{z}{2} \text{erfc}\left(\frac{z^2}{4\alpha\tau}\right)^{\frac{1}{2}} \right] \quad (13\text{-}34)$$

where the complementary error function is defined by

$$\text{erfc } x = 1 - \text{erf } x$$

and

$$\text{erf } x = \frac{2}{\sqrt{\pi}} \int_0^x e^{-n^2} \, d\eta$$

The solution for $T_i$ is

$$T_i = \frac{(1 - \beta)I_s}{\rho c \alpha a} \left[ \text{erfc}\left(\frac{z}{2\sqrt{\alpha\tau}}\right) - e^{-az} \right]$$
$$+ \frac{(1 - \beta)I_s}{2\rho c \alpha a} \exp\left(\alpha\tau a^2\right) \left\{ e^{-az} \text{erfc}\left[ a\sqrt{\alpha\tau} - \frac{z}{2\sqrt{\alpha\tau}} \right] \right.$$
$$\left. - e^{az} \text{erfc}\left[ a\sqrt{\alpha\tau} + \frac{z}{2\sqrt{\alpha\tau}} \right] \right\} \quad (13\text{-}35)$$

These equations are in good agreement with both laboratory and field measurements [16, 17].

## REVIEW QUESTIONS
1. Why is the sky blue?
2. Define albedo.
3. What is meant by the atmospheric greenhouse effect?
4. What is the dry adiabatic lapse rate?
5. What are the conditions for stable and unstable atmospheres in terms of the ambient lapse rates?
6. What is meant by a temperature inversion? How can such a condition develop? What is the significance of the temperature inversion in terms of atmospheric pollution?
7. What is meant by a pseudoadiabatic process?
8. What is meant by the potential temperature of a parcel of air?
9. Why cannot atmospheric evaporation rates be calculated with ordinary molecular diffusion equations?
10. What is meant by a convectively stable atmosphere?
11. What do the following terms mean: epilimnion, metalimnion, hypolimnion, thermocline?
12. What is meant by the phrase "the lake turns over"?

13.    What is meant by the turbidity factor?

## PROBLEMS

13-1.    Calculate the absorption rate for solar radiation on bright fine sand for a solar altitude angle of 50° and a turbidity factor of 3.5.

13-2.    Calculate the ground insolation for a solar altitude of 30° in a smoggy environment with a turbidity factor of 4.0. How does this compare with the ground insolation for clear air at a solar altitude of 90°?

13-3.    Plot the ground solar insolation as a function of turbidity factor for a solar altitude of 80°.

13-4.    Using the spectrum of Fig. 13-2, estimate the fraction of earth-bound radiation that may be transmitted by the atmosphere. What is the magnitude of this transmitted energy?

13-5.    Using the spectrum of Fig. 13-2, estimate the magnitude of the solar radiation flux at the earth's surface. How does this calculation compare with one based on Eq. (13-3)?

13-6.    A pollution-control enthusiast claims that a certain metropolitan area has such a high concentration of atmospheric contaminants that the solar insolation is attenuated by 50 percent. How do you evaluate this claim? Assuming a turbidity factor of 4.5, what solar altitude would be necessary to produce the 50 percent attenuation factor?

13-7.    A light breeze at 5 mph blows across a standard evaporation pan. The atmospheric conditions are 70°F and 40 percent relative humidity. What is the evaporation rate for a land pan in $lb_m$/hr-ft$^2$? What would be the evaporation rate for zero velocity?

13-8.    An evaporation rate of 0.2 $lb_m$/hr-ft$^2$ is experienced for a 10 mph breeze blowing across a standard land pan. What is the relative humidity if the dry bulb (ambient) air temperature is 105°F?

13-9.    Solar radiation arrives at the surface of a lake during the time when the solar altitude is 90°. The lake is quite turbid with an extinction coefficient of 0.25 and the atmosphere has a turbidity factor of 2.8. What fraction of the incident radiation has been absorbed at depths of (a) 0 ft, (b) 0.25 ft, (c) 0.5 ft, (d) 1.0 ft?

13-10.    Calculate the heat-generation rates for the depths given in Prob. 13-9.

13-11.    A 10-mph breeze blows across a lake at 70°F and 50 percent relative humidity. The lake surface temperature is 65°F and $\beta = 0.4$, $a = 0.10$. Calculate the value of $k_L$ for a clear day with a solar altitude of 90°. Assume that the water surface radiates to the sky as if the sky were at −100°F.

13-12.    Construct a plot of temperature and pressure vs. altitude from sea level to 4000 ft for a dry adiabatic lapse rate and an air temperature at the ground of 70°F. Using this pressure profile, plot the temperature a saturated parcel of air would assume if it were to follow a pseudoadiabatic process from ground level to 4,000 ft. How much moisture would have been condensed when the parcel reaches 4,000 ft?

# REFERENCES

1. Kondratyev, K. Y.: "Radiative Heat Exchange in the Atmosphere," Pergamon Press, New York, 1965.
2. Geiger, R.: "The Climate Near the Ground," rev. ed., Harvard University Press, Cambridge, Mass., 1965.
3. Eagleson, P. S.: "Dynamic Hydrology," McGraw-Hill Book Company, New York, 1970.
4. Raphael, J. M.: Prediction of Temperature in Rivers and Reservoirs, *Proc. ASCE Power Div.*, no. PO2, Paper 3200, July, 1962.
5. Briggs, G. A.: Plume Rise: A Critical Survey, Air Resources Atmospheric Turbulence and Diffusion Laboratory, Environmental Science Services Administration, Oak Ridge, Tennessee, May, 1969.
6. Scorer, R. S.: "Air Pollution," Pergamon Press, New York, 1968.
7. Pasquill, F.: "Atmospheric Diffusion," D. Van Nostrand Company, Inc., Princeton, N.J., 1962.
8. Holman, J. P.: "Thermodynamics," McGraw-Hill Book Company, New York, 1969.
9. Magill, P. L., ed.: "Air Pollution Handbook," McGraw-Hill Book Company, New York, 1956.
10. Instructions for Climatological Observers, *U.S. Dept. Commerce Weather Bur. Circ. B.*, 10th ed. rev., October, 1955.
11. Water-loss Investigations: vol. 1, *Lake Hefner Studies Tech. Rept., U.S. Geol. Surv. Circ.* 229, 1952.
12. Nordenson, T. J., and D. R. Baker: Comparative Evaluation of Evaporation Instruments, *J. Geophys. Res.*, vol. 67, no. 2, p. 671, February, 1962.
13. Kohler, M. A., T. J. Nordenson, and W. E. Fox: Evaporation from Pans and Lakes, *U.S. Dept. Commerce Weather Bur. Res. Paper* 38, May, 1955.
14. Kraus, E. B., and C. Rooth: Temperature and Steady State Vertical Heat Flux in the Ocean Surface Layers, *Tellus*, vol. 13, pp. 231–239, 1961.
15. Hutchinson, G. E.: "A Treatise on Limnology," vol. 1, John Wiley & Sons, Inc., 1957.
16. Dake, J. M. K., and D. R. F. Harleman: Thermal Stratification in Lakes: Analytical and Laboratory Studies, *Water Resources Research*, vol. 5, no. 2, p. 484, April, 1969.
17. Goldman, C. R., and C. R. Carter: An Investigation by Rapid Carbon—14 Bioassay of Factors Affecting the Cultural Entrophication of Lake Tahoe, California, *J. Water Pollution Control Federation*, p. 1044, July, 1965.
18. Bachmann, R. W., and C. R. Goldman: Hypolimnetic Heating in Castle Lake, California, *Limnol. Oceanog.*, vol. 10, p. 2, April, 1965.
19. Sutton, O. G.: "Micrometeorology," McGraw-Hill Book Company, New York, 1956.
20. Haltiner, G. J., and F. L. Martin: "Dynamical and Physical Meteorology," McGraw-Hill Book Company, New York, 1957.

# APPENDIX A

## Table A-1   The Error Function

| $\dfrac{x}{2\sqrt{\alpha\tau}}$ | $\text{erf}\,\dfrac{x}{2\sqrt{\alpha\tau}}$ | $\dfrac{x}{2\sqrt{\alpha\tau}}$ | $\text{erf}\,\dfrac{x}{2\sqrt{\alpha\tau}}$ | $\dfrac{x}{2\sqrt{\alpha\tau}}$ | $\text{erf}\,\dfrac{x}{2\sqrt{\alpha\tau}}$ |
|---|---|---|---|---|---|
| 0.00 | 0.00000 | 0.76 | 0.71754 | 1.52 | 0.96841 |
| 0.02 | 0.02256 | 0.78 | 0.73001 | 1.54 | 0.97059 |
| 0.04 | 0.04511 | 0.80 | 0.74210 | 1.56 | 0.97263 |
| 0.06 | 0.06762 | 0.82 | 0.75381 | 1.58 | 0.97455 |
| 0.08 | 0.09008 | 0.84 | 0.76514 | 1.60 | 0.97635 |
| 0.10 | 0.11246 | 0.86 | 0.77610 | 1.62 | 0.97804 |
| 0.12 | 0.13476 | 0.88 | 0.78669 | 1.64 | 0.97962 |
| 0.14 | 0.15695 | 0.90 | 0.79691 | 1.66 | 0.98110 |
| 0.16 | 0.17901 | 0.92 | 0.80677 | 1.68 | 0.98249 |
| 0.18 | 0.20094 | 0.94 | 0.81627 | 1.70 | 0.98379 |
| 0.20 | 0.22270 | 0.96 | 0.82542 | 1.72 | 0.98500 |
| 0.22 | 0.24430 | 0.98 | 0.83423 | 1.74 | 0.98613 |
| 0.24 | 0.26570 | 1.00 | 0.84270 | 1.76 | 0.98719 |
| 0.26 | 0.28690 | 1.02 | 0.85084 | 1.78 | 0.98817 |
| 0.28 | 0.30788 | 1.04 | 0.85865 | 1.80 | 0.98909 |
| 0.30 | 0.32863 | 1.06 | 0.86614 | 1.82 | 0.98994 |
| 0.32 | 0.34913 | 1.08 | 0.87333 | 1.84 | 0.99074 |
| 0.34 | 0.36936 | 1.10 | 0.88020 | 1.86 | 0.99147 |
| 0.36 | 0.38933 | 1.12 | 0.88079 | 1.88 | 0.99216 |
| 0.38 | 0.40901 | 1.14 | 0.89308 | 1.90 | 0.99279 |
| 0.40 | 0.42839 | 1.16 | 0.89910 | 1.92 | 0.99338 |
| 0.42 | 0.44749 | 1.18 | 0.90484 | 1.94 | 0.99392 |
| 0.44 | 0.46622 | 1.20 | 0.91031 | 1.96 | 0.99443 |
| 0.46 | 0.48466 | 1.22 | 0.91553 | 1.98 | 0.99489 |
| 0.48 | 0.50275 | 1.24 | 0.92050 | 2.00 | 0.995322 |
| 0.50 | 0.52050 | 1.26 | 0.92524 | 2.10 | 0.997020 |
| 0.52 | 0.53790 | 1.28 | 0.92973 | 2.20 | 0.998137 |
| 0.54 | 0.55494 | 1.30 | 0.93401 | 2.30 | 0.998857 |
| 0.56 | 0.57162 | 1.32 | 0.93806 | 2.40 | 0.999311 |
| 0.58 | 0.58792 | 1.34 | 0.94191 | 2.50 | 0.999593 |
| 0.60 | 0.60386 | 1.36 | 0.94556 | 2.60 | 0.999764 |
| 0.62 | 0.61941 | 1.38 | 0.94902 | 2.70 | 0.999866 |
| 0.64 | 0.63459 | 1.40 | 0.95228 | 2.80 | 0.999925 |
| 0.66 | 0.64938 | 1.42 | 0.95538 | 2.90 | 0.999959 |
| 0.68 | 0.66278 | 1.44 | 0.95830 | 3.00 | 0.999978 |
| 0.70 | 0.67780 | 1.46 | 0.96105 | 3.20 | 0.999994 |
| 0.72 | 0.69143 | 1.48 | 0.96365 | 3.40 | 0.999998 |
| 0.74 | 0.70468 | 1.50 | 0.96610 | 3.60 | 1.000000 |

# Table A-2  Property Values for Metals†

| Metals | Properties at 68°F | | | | $k$, thermal conductivity, Btu/hr-ft-°F | | | | | | | | | |
|---|---|---|---|---|---|---|---|---|---|---|---|---|---|---|
| | $\rho$, lb$_m$/ft$^3$ | $c_p$, Btu/lb$_m$-°F | $k$, Btu/hr-ft-°F | $\alpha$, ft$^2$/hr | −148°F −100°C | 32°F 0°C | 212°F 100°C | 392°F 200°C | 572°F 300°C | 752°F 400°C | 1112°F 600°C | 1472°F 800°C | 1832°F 1000°C | 2192°F 1200°C |
| *Aluminum* | | | | | | | | | | | | | | |
| Pure | 169 | 0.214 | 118 | 3.665 | 124 | 117 | 119 | 124 | 132 | 144 | | | | |
| Al-Cu (duralumin), 94–96 % Al, 3–5 % Cu, trace Mg | | | | | | | | | | | | | | |
| Al-Si (silumin, copper-bearing), 86.5 % Al, 12.5 % Si, 1 % Cu | 174 | 0.211 | 95 | 2.580 | 73 | 92 | 105 | 112 | | | | | | |
| Cu | 166 | 0.207 | 79 | 2.311 | 69 | 79 | 83 | 88 | 93 | | | | | |
| *Lead* | 710 | 0.031 | 20 | 0.924 | 21.3 | 20.3 | 19.3 | 18.2 | 17.2 | | | | | |
| *Iron* | | | | | | | | | | | | | | |
| Pure | 493 | 0.108 | 42 | 0.785 | 50 | 42 | 39 | 36 | 32 | 28 | 23 | 21 | 20 | 21 |
| Wrought iron 0.5 % C | 490 | 0.11 | 34 | 0.634 | | 34 | 33 | 30 | 28 | 26 | 21 | 19 | 19 | 19 |
| Steel (C max 1.5 %): | | | | | | | | | | | | | | |
| Carbon steel: | | | | | | | | | | | | | | |
| 0.5 % C | 489 | 0.111 | 31 | 0.570 | | 32 | 30 | 28 | 26 | 24 | 20 | 18 | 17 | 18 |
| 1.0 % C | 487 | 0.113 | 25 | 0.452 | | 25 | 25 | 24 | 23 | 21 | 19 | 17 | 16 | 17 |
| 1.5 % C | 484 | 0.116 | 21 | 0.376 | | 21 | 21 | 21 | 20 | 19 | 18 | 16 | 16 | 17 |
| Nickel steel: | | | | | | | | | | | | | | |
| 0 % Ni | 493 | 0.108 | 42 | 0.785 | | | | | | | | | | |
| 20 % Ni | 499 | 0.11 | 11 | 0.204 | | | | | | | | | | |
| 40 % Ni | 510 | 0.11 | 6 | 0.108 | | | | | | | | | | |
| 80 % Ni | 538 | 0.11 | 20 | 0.344 | | | | | | | | | | |
| Invar: | | | | | | | | | | | | | | |
| 36 % Ni | 508 | 0.11 | 6.2 | 0.108 | | | | | | | | | | |
| Chrome steel: | | | | | | | | | | | | | | |
| 0 % Cr | 493 | 0.108 | 42 | 0.785 | 50 | 42 | 39 | 36 | 32 | 28 | 23 | 21 | 20 | 21 |
| 1 % Cr | 491 | 0.11 | 35 | 0.645 | | 36 | 32 | 30 | 27 | 24 | 21 | 19 | 19 | |
| 5 % Cr | 489 | 0.11 | 23 | 0.430 | | 23 | 22 | 21 | 21 | 19 | 17 | 17 | 17 | 17 |
| 20 % Cr | 480 | 0.11 | 13 | 0.258 | | 13 | 13 | 13 | 13 | 14 | 14 | 15 | 17 | 17 |

| Metal | ρ | c | k | α | | | | | | | | | |
|---|---|---|---|---|---|---|---|---|---|---|---|---|---|
| **Chrome nickel:** | | | | | | | | | | | | | |
| 15 % Cr, 10 % Ni | 491 | 0.11 | 11 | 0.204 | 9.4 | 10 | 10 | 11 | 11 | 13 | 15 | 18 | |
| 18 % Cr, 8 % Ni (V2A) | 488 | 0.11 | 9.4 | 0.172 | | | | | | | | | |
| 25 % Cr, 20 % Ni | 491 | 0.11 | 7.4 | 0.140 | | | | | | | | | |
| **Tungsten steel:** | | | | | | | | | | | | | |
| 0 % W | 493 | 0.108 | 42 | 0.785 | | | | | | | | | |
| 1 % W | 494 | 0.107 | 38 | 0.720 | | | | | | | | | |
| 5 % W | 504 | 0.104 | 31 | 0.591 | | | | | | | | | |
| 10 % W | 519 | 0.100 | 28 | 0.527 | | | | | | | | | |
| *Copper* | | | | | | | | | | | | | |
| Pure | 559 | 0.0915 | 223 | 4.353 | 235 | 223 | 219 | 216 | 213 | 210 | 204 | | |
| Aluminum bronze, 95 % Cu, 5 % Al | 541 | 0.098 | 48 | 0.903 | | | | | | | | | |
| Bronze, 75 % Cu, 25 % Sn | 541 | 0.082 | 15 | 0.333 | | | | | | | | | |
| Red brass, 85 % Cu, 9 % Sn, 6 % Zn | 544 | 0.092 | 35 | 0.699 | 34 | 41 | | | | | | | |
| Brass, 70 % Cu, 30 % Zn | 532 | 0.092 | 64 | 1.322 | 51 | 74 | 83 | 85 | 85 | | | | |
| German silver, 62 % Cu, 15 % Ni, 22 % Zn | 538 | 0.094 | 14.4 | 0.290 | 11.1 | 18 | 23 | 26 | 28 | | | | |
| Constantan, 60 % Cu, 40 % Ni | 557 | 0.098 | 13.1 | 0.237 | 12 | 12.8 | 15 | | | | | | |
| *Magnesium* | | | | | | | | | | | | | |
| Pure | 109 | 0.242 | 99 | 3.762 | 103 | 99 | 97 | 94 | 91 | 85 | 85 | | |
| Mg-Al (electrolytic), 6–8 % Al, 1–2 % Zn | 113 | 0.24 | 38 | 1.397 | 30 | 36 | 43 | 48 | | | | | |
| *Molybdenum* | 638 | 0.060 | 71 | 2.074 | 80 | 72 | 68 | 66 | 63 | 61 | 59 | 57 | 53 |
| *Nickel* | | | | | | | | | | | | | |
| Pure (99.9 %) | 556 | 0.1065 | 52 | 0.882 | 60 | 54 | 48 | 42 | 37 | 34 | | | |
| Nickel chrome, 90 % Ni, 10 % Cr | 541 | 0.106 | 10 | 0.172 | 9.9 | 10.9 | 12.1 | 13.2 | 14.2 | | | | |
| *Silver*, pure | 657 | 0.0559 | 235 | 6.418 | 242 | 237 | 240 | 216 | 209 | 208 | | | |
| *Tungsten* | 1208 | 0.0321 | 94 | 2.430 | 96 | 87 | 82 | 77 | 73 | 65 | | | |
| *Zinc*, pure | 446 | 0.0918 | 64.8 | 1.591 | 66 | 65 | 63 | 61 | 58 | 54 | | | |
| *Tin*, pure | 456 | 0.0541 | 37 | 1.505 | 43 | 41 | 38.1 | 34 | 33 | | | | |

† From E. R. G. Eckert and R. M. Drake, "Heat and Mass Transfer," 2d ed., McGraw-Hill Book Company, New York, 1959.

## Table A-3    Properties of Nonmetals†

| Substance | Temperature, °F | $k$, Btu/hr-ft-°F |
|---|---|---|
| *Structural and heat-resistant materials* | | |
| Asphalt | 68–132 | 0.43–0.44 |
| Brick: | | |
|    Building brick, common | 68 | 0.40 |
|    Building brick, face | | 0.76 |
|    Carborundum brick | 1110 | 10.7 |
| | 2550 | 6.4 |
|    Chrome brick | 392 | 1.34 |
| | 1022 | 1.43 |
| | 1652 | 1.15 |
|    Diatomaceous earth, molded and fired | 400 | 0.14 |
| | 1600 | 0.18 |
|    Fireclay brick (burnt 2426°F) | 932 | 0.60 |
| | 1472 | 0.62 |
| | 2012 | 0.63 |
|    Fireclay brick (burnt 2642°F) | 932 | 0.74 |
| | 1472 | 0.79 |
| | 2012 | 0.81 |
|    Fireclay brick (Missouri) | 392 | 0.58 |
| | 1112 | 0.85 |
| | 2552 | 1.02 |
|    Magnesite | 400 | 2.2 |
| | 1200 | 1.6 |
| | 2200 | 1.1 |
| Cement, portland | | 0.17 |
| Cement, mortar | 75 | 0.67 |
| Concrete, cinder | 75 | 0.44 |
| Concrete, stone 1-2-4 mix | 69 | 0.79 |
| Glass, window | 68 | 0.45 (avg) |
| Glass, borosilicate | 86–167 | 0.63 |
| Plaster, gypsum | 70 | 0.28 |
| Plaster, metal lath | 70 | 0.27 |
| Plaster, wood lath | 70 | 0.16 |
| *Stone* | | |
| Granite | | 1.0–2.3 |
| Limestone | 210–570 | 0.73–0.77 |
| Marble | | 1.20–1.70 |
| Sandstone | 104 | 1.06 |
| *Wood (across the grain)* | | |
| Balsa, 8.8 lb/ft³ | 86 | 0.032 |
| Cypress | 86 | 0.056 |
| Fir | 75 | 0.063 |
| Maple or oak | 86 | 0.096 |
| Yellow pine | 75 | 0.085 |
| White pine | 86 | 0.065 |

## Table A-3    Properties of Nonmetals† (*Continued*)

| Substance | Temperature, °F | $k$, Btu/hr-ft-°F |
|---|---|---|
| *Insulating material* | | |
| Asbestos: | | |
| Asbestos-cement boards | 68 | 0.43 |
| Asbestos sheets | 124 | 0.096 |
| Asbestos felt, 40 laminations/in. | 100 | 0.033 |
| | 300 | 0.040 |
| | 500 | 0.048 |
| Asbestos felt, 20 laminations/in. | 100 | 0.045 |
| | 300 | 0.055 |
| | 500 | 0.065 |
| Asbestos, corrugated, 4 plies/in. | 100 | 0.05 |
| | 200 | 0.058 |
| | 300 | 0.069 |
| Asbestos cement | | 1.2 |
| Asbestos, loosely packed | −50 | 0.086 |
| | 32 | 0.089 |
| | 210 | 0.093 |
| Balsam wool, 2.2 lb/ft³ | 90 | 0.023 |
| Cardboard, corrugated | | 0.037 |
| Celotex | 90 | 0.028 |
| Corkboard, 10 lb/ft³ | 86 | 0.025 |
| Cork, regranulated | 90 | 0.026 |
| Cork, ground | 90 | 0.025 |
| Diatomaceous earth (Sil-o-cel) | 32 | 0.035 |
| Felt, hair | 86 | 0.021 |
| Felt, wool | 86 | 0.03 |
| Fiber insulating board | 70 | 0.028 |
| Glass wool, 1.5 lb/ft³ | 75 | 0.022 |
| Insulex, dry | 90 | 0.037 |
| | | 0.083 |
| Kapok | 86 | 0.020 |
| Magnesia, 85% | 100 | 0.039 |
| | 200 | 0.041 |
| | 300 | 0.043 |
| | 400 | 0.046 |
| Rock wool, 10 lb/ft³ | 90 | 0.023 |
| Rock wool, loosely packed | 300 | 0.039 |
| | 500 | 0.050 |
| Sawdust | 75 | 0.034 |
| Silica aerogel | 90 | 0.014 |
| Wood shavings | 75 | 0.034 |

† From A. I. Brown and S. M. Marco, "Introduction to Heat Transfer," 3d ed., McGraw-Hill Book Company, New York, 1958.

## Table A-4  Properties of Saturated Liquids†

| $T,$ °F | $\rho,$ $\dfrac{\text{lb}_m}{\text{ft}^3}$ | $c_p,$ $\dfrac{\text{Btu}}{\text{lb}_m\text{-}°\text{F}}$ | $\nu, \dfrac{\text{ft}^2}{\text{sec}}$ | $k,$ $\dfrac{\text{Btu}}{\text{hr-ft-}°\text{F}}$ | $\alpha, \dfrac{\text{ft}^2}{\text{hr}}$ | Pr | $\beta, \dfrac{1}{R}$ |
|---|---|---|---|---|---|---|---|
| **Ammonia, NH₃** | | | | | | | |
| −58 | 43.93 | 1.066 | $0.468 \times 10^{-5}$ | 0.316 | $6.75 \times 10^{-3}$ | 2.60 | |
| −40 | 43.18 | 1.067 | 0.437 | 0.316 | 6.88 | 2.28 | |
| −22 | 42.41 | 1.069 | 0.417 | 0.317 | 6.98 | 2.15 | |
| −4 | 41.62 | 1.077 | 0.410 | 0.316 | 7.05 | 2.09 | |
| 14 | 40.80 | 1.090 | 0.407 | 0.314 | 7.07 | 2.07 | |
| 32 | 39.96 | 1.107 | 0.402 | 0.312 | 7.05 | 2.05 | |
| 50 | 39.09 | 1.126 | 0.396 | 0.307 | 6.98 | 2.04 | |
| 68 | 38.19 | 1.146 | 0.386 | 0.301 | 6.88 | 2.02 | $1.36 \times 10^{-3}$ |
| 86 | 37.23 | 1.168 | 0.376 | 0.293 | 6.75 | 2.01 | |
| 104 | 36.27 | 1.194 | 0.366 | 0.285 | 6.59 | 2.00 | |
| 122 | 35.23 | 1.222 | 0.355 | 0.275 | 6.41 | 1.99 | |
| **Carbon dioxide, CO₂** | | | | | | | |
| −58 | 72.19 | 0.44 | $0.128 \times 10^{-5}$ | 0.0494 | $1.558 \times 10^{-3}$ | 2.96 | |
| −40 | 69.78 | 0.45 | 0.127 | 0.0584 | 1.864 | 2.46 | |
| −22 | 67.22 | 0.47 | 0.126 | 0.0645 | 2.043 | 2.22 | |
| −4 | 64.45 | 0.49 | 0.124 | 0.0665 | 2.110 | 2.12 | |
| 14 | 61.39 | 0.52 | 0.122 | 0.0635 | 1.989 | 2.20 | |
| 32 | 57.87 | 0.59 | 0.117 | 0.0604 | 1.774 | 2.38 | |
| 50 | 53.69 | 0.75 | 0.109 | 0.0561 | 1.398 | 2.80 | |
| 68 | 48.23 | 1.2 | 0.098 | 0.0504 | 0.860 | 4.10 | $3.67 \times 10^{-3}$ |
| 86 | 37.32 | 8.7 | 0.086 | 0.0406 | 0.108 | 28.7 | |
| **Sulfur dioxide, SO₂** | | | | | | | |
| −58 | 97.44 | 0.3247 | $0.521 \times 10^{-5}$ | 0.140 | $4.42 \times 10^{-3}$ | 4.24 | |
| −40 | 95.94 | 0.3250 | 0.456 | 0.136 | 4.38 | 3.74 | |
| −22 | 94.43 | 0.3252 | 0.399 | 0.133 | 4.33 | 3.31 | |
| −4 | 92.93 | 0.3254 | 0.349 | 0.130 | 4.29 | 2.93 | |
| 14 | 91.37 | 0.3255 | 0.310 | 0.126 | 4.25 | 2.62 | |
| 32 | 89.80 | 0.3257 | 0.277 | 0.122 | 4.19 | 2.38 | |
| 50 | 88.18 | 0.3259 | 0.250 | 0.118 | 4.13 | 2.18 | |
| 68 | 86.55 | 0.3261 | 0.226 | 0.115 | 4.07 | 2.00 | $1.08 \times 10^{-3}$ |
| 86 | 84.86 | 0.3263 | 0.204 | 0.111 | 4.01 | 1.83 | |
| 104 | 82.98 | 0.3266 | 0.186 | 0.107 | 3.95 | 1.70 | |
| 122 | 81.10 | 0.3268 | 0.174 | 0.102 | 3.87 | 1.61 | |

## Table A-4    Properties of Saturated Liquids† (*Continued*)

| $T$, °F | $\rho$, $\dfrac{lb_m}{ft^3}$ | $c_p$, $\dfrac{Btu}{lb_m\text{-}°F}$ | $\nu$, $\dfrac{ft^2}{sec}$ | $k$, $\dfrac{Btu}{hr\text{-}ft\text{-}°F}$ | $\alpha$, $\dfrac{ft^2}{hr}$ | Pr | $\beta$, $\dfrac{1}{R}$ |
|---|---|---|---|---|---|---|---|
| \multicolumn{8}{l}{Dichlorodifluoromethane (Freon), $CCl_2F_2$} |

Dichlorodifluoromethane (Freon), $CCl_2F_2$

| $T$, °F | $\rho$ | $c_p$ | $\nu$ | $k$ | $\alpha$ | Pr | $\beta$ |
|---|---|---|---|---|---|---|---|
| −58 | 96.56 | 0.2090 | $0.334 \times 10^{-5}$ | 0.039 | $1.94 \times 10^{-3}$ | 6.2 | $1.4 \times 10^{-4}$ |
| −40 | 94.81 | 0.2113 | 0.300 | 0.040 | 1.99 | 5.4 | |
| −22 | 92.99 | 0.2139 | 0.272 | 0.040 | 2.04 | 4.8 | |
| −4 | 91.18 | 0.2167 | 0.253 | 0.041 | 2.09 | 4.4 | |
| 14 | 89.24 | 0.2198 | 0.238 | 0.042 | 2.13 | 4.0 | |
| 32 | 87.24 | 0.2232 | 0.230 | 0.042 | 2.16 | 3.8 | |
| 50 | 85.17 | 0.2268 | 0.219 | 0.042 | 2.17 | 3.6 | |
| 68 | 83.04 | 0.2307 | 0.213 | 0.042 | 2.17 | 3.5 | |
| 86 | 80.85 | 0.2349 | 0.209 | 0.041 | 2.17 | 3.5 | |
| 104 | 78.48 | 0.2393 | 0.206 | 0.040 | 2.15 | 3.5 | |
| 122 | 75.91 | 0.2440 | 0.204 | 0.039 | 2.11 | 3.5 | |

Glycerin, $C_2H_3(OH)_3$

| $T$, °F | $\rho$ | $c_p$ | $\nu$ | $k$ | $\alpha$ | Pr | $\beta$ |
|---|---|---|---|---|---|---|---|
| 32 | 79.66 | 0.540 | 0.0895 | 0.163 | $3.84 \times 10^{-3}$ | $84.7 \times 10^3$ | |
| 50 | 79.29 | 0.554 | 0.0323 | 0.164 | 3.74 | 31.0 | |
| 68 | 78.91 | 0.570 | 0.0127 | 0.165 | 3.67 | 12.5 | $0.28 \times 10^{-3}$ |
| 86 | 78.54 | 0.584 | 0.0054 | 0.165 | 3.60 | 5.38 | |
| 104 | 78.16 | 0.600 | 0.0024 | 0.165 | 3.54 | 2.45 | |
| 122 | 77.72 | 0.617 | 0.0016 | 0.166 | 3.46 | 1.63 | |

Ethylene glycol, $C_2H_4(OH_2)$

| $T$, °F | $\rho$ | $c_p$ | $\nu$ | $k$ | $\alpha$ | Pr | $\beta$ |
|---|---|---|---|---|---|---|---|
| 32 | 70.59 | 0.548 | $61.92 \times 10^{-5}$ | 0.140 | $3.62 \times 10^{-3}$ | 615 | |
| 68 | 69.71 | 0.569 | 20.64 | 0.144 | 3.64 | 204 | $0.36 \times 10^{-3}$ |
| 104 | 68.76 | 0.591 | 9.35 | 0.148 | 3.64 | 93 | |
| 140 | 67.90 | 0.612 | 5.11 | 0.150 | 3.61 | 51 | |
| 176 | 67.27 | 0.633 | 3.21 | 0.151 | 3.57 | 32.4 | |
| 212 | 66.08 | 0.655 | 2.18 | 0.152 | 3.52 | 22.4 | |

Engine oil (unused)

| $T$, °F | $\rho$ | $c_p$ | $\nu$ | $k$ | $\alpha$ | Pr | $\beta$ |
|---|---|---|---|---|---|---|---|
| 32 | 56.13 | 0.429 | 0.0461 | 0.085 | $3.53 \times 10^{-3}$ | 47,100 | |
| 68 | 55.45 | 0.449 | 0.0097 | 0.084 | 3.38 | 10,400 | $0.39 \times 10^{-3}$ |
| 104 | 54.69 | 0.469 | 0.0026 | 0.083 | 3.23 | 2,870 | |
| 140 | 53.94 | 0.489 | $0.803 \times 10^{-3}$ | 0.081 | 3.10 | 1,050 | |
| 176 | 53.19 | 0.509 | 0.404 | 0.080 | 2.98 | 490 | |
| 212 | 52.44 | 0.530 | 0.219 | 0.079 | 2.86 | 276 | |
| 248 | 51.75 | 0.551 | 0.133 | 0.078 | 2.75 | 175 | |
| 284 | 51.00 | 0.572 | 0.086 | 0.077 | 2.66 | 116 | |
| 320 | 50.31 | 0.593 | 0.060 | 0.076 | 2.57 | 84 | |

## Table A-4   Properties of Saturated Liquids† (Continued)

| $T,$ °F | $\rho,$ $\dfrac{lb_m}{ft^3}$ | $c_p,$ $\dfrac{Btu}{lb_m\text{-}°F}$ | $\nu, \dfrac{ft^2}{sec}$ | $k,$ $\dfrac{Btu}{hr\text{-}ft\text{-}°F}$ | $\alpha, \dfrac{ft^2}{hr}$ | Pr | $\beta, \dfrac{1}{R}$ |
|---|---|---|---|---|---|---|---|

Mercury, Hg

| 32 | 850.78 | 0.0335 | $0.133 \times 10^{-5}$ | 4.74 | $166.6 \times 10^{-3}$ | 0.0288 | |
| 68 | 847.71 | 0.0333 | 0.123 | 5.02 | 178.5 | 0.0249 | $1.01 \times 10^{-4}$ |
| 122 | 843.14 | 0.0331 | 0.112 | 5.43 | 194.6 | 0.0207 | |
| 212 | 835.57 | 0.0328 | 0.0999 | 6.07 | 221.5 | 0.0162 | |
| 302 | 828.06 | 0.0326 | 0.0918 | 6.64 | 246.2 | 0.0134 | |
| 392 | 820.61 | 0.0375 | 0.0863 | 7.13 | 267.7 | 0.0116 | |
| 482 | 813.16 | 0.0324 | 0.0823 | 7.55 | 287.0 | 0.0103 | |
| 600 | 802 | 0.032 | 0.0724 | 8.10 | 316 | 0.0083 | |

† From E. R. G. Eckert and R. M. Drake, "Heat and Mass Transfer," 2d ed., McGraw-Hill Book Company, New York, 1959.

## Table A-5   Properties of Air at Atmospheric Pressure†

The listed properties are not strongly pressure-dependent and may be used over a fairly wide range of pressures.

| $T,$ °F | $\mu, \dfrac{lb_m}{hr\text{-}ft}$ | $k, \dfrac{Btu}{hr\text{-}ft\text{-}°F}$ | $c_p, \dfrac{Btu}{lb_m\text{-}°F}$ | Pr |
|---|---|---|---|---|
| −100 | 0.0319 | 0.0104 | 0.239 | 0.739 |
| −50 | 0.0358 | 0.0118 | 0.239 | 0.729 |
| 0 | 0.0394 | 0.0131 | 0.240 | 0.718 |
| 50 | 0.0427 | 0.0143 | 0.240 | 0.712 |
| 100 | 0.0459 | 0.0157 | 0.240 | 0.706 |
| 150 | 0.0484 | 0.0167 | 0.241 | 0.699 |
| 200 | 0.0519 | 0.0181 | 0.241 | 0.693 |
| 250 | 0.0547 | 0.0192 | 0.242 | 0.690 |
| 300 | 0.0574 | 0.0203 | 0.243 | 0.686 |
| 400 | 0.0626 | 0.0225 | 0.245 | 0.681 |
| 500 | 0.0675 | 0.0246 | 0.248 | 0.680 |
| 600 | 0.0721 | 0.0265 | 0.250 | 0.680 |
| 700 | 0.0765 | 0.0284 | 0.254 | 0.682 |
| 800 | 0.0806 | 0.0303 | 0.257 | 0.684 |
| 900 | 0.0846 | 0.0320 | 0.260 | 0.687 |
| 1000 | 0.0884 | 0.0337 | 0.263 | 0.690 |

† From *Natl. Bur. Std.* (*U.S.*) *Circ.* 564, 1955.

## Table A-6    Properties of Gases at Atmospheric Pressure†

| $T$, °F | $\rho$, $\dfrac{lb_m}{ft^3}$ | $c_p$, $\dfrac{Btu}{lb_m\text{-}°F}$ | $\mu$, $\dfrac{lb_m}{sec\text{-}ft}$ | $\nu$, $\dfrac{ft^2}{sec}$ | $k$, $\dfrac{Btu}{hr\text{-}ft\text{-}°F}$ | $\alpha$, $\dfrac{ft^2}{hr}$ | Pr |
|---|---|---|---|---|---|---|---|
| **Helium** | | | | | | | |
| −200 | 0.211 | 1.242 | $84.3 \times 10^{-7}$ | $39.95 \times 10^{-5}$ | 0.0536 | 2.044 | 0.70 |
| −100 | 0.0152 | 1.242 | 105.2 | 69.30 | 0.0680 | 3.599 | 0.694 |
| 0 | 0.0119 | 1.242 | 122.1 | 102.8 | 0.0784 | 5.299 | 0.70 |
| 200 | 0.00829 | 1.242 | 154.9 | 186.9 | 0.0977 | 9.490 | 0.71 |
| 400 | 0.00637 | 1.242 | 184.8 | 289.9 | 0.114 | 14.40 | 0.72 |
| 600 | 0.00517 | 1.242 | 209.2 | 404.5 | 0.130 | 20.21 | 0.72 |
| 800 | 0.00439 | 1.242 | 233.5 | 531.9 | 0.145 | 25.81 | 0.72 |
| 1000 | 0.00376 | 1.242 | 256.5 | 682.5 | 0.159 | 34.00 | 0.72 |
| **Hydrogen** | | | | | | | |
| −190 | 0.01022 | 3.010 | $3.760 \times 10^{-6}$ | $36.79 \times 10^{-5}$ | 0.0567 | 1.84 | 0.718 |
| −100 | 0.00766 | 3.234 | 4.578 | 59.77 | 0.741 | 2.99 | 0.719 |
| −10 | 0.00613 | 3.358 | 5.321 | 86.80 | 0.0902 | 4.38 | 0.713 |
| 80 | 0.00511 | 3.419 | 6.023 | 117.9 | 0.105 | 6.02 | 0.706 |
| 170 | 0.00438 | 3.448 | 6.689 | 152.7 | 0.119 | 7.87 | 0.697 |
| 260 | 0.00383 | 3.461 | 7.300 | 190.6 | 0.132 | 9.95 | 0.690 |
| 350 | 0.00341 | 3.463 | 7.915 | 232.1 | 0.145 | 12.26 | 0.682 |
| 440 | 0.00307 | 3.465 | 8.491 | 276.6 | 0.157 | 14.79 | 0.675 |
| 530 | 0.00279 | 3.471 | 9.055 | 324.6 | 0.169 | 17.50 | 0.668 |
| 620 | 0.00255 | 3.472 | 9.599 | 376.4 | 0.182 | 20.56 | 0.664 |
| 800 | 0.00218 | 3.481 | 10.68 | 489.9 | 0.203 | 26.75 | 0.659 |
| 980 | 0.00191 | 3.505 | 11.69 | 612 | 0.222 | 33.18 | 0.664 |
| 1160 | 0.00170 | 3.540 | 12.62 | 743 | 0.238 | 39.59 | 0.676 |
| **Oxygen** | | | | | | | |
| −190 | 0.1635 | 0.2192 | $7.721 \times 10^{-6}$ | $4.722 \times 10^{-5}$ | 0.00790 | 0.2204 | 0.773 |
| −100 | 0.1221 | 0.2181 | 9.979 | 9.173 | 0.01054 | 0.3958 | 0.745 |
| −10 | 0.0975 | 0.2187 | 12.01 | 12.32 | 0.01305 | 0.6120 | 0.725 |
| 80 | 0.0812 | 0.2198 | 13.86 | 17.07 | 0.01546 | 0.8662 | 0.709 |
| 170 | 0.0695 | 0.2219 | 15.56 | 22.39 | 0.01774 | 1.150 | 0.702 |
| 260 | 0.0609 | 0.2250 | 17.16 | 28.18 | 0.02000 | 1.460 | 0.695 |
| 350 | 0.0542 | 0.2285 | 18.66 | 34.43 | 0.02212 | 1.786 | 0.694 |
| 440 | 0.0487 | 0.2322 | 20.10 | 41.27 | 0.02411 | 2.132 | 0.697 |
| 530 | 0.0443 | 0.2360 | 21.48 | 48.49 | 0.02610 | 2.496 | 0.700 |
| 620 | 0.0406 | 0.2399 | 22.79 | 56.13 | 0.02792 | 2.867 | 0.704 |

## Table A-6    Properties of Gases at Atmospheric Pressure† *(Continued)*

| $T$, °F | $\rho$, $\dfrac{\text{lb}_m}{\text{ft}^3}$ | $c_p$, $\dfrac{\text{Btu}}{\text{lb}_m\text{-°F}}$ | $\mu$, $\dfrac{\text{lb}_m}{\text{sec-ft}}$ | $\nu$, $\dfrac{\text{ft}^2}{\text{sec}}$ | $k$, $\dfrac{\text{Btu}}{\text{hr-ft-°F}}$ | $\alpha$, $\dfrac{\text{ft}^2}{\text{hr}}$ | Pr |
|---|---|---|---|---|---|---|---|
| **Nitrogen** | | | | | | | |
| −100 | 0.1068 | 0.2491 | $8.700 \times 10^{-6}$ | $8.146 \times 10^{-5}$ | 0.01054 | 0.3962 | 0.747 |
| 80 | 0.0713 | 0.2486 | 11.99 | 16.82 | 0.01514 | 0.8542 | 0.713 |
| 260 | 0.533 | 0.2498 | 14.77 | 27.71 | 0.01927 | 1.447 | 0.691 |
| 440 | 0.0426 | 0.2521 | 17.27 | 40.54 | 0.02302 | 2.143 | 0.684 |
| 620 | 0.0355 | 0.2569 | 19.56 | 55.10 | 0.02646 | 2.901 | 0.686 |
| 800 | 0.0308 | 0.2620 | 21.59 | 70.10 | 0.02960 | 3.668 | 0.691 |
| 980 | 0.0267 | 0.2681 | 23.41 | 87.68 | 0.03241 | 4.528 | 0.700 |
| 1160 | 0.0237 | 0.2738 | 25.19 | 98.02 | 0.03507 | 5.404 | 0.711 |
| 1340 | 0.0213 | 0.2789 | 26.88 | 126.2 | 0.03741 | 6.297 | 0.724 |
| 1520 | 0.0194 | 0.2832 | 28.41 | 146.4 | 0.03958 | 7.204 | 0.736 |
| 1700 | 0.0178 | 0.2875 | 29.90 | 168.0 | 0.04151 | 8.111 | 0.748 |
| **Carbon dioxide** | | | | | | | |
| −64 | 0.1544 | 0.187 | $7.462 \times 10^{-6}$ | $4.833 \times 10^{-5}$ | 0.006243 | 0.2294 | 0.818 |
| −10 | 0.1352 | 0.192 | 8.460 | 6.257 | 0.007444 | 0.2868 | 0.793 |
| 80 | 0.1122 | 0.208 | 10.051 | 8.957 | 0.009575 | 0.4103 | 0.770 |
| 170 | 0.0959 | 0.215 | 11.561 | 12.05 | 0.01183 | 0.5738 | 0.755 |
| 260 | 0.0838 | 0.225 | 12.98 | 15.49 | 0.01422 | 0.7542 | 0.738 |
| 350 | 0.0744 | 0.234 | 14.34 | 19.27 | 0.01674 | 0.9615 | 0.721 |
| 440 | 0.0670 | 0.242 | 15.63 | 23.33 | 0.01937 | 1.195 | 0.702 |
| 530 | 0.0608 | 0.250 | 16.85 | 27.71 | 0.02208 | 1.453 | 0.685 |
| 620 | 0.0558 | 0.257 | 18.03 | 32.31 | 0.02491 | 1.737 | 0.608 |
| **Ammonia, $NH_3$** | | | | | | | |
| 32 | 0.495 | 0.520 | $6.285 \times 10^{-6}$ | $1.27 \times 10^{-4}$ | 0.0127 | 0.507 | 0.90 |
| 122 | 0.405 | 0.520 | 7.415 | 1.83 | 0.0156 | 0.744 | 0.88 |
| 212 | 0.0349 | 0.534 | 8.659 | 2.48 | 0.0189 | 1.015 | 0.87 |
| 302 | 0.0308 | 0.553 | 9.859 | 3.20 | 0.0226 | 1.330 | 0.87 |
| 392 | 0.0275 | 0.572 | 11.08 | 4.03 | 0.0270 | 1.713 | 0.84 |

† From E. R. G. Eckert and R. M. Drake, "Heat and Mass Transfer," 2d ed., McGraw-Hill Book Company, New York, 1959.

The values of $C_p$, $\mu$, $k$, and Pr are not strongly pressure-dependent and may be used over a fairly wide range of pressures.

# Table A-7  Physical Properties of Some Common Low-melting-point Metals[†]

| Metal | Melting point, °F | Normal boiling point, °F | Temperature, °F | Density, $\dfrac{lb_m}{ft^3}$ | Viscosity $\times 10^3$, $\dfrac{lb_m}{ft\text{-}sec}$ | Heat capacity, $\dfrac{Btu}{lb_m\text{-}°F}$ | Thermal conductivity, $\dfrac{Btu}{hr\text{-}ft\text{-}°F}$ | Prandtl number |
|---|---|---|---|---|---|---|---|---|
| Bismuth | 520 | 2691 | 600 | 625 | 1.09 | 0.0345 | 9.5 | 0.014 |
|  |  |  | 1400 | 591 | 0.53 | 0.0393 | 9.0 | 0.0084 |
| Lead | 621 | 3159 | 700 | 658 | 1.61 | 0.038 | 9.3 | 0.024 |
|  |  |  | 1300 | 633 | 0.92 | 0.037 | 8.6 | 0.016 |
| Lithium | 354 | 2403 | 400 | 31.6 | 0.40 | 1.0 | 22.0 | 0.065 |
|  |  |  | 1800 | 27.6 | 0.28 | 1.0 |  |  |
| Mercury | −38.0 | 675 | 50 | 847 | 1.07 | 0.033 | 4.7 | 0.027 |
|  |  |  | 600 | 802 | 0.58 | 0.032 | 8.1 | 0.0084 |
| Potassium | 147 | 1400 | 300 | 50.4 | 0.25 | 0.19 | 26.0 | 0.0066 |
|  |  |  | 1300 | 42.1 | 0.09 | 0.18 | 19.1 | 0.0031 |
| Sodium | 208 | 1621 | 400 | 56.3 | 0.29 | 0.32 | 46.4 | 0.0072 |
|  |  |  | 1300 | 48.6 | 0.12 | 0.30 | 34.5 | 0.0038 |
| Sodium potassium: |  |  |  |  |  |  |  |  |
| 22% Na | 66.2 | 1518 | 200 | 53.0 | 0.330 | 0.226 | 14.1 | 0.019 |
|  |  |  | 1400 | 43.1 | 0.0981 | 0.211 |  |  |
| 56% Na | 12 | 1443 | 200 | 55.4 | 0.390 | 0.270 | 14.8 | 0.026 |
|  |  |  | 1400 | 46.2 | 0.108 | 0.249 | 16.7 | 0.058 |
| Lead bismuth, 44.5% Pb | 257 | 3038 | 550 | 646 | 1.18 | 0.035 | 6.20 | 0.024 |
|  |  |  | 1200 | 614 | 0.77 |  |  |  |

[†] From J. G. Knudsen and D. L. Katz, "Fluid Dynamics and Heat Transfer," McGraw-Hill Book Company, New York, 1958.

**Table A-8  Diffusion Coefficients of Gases and Vapors in Air (at 25°C and 1 atm)†**

| Substance | $D, \dfrac{cm^2}{sec}$ | $Sc = \dfrac{\nu}{D}$ |
|---|---|---|
| Ammonia | 0.28 | 0.78 |
| Carbon dioxide | 0.164 | 0.94 |
| Hydrogen | 0.410 | 0.22 |
| Oxygen | 0.206 | 0.75 |
| Water | 0.256 | 0.60 |
| Ethyl ether | 0.093 | 1.66 |
| Methanol | 0.159 | 0.97 |
| Ethyl alcohol | 0.119 | 1.30 |
| Formic acid | 0.159 | 0.97 |
| Acetic acid | 0.133 | 1.16 |
| Aniline | 0.073 | 2.14 |
| Benzene | 0.088 | 1.76 |
| Toluene | 0.084 | 1.84 |
| Ethyl benzene | 0.077 | 2.01 |
| Propyl benzene | 0.059 | 2.62 |

† From J. H. Perry (ed.), "Chemical Engineers' Handbook," 4th ed., McGraw-Hill Book Company, New York, 1963.

## Table A-9   Properties of Water (Saturated Liquid)†

| °F | $c_p$, $\dfrac{\text{Btu}}{\text{lb}_m\text{-}°\text{F}}$ | $\rho$, $\dfrac{\text{lb}_m}{\text{ft}^3}$ | $\mu$, $\dfrac{\text{lb}_m}{\text{ft-hr}}$ | $k$, $\dfrac{\text{Btu}}{\text{hr-ft-}°\text{F}}$ | Pr, $\dfrac{c_p\mu}{k}$ | $\dfrac{g\beta\rho^2 c_p}{\mu k}$, $\dfrac{1}{\text{ft}^3\text{-}°\text{F}}$ |
|---|---|---|---|---|---|---|
| 32  | 1.009 | 62.42 | 4.33 | 0.327 | 13.35 | |
| 40  | 1.005 | 62.42 | 3.75 | 0.332 | 11.35 | $0.3 \times 10^8$ |
| 50  | 1.002 | 62.38 | 3.17 | 0.338 | 9.40  | $1.0 \times 10^8$ |
| 60  | 1.000 | 62.34 | 2.71 | 0.344 | 7.88  | $1.7 \times 10^8$ |
| 70  | 0.998 | 62.27 | 2.37 | 0.349 | 6.78  | $2.3 \times 10^8$ |
| 80  | 0.998 | 62.17 | 2.08 | 0.355 | 5.85  | $3.0 \times 10^8$ |
| 90  | 0.997 | 62.11 | 1.85 | 0.360 | 5.12  | $3.9 \times 10^8$ |
| 100 | 0.997 | 61.99 | 1.65 | 0.364 | 4.53  | $5.2 \times 10^8$ |
| 110 | 0.997 | 61.84 | 1.49 | 0.368 | 4.04  | $6.6 \times 10^8$ |
| 120 | 0.997 | 61.73 | 1.36 | 0.372 | 3.64  | $7.7 \times 10^8$ |
| 130 | 0.998 | 61.54 | 1.24 | 0.375 | 3.30  | $8.9 \times 10^8$ |
| 140 | 0.998 | 61.39 | 1.14 | 0.378 | 3.01  | $10.2 \times 10^8$ |
| 150 | 0.999 | 61.20 | 1.04 | 0.381 | 2.73  | $12.0 \times 10^8$ |
| 160 | 1.000 | 61.01 | 0.97 | 0.384 | 2.53  | $13.9 \times 10^8$ |
| 170 | 1.001 | 60.79 | 0.90 | 0.386 | 2.33  | $15.5 \times 10^8$ |
| 180 | 1.002 | 60.57 | 0.84 | 0.389 | 2.16  | $17.1 \times 10^8$ |
| 190 | 1.003 | 60.35 | 0.79 | 0.390 | 2.03  | |
| 200 | 1.004 | 60.13 | 0.74 | 0.392 | 1.90  | |
| 220 | 1.007 | 59.63 | 0.65 | 0.395 | 1.66  | |
| 240 | 1.010 | 59.10 | 0.59 | 0.396 | 1.51  | |
| 260 | 1.015 | 58.51 | 0.53 | 0.396 | 1.36  | |
| 280 | 1.020 | 57.94 | 0.48 | 0.396 | 1.24  | |
| 300 | 1.026 | 57.31 | 0.45 | 0.395 | 1.17  | |
| 350 | 1.044 | 55.59 | 0.38 | 0.391 | 1.02  | |
| 400 | 1.067 | 53.65 | 0.33 | 0.384 | 1.00  | |
| 450 | 1.095 | 51.55 | 0.29 | 0.373 | 0.85  | |
| 500 | 1.130 | 49.02 | 0.26 | 0.356 | 0.83  | |
| 550 | 1.200 | 45.92 | 0.23 | | | |
| 600 | 1.362 | 42.37 | 0.21 | | | |

† From A. I. Brown and S. M. Marco, "Introduction to Heat Transfer," 3d ed., McGraw-Hill Book Company, New York, 1958.

## Table A-10   Normal Total Emissivity of Various Surfaces†

| Surface | $T$, °F | Emissivity, $\epsilon$ |
|---|---|---|
| Metals and their oxides | | |
| | | |
| Aluminum: | | |
|   Highly polished plate, 98.3% pure | 440–1070 | 0.039–0.057 |
|   Commercial sheet | 212 | 0.09 |
|   Heavily oxidized | 299–940 | 0.20–0.31 |
|   Al-surfaced roofing | 100 | 0.216 |
| Brass: | | |
|   Highly polished: | | |
|     73.2% Cu, 26.7% Zn | 476–674 | 0.028–0.031 |
|     62.4% Cu, 36.8% Zn, 0.4% Pb, 0.3% Al | 494–710 | 0.033–0.037 |
|     82.9% Cu, 17.0% Zn | 530 | 0.030 |
|   Hard-rolled, polished, but direction of polishing visible | 70 | 0.038 |
|   Dull plate | 120–660 | 0.22 |
|   Chromium (see nickel alloys for Ni-Cr steels), polished | 100–2000 | 0.08–0.36 |
| Copper: | | |
|   Polished | 242 | 0.023 |
| | 212 | 0.052 |
|   Plate, heated long time, covered with thick oxide layer | 77 | 0.78 |
| Gold, pure, highly polished | 440–1160 | 0.018–0.035 |
| Iron and steel (not including stainless): | | |
|   Steel, polished | 212 | 0.066 |
|   Iron, polished | 800–1880 | 0.14–0.38 |
|   Cast iron, newly turned | 72 | 0.44 |
|   Cast iron, turned and heated | 1620–1810 | 0.60–0.70 |
|   Mild steel; A | 450–1950 | 0.20–0.32 |
| Oxidized surfaces: | | |
|   Iron plate, pickled, then rusted red | 68 | 0.61 |
|   Iron, dark-gray surface | 212 | 0.31 |
|   Rough ingot iron | 1700–2040 | 0.87–0.95 |
|   Sheet steel with strong, rough oxide layer | 75 | 0.80 |
| Lead: | | |
|   Unoxidized, 99.96% pure | 260–440 | 0.057–0.075 |
|   Gray oxidized | 75 | 0.28 |
|   Oxidized at 300°F | 390 | 0.63 |
| Magnesium, magnesium oxide | 530–1520 | 0.55–0.20 |
| Molybdenum: | | |
|   Filament | 1340–4700 | 0.096–0.202 |
|   Massive, polished | 212 | 0.071 |
| Monel metal, oxidized at 1110°F | 390–1110 | 0.41–0.46 |
| Nickel: | | |
|   Polished | 212 | 0.072 |
|   Nickel oxide | 1200–2290 | 0.59–0.86 |

## Table A-10    Normal Total Emissivity of Various Surfaces† (*Continued*)

| Surface | T, °F | Emissivity, ε |
|---|---|---|
| Nickel alloys: | | |
| Copper nickel, polished | 212 | 0.059 |
| Nichrome wire, bright | 120–1830 | 0.65–0.79 |
| Nichrome wire, oxidized | 120–930 | 0.95–0.98 |
| Platinum; polished plate, pure | 440–1160 | 0.054–0.104 |
| Silver: | | |
| Polished, pure | 440–1160 | 0.020–0.032 |
| Polished | 100–700 | 0.022–0.031 |
| Stainless steels: | | |
| Polished | 212 | 0.074 |
| Type 301; B | 450–1725 | 0.54–0.63 |
| Tin, bright tinned iron | 76 | 0.043 and 0.064 |
| Tungsten, filament | 6000 | 0.39 |
| Zinc, galvanized sheet iron, fairly bright | 82 | 0.23 |
| Refractories, building materials, paints, and miscellaneous | | |
| Alumina (85–99.5% $Al_2O_3$, 0–12% $SiO_2$, 0–1% $Ge_2O_3$); | | |
| effect of mean grain size, microns ($\mu$): | | |
| 10 $\mu$ | | 0.30–0.18 |
| 50 $\mu$ | | 0.39–0.28 |
| 100 $\mu$ | | 0.50–0.40 |
| Asbestos, board | 74 | 0.96 |
| Brick: | | |
| Red, rough, but no gross irregularities | 70 | 0.93 |
| Fireclay | 1832 | 0.75 |
| Carbon: | | |
| T-carbon (Gebruder Siemens) 0.9% ash, started with emissivity of 0.72 at 260°F but on heating changed to values given | 260–1160 | 0.81–0.79 |
| Filament | 1900–2560 | 0.526 |
| Rough plate | 212–608 | 0.77 |
| Lampblack, rough deposit | 212–932 | 0.84–0.78 |
| Concrete tiles | 1832 | 0.63 |
| Enamel, white fused, on iron | 66 | 0.90 |
| Glass: | | |
| Smooth | 72 | 0.94 |
| Pyrex, lead, and soda | 500–1000 | 0.95–0.85 |
| Paints, lacquers, varnishes: | | |
| Snow-white enamel varnish on rough iron plate | 73 | 0.906 |
| Black shiny lacquer, sprayed on iron | 76 | 0.875 |
| Black shiny shellac on tinned iron sheet | 70 | 0.821 |
| Black matte shellac | 170–295 | 0.91 |
| Black or white lacquer | 100–200 | 0.80–0.95 |
| Flat black lacquer | 100–200 | 0.96–0.98 |

**Table A-10    Normal Total Emissivity of Various Surfaces**† (*Continued*)

| Surface | $T$, °F | Emissivity, $\epsilon$ |
|---|---|---|
| Paints, lacquers, varnishes: | | |
|   Aluminum paints and lacquers: | | |
|     10% Al, 22% lacquer body, on rough or smooth | | |
|       surface | 212 | 0.52 |
|     Other Al paints, varying age and Al content | 212 | 0.27–0.67 |
| Porcelain, glazed | 72 | 0.92 |
| Quartz, rough, fused | 70 | 0.93 |
| Roofing paper | 69 | 0.91 |
| Rubber, hard, glossy plate | 74 | 0.94 |
| Water | 32–212 | 0.95–0.963 |

† Courtesy of H. C. Hottel, from W. H. McAdams, "Heat Transmission," 3d ed., McGraw-Hill Book Company, New York, 1954.

## Table A-11    Steel-pipe Dimensions

| Nominal pipe size, in. | Outside diam., in. | Sched- ule no. | Wall thick- ness, in. | Inside diam., in. | Metal sectional area, in.$^2$ | Inside cross- sectional area, ft$^2$ |
|---|---|---|---|---|---|---|
| $\frac{1}{8}$ | 0.405 | 40 | 0.068 | 0.269 | 0.072 | 0.00040 |
|  |  | 80 | 0.095 | 0.215 | 0.093 | 0.00025 |
| $\frac{1}{4}$ | 0.540 | 40 | 0.088 | 0.364 | 0.125 | 0.00072 |
|  |  | 80 | 0.119 | 0.302 | 0.157 | 0.00050 |
| $\frac{3}{8}$ | 0.675 | 40 | 0.091 | 0.493 | 0.167 | 0.00133 |
|  |  | 80 | 0.126 | 0.423 | 0.217 | 0.00098 |
| $\frac{1}{2}$ | 0.840 | 40 | 0.109 | 0.622 | 0.250 | 0.00211 |
|  |  | 80 | 0.147 | 0.546 | 0.320 | 0.00163 |
| $\frac{3}{4}$ | 1.050 | 40 | 0.113 | 0.824 | 0.333 | 0.00371 |
|  |  | 80 | 0.154 | 0.742 | 0.433 | 0.00300 |
| 1 | 1.315 | 40 | 0.133 | 1.049 | 0.494 | 0.00600 |
|  |  | 80 | 0.179 | 0.957 | 0.639 | 0.00499 |
| $1\frac{1}{2}$ | 1.900 | 40 | 0.145 | 1.610 | 0.799 | 0.01414 |
|  |  | 80 | 0.200 | 1.500 | 1.068 | 0.01225 |
|  |  | 160 | 0.281 | 1.338 | 1.429 | 0.00976 |
| 2 | 2.375 | 40 | 0.154 | 2.067 | 1.075 | 0.02330 |
|  |  | 80 | 0.218 | 1.939 | 1.477 | 0.02050 |
| 3 | 3.500 | 40 | 0.216 | 3.068 | 2.228 | 0.05130 |
|  |  | 80 | 0.300 | 2.900 | 3.016 | 0.04587 |
| 4 | 4.500 | 40 | 0.237 | 4.026 | 3.173 | 0.08840 |
|  |  | 80 | 0.337 | 3.826 | 4.407 | 0.07986 |
| 5 | 5.563 | 40 | 0.258 | 5.047 | 4.304 | 0.1390 |
|  |  | 80 | 0.375 | 4.813 | 6.112 | 0.1263 |
|  |  | 120 | 0.500 | 4.563 | 7.953 | 0.1136 |
|  |  | 160 | 0.625 | 4.313 | 9.696 | 0.1015 |
| 6 | 6.625 | 40 | 0.280 | 6.065 | 5.584 | 0.2006 |
|  |  | 80 | 0.432 | 5.761 | 8.405 | 0.1810 |
| 10 | 10.75 | 40 | 0.365 | 10.020 | 11.90 | 0.5475 |
|  |  | 60 | 0.500 | 9.750 | 16.10 | 0.5185 |

## Table A-12   Conversion Factors

Length:
  12 in. = 1 ft
  2.54 cm = 1 in.
  $1 \mu = 10^{-6}$ m $= 10^{-4}$ cm
Mass:
  1 kg = 2.205 $lb_m$
  1 slug = 32.16 $lb_m$
  454 g = 1 $lb_m$
Force:
  1 dyne $= 2.248 \times 10^{-6}$ $lb_f$
  1 $lb_f$ = 4.448 newtons
  $10^5$ dynes = 1 newton
Energy:
  1 ft-$lb_f$ = 1.356 joules
  1 kwhr = 3413 Btu
  1 hp-hr = 2545 Btu
  1 Btu = 252 cal
  1 Btu = 778 ft-$lb_f$
Pressure:
  1 atm = 14.696 $lb_f$/in.$^2$ = 2116 $lb_f$/ft$^2$
  1 atm $= 1.01325 \times 10^5$ newtons/m$^2$
  1 in. Hg = 70.73 $lb_f$/ft$^2$
Viscosity:
  1 centipoise = 2.42 $lb_m$/hr-ft
  1 $lb_f$-sec/ft$^2$ = 32.16 $lb_m$/sec-ft
Thermal conductivity:
  1 cal/sec-cm-°C = 242 Btu/hr-ft-°F
  1 watt/cm-°C = 57.79 Btu/hr-ft-°F

Useful conversions to SI Units

Length:
  1 in. = 0.0254 m
  1 ft = 0.3048 m
  1 mile = 1.60934 km
Area:
  1 in.$^2$ = 645.16 mm$^2$
  1 ft$^2$ = 0.092903 m$^2$
  1 sq mile = 2.58999 km$^2$
Volume:
  1 in.$^3$ $= 1.63871 \times 10^{-5}$ m$^3$
  1 ft$^3$ = 0.0283168 m$^3$
  1 gal = 231 in.$^3$ = 0.004546092 m$^3$
Mass:
  1 $lb_m$ = 0.45359237 kg
Density:
  1 $lb_m$/in.$^3$ $= 2.76799 \times 10^4$ kg/m$^3$
  1 $lb_m$/ft$^3$ = 16.0185 kg/m$^3$
Force:
  1 dyne $= 10^{-5}$ N
  1 $lb_f$ = 4.44822 N

Pressure:

   1 atm $= 1.01325 \times 10^5$ N/m$^2$

   1 lb$_f$/in.$^2$ $= 6894.76$ N/m$^2$

Energy:

   1 erg $= 10^{-7}$ J

   1 Btu $= 1055.06$ J

   1 ft-lb$_f$ $= 1.35582$ J

   1 cal (15°C) $= 4.1855$ J

Power:

   1 hp $= 745.7$ W

   1 Btu/hr $= 0.293$ W

Heat flux:

   1 Btu/hr-ft$^2$ $= 3.15372$ W/m$^2$

   1 Btu/hr-ft $= 0.96128$ W/m

Thermal conductivity:

   1 Btu/hr-ft-°F $= 1.730278$ W/m$^2$-°C

Heat-transfer coefficient

   1 Btu/hr-ft$^2$-°F $= 5.67683$ W/m$^2$-°C

# APPENDIX B

# EXACT SOLUTIONS TO THE LAMINAR-BOUNDARY-LAYER EQUATIONS

## EXACT SOLUTIONS TO THE LAMINAR - BOUNDARY - LAYER EQUATIONS

We wish to obtain a solution to the laminar-boundary-layer momentum and energy equations, assuming constant fluid properties and zero pressure gradient. We have

Continuity:
$$\frac{\partial u}{\partial x} + \frac{\partial v}{\partial y} = 0 \qquad \text{(B-1)}$$

Momentum:
$$u \frac{\partial u}{\partial x} + v \frac{\partial u}{\partial y} = \nu \frac{\partial^2 u}{\partial y^2} \qquad \text{(B-2)}$$

Energy:
$$u \frac{\partial T}{\partial x} + v \frac{\partial T}{\partial y} = \alpha \frac{\partial^2 T}{\partial y^2} \qquad \text{(B-3)}$$

It will be noted that the viscous-dissipation term is omitted from the energy equation for the present. In accordance with the order-of-magnitude analysis of Sec. 6-1,

$$\delta \sim \sqrt{\frac{\nu x}{u_\infty}} \qquad \text{(B-4)}$$

The assumption is now made that the velocity profiles have similar shapes at various distances from the leading edge of the flat plate. The significant variable is then $y/\delta$, and we assume that the velocity may be expressed as a function of this variable. We then have

$$\frac{u}{u_\infty} = g\left(\frac{y}{\delta}\right)$$

Introducing the order-of-magnitude estimate for $\delta$ from (B-4),

$$\frac{u}{u_\infty} = g(\eta) \qquad \text{(B-5)}$$

where
$$\eta = \frac{y}{\sqrt{\nu x/u_\infty}} = y \sqrt{\frac{u_\infty}{\nu x}} \qquad \text{(B-6)}$$

$\eta$ is called the similarity variable, and $g(\eta)$ is the function we seek as a solution. In accordance with the continuity equation, a stream function $\psi$ may be defined so that

$$u = \frac{\partial \psi}{\partial y} \tag{B-7}$$

$$v = -\frac{\partial \psi}{\partial x} \tag{B-8}$$

Inserting (B-7) in (B-5) gives

$$\psi = \int u_\infty g(\eta)\, dy = \int u_\infty \sqrt{\frac{\nu x}{u_\infty}}\, g(\eta)\, d\eta$$

or

$$\psi = u_\infty \sqrt{\frac{\nu x}{u_\infty}}\, f(\eta) \tag{B-9}$$

where $f(\eta) = \int g(\eta)\, d\eta$.

From (B-8) and (B-9) we obtain

$$v = \frac{1}{2} \sqrt{\frac{\nu u_\infty}{x}} \left( \eta \frac{df}{d\eta} - f \right) \tag{B-10}$$

Making similar transformations on the other terms in Eq. (B-2), we obtain

$$f \frac{d^2 f}{d\eta^2} + 2 \frac{d^3 f}{d\eta^3} = 0 \tag{B-11}$$

This is an ordinary differential equation, which may be solved numerically for the function $f(\eta)$. The boundary conditions are

|  Physical coordinates: |  | Similarity coordinates: |  |
| --- | --- | --- | --- |
| $u = 0$ | at $y = 0$ | $\dfrac{df}{d\eta} = 0$ | at $\eta = 0$ |
| $v = 0$ | at $y = 0$ | $f = 0$ | at $\eta = 0$ |
| $\dfrac{\partial u}{\partial y} = 0$ | at $y \to \infty$ | $\dfrac{df}{d\eta} = 1.0$ | at $\eta \to \infty$ |

The first solution to Eq. (B-11) was obtained by Blasius.[†] The values of $u$ and $v$ as obtained from this solution are presented in Fig. B-1.

The energy equation is solved in a similar manner by first defining a dimensionless temperature variable as

$$\theta(\eta) = \frac{T(\eta) - T_w}{T_\infty - T_w} \tag{B-12}$$

where it is also assumed that $\theta$ and $T$ may be expressed as functions of the similarity variable $\eta$. Equation (B-3) then becomes

$$\frac{d^2\theta}{d\eta^2} + \tfrac{1}{2}\,\mathrm{Pr}\, f \frac{d\theta}{d\eta} = 0 \tag{B-13}$$

with the boundary conditions

$$\theta = \begin{cases} 0 & \text{at } y = 0,\ \eta = 0 \\ 1.0 & \text{at } y = \infty,\ \eta = \infty \end{cases}$$

[†] H. Blasius, *Z. Math. Phys.*, vol. 56, p. 1, 1908.

**Fig. B-1** Velocity profiles in laminar boundary layer.

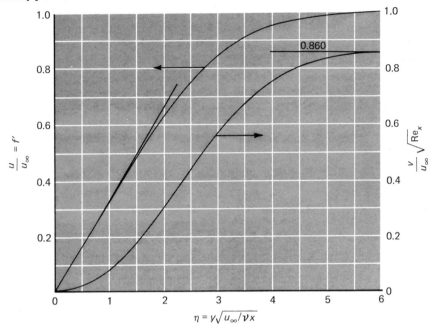

Given the function $f(\eta)$, the solution to Eq. (B-13) may be obtained as

$$\theta(\eta) = \frac{\displaystyle\int_0^\eta \exp\left(-\frac{Pr}{2}\int_0^\eta f\,d\eta\right)d\eta}{\displaystyle\int_0^\infty \exp\left(-\frac{Pr}{2}\int_0^\eta f\,d\eta\right)d\eta} \tag{B-14}$$

This solution is given by Pohlhausen[†] and is shown in Fig. B-2. For Prandtl numbers between 0.6 and 15 it was found that the dimensionless temperature gradient at the surface could be represented satisfactorily by

$$\frac{d\theta(\eta)}{d\eta}\Big)_{\eta=0} = 0.332\ Pr^{\frac{1}{3}} \tag{B-15}$$

The heat-transfer coefficient may subsequently be expressed by

$$Nu_x = 0.332\ Re_x^{\frac{1}{2}}\ Pr^{\frac{1}{3}} \tag{B-16}$$

in agreement with the results of Chap. 5.

Now let us consider a solution of the complete energy equation, including the viscous-dissipation term. We have

$$u\frac{\partial T}{\partial x} + v\frac{\partial T}{\partial x} = \alpha\frac{\partial^2 T}{\partial y^2} + \frac{\mu}{\rho c_p}\left(\frac{\partial u}{\partial y}\right)^2 \tag{B-17}$$

[†] E. Pohlhausen, Z. Angew. Math. Mech., vol. 1, p. 115, 1921.

Fig. B-2    Temperature profiles in laminar boundary layer with isothermal wall.

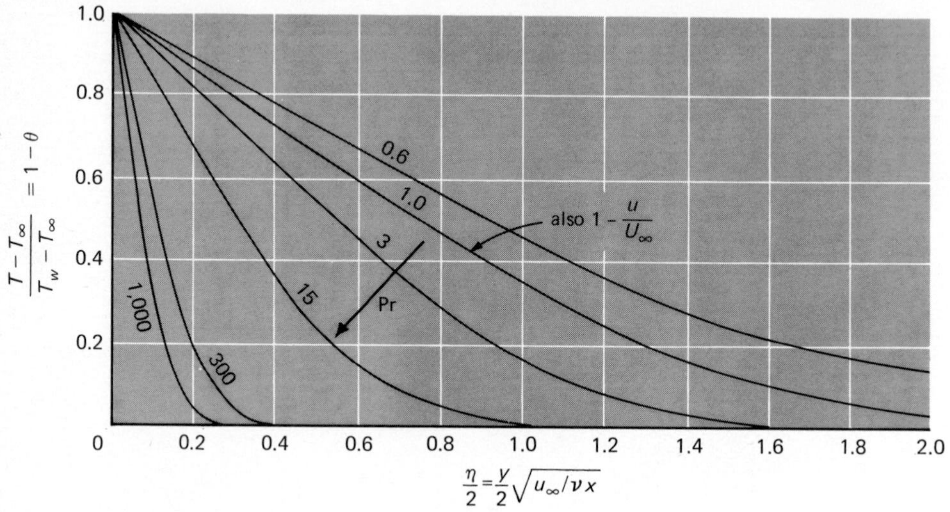

The solution to this equation is first obtained for the case of an adiabatic plate. Introducing a new dimensionless temperature profile in terms of the stagnation temperature $T_0$,

$$\theta(\eta) = \frac{T(\eta) - T_\infty}{T_0 - T_\infty} = \frac{T(\eta) - T_\infty}{u_\infty^2/2c_p}$$

Eq. (B-17) becomes

$$\frac{d^2\theta}{d\eta^2} + \tfrac{1}{2}\operatorname{Pr} f \frac{d\theta}{d\eta} + 2\operatorname{Pr}\left(\frac{d^2f}{d\eta^2}\right)^2 = 0 \qquad (\text{B-18})$$

For the adiabatic-wall case the boundary conditions are

$$\frac{d\theta}{d\eta} = 0 \qquad \text{at } y = 0, \ \eta = 0$$
$$\theta = 0 \qquad \text{at } y = \infty, \eta = \infty$$

The solution to Eq. (B-18) is given by Pohlhausen as

$$\theta_a(\eta,\operatorname{Pr}) = 2\operatorname{Pr}\int_\eta^\infty \left(\frac{d^2f}{d\eta^2}\right)^{\operatorname{Pr}}\left[\int_0^\eta\left(\frac{d^2f}{d\eta^2}\right)^{2-\operatorname{Pr}} d\eta\right] d\eta \qquad (\text{B-19})$$

where the symbol $\theta_a$ has been used to indicate the adiabatic-wall solution. A graphical plot of the solution is given in Fig. B-3. The recovery factor is given as

$$r = \theta_a(0,\operatorname{Pr})$$

For Prandtl numbers near unity this reduces to the relation given in Eq. (5-80),

$$r = \operatorname{Pr}^{\frac{1}{2}} \qquad (5\text{-}80)$$

Now consider the case where the wall is maintained at some temperature other than

**Fig. B-3**   Temperature profiles in laminar boundary layer with adiabatic wall.

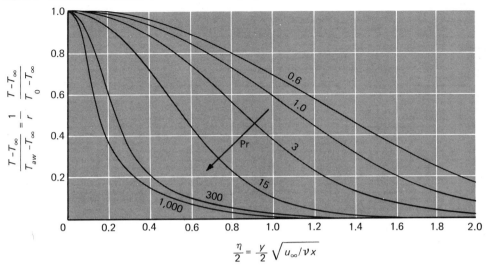

$T_{aw}$; that is, there is heat transfer either to or from the fluid. The boundary conditions are now expressed as

$$T = \begin{cases} T_w & \text{at } y = 0, \ \eta = 0 \\ T_\infty & \text{at } y = \infty, \ \eta = \infty \end{cases}$$

We observe that the viscous-heating term in Eq. (B-18) contributes a particular solution to the equation. If there were no viscous heating, the adiabatic-wall solution would yield a uniform temperature profile throughout the boundary layer. We now assume that the temperature profile for the combined case of a heated wall and viscous dissipation can be represented by a linear combination of the solutions given in Eqs. (B-14) and (B-19). This assumption is justified in view of the fact that Eq. (B-18) is linear in the dependent variable $\theta$. We then write

$$T - T_\infty = (T_a - T_\infty) + (T_c - T_{aw}) \tag{B-20}$$

where $T_a - T_\infty$ is the temperature distribution from Eq. (B-19) and $T_c - T_{aw}$ is the solution from Eq. (B-14), $T_{aw}$ taking the same role as $T_\infty$ in that solution. Equation (B-20) may be written

$$T - T_\infty = \theta_a(T_0 - T_\infty) + \theta(T_{aw} - T_w) + T_w - T_{aw} \tag{B-21}$$

This solution may be tested by inserting it in Eq. (B-18). There results

$$(T_0 - T_\infty)\left[\frac{d^2\theta_a}{d\eta^2} + \tfrac{1}{2}\Pr f \frac{d\theta_a}{d\eta} + 2\Pr\left(\frac{d^2f}{d\eta^2}\right)^2\right] + (T_{aw} - T_w)\left(\frac{d^2\theta}{d\eta^2} + \tfrac{1}{2}\Pr f \frac{d\theta}{d\eta}\right) = 0$$

An inspection of this relation indicates that Eq. (B-21) is a valid solution for the actual boundary conditions of the heated wall. The temperature gradient at the wall may thus be expressed as

$$\frac{\partial T}{\partial \eta}\Big)_{\eta=0} = (T_0 - T_\infty)\frac{\partial \theta_a}{\partial \eta}\Big)_{\eta=0} + (T_{aw} - T_w)\frac{\partial \theta}{\partial \eta}\Big)_{\eta=0}$$

The first term is zero, and Eq. (B-15) may be used to evaluate the second term. There results

$$\frac{\partial T}{\partial \eta}\Big)_{\eta=0} = 0.332(T_{aw} - T_w)\, \mathrm{Pr}^{\frac{1}{3}} \tag{B-22}$$

This relation immediately suggests the definition of the heat-transfer coefficient for the case where viscous heating is important by the relation

$$\frac{q}{A} = h(T_w - T_{aw}) \tag{B-23}$$

The analysis then proceeds as discussed in Sec. 5-11.

# APPENDIX C

# ANSWERS
# TO SELECTED
# PROBLEMS

1-5.  694°F

1-7.  7.2 Btu/hr

1-9.  2.38 × 10⁴ Btu/hr-ft²

1-11.  98.75°F

1-15.  450 Btu/hr-ft²

2-1.  0.586 ft

2-3.  128 Btu/hr-ft²

2-5.  1,425 ft²

2-11.  353°F

2-21.  70.9 Btu/hr

2-23.  78.3 Btu/hr

2-25.  830 Btu/hr

2-39.  41.5 Btu/hr

3-2.  8 percent

3-5.  14,700 Btu/hr

3-12.  237 Btu/hr

3-14.  6692 Btu/hr-ft

3-22.  522.5 Btu/hr-ft

4-3.  0.387 hr

4-7.  31.2 Btu/hr-ft²

4-9.  3.32 hr

4-11.  0.039 hr

4-15.  370°F, 337°F

4-17.  295°F

4-18.  1.1 hr

4-29.  0.394 hr

5-3.  0.00178 ft

5-5.  0.0352 ft/sec; 0.0249 ft/sec

5-7.  161.5 Btu/hr-ft

5-9.  0.56 ft

5-13.  1470 Btu/hr-ft²

5-15.  13,950 Btu/hr

5-19.  64/Re_d

5-25.  117,250 Btu/hr

5-27.  450 Btu/hr

5-34.  0.0073 ft

6-3.  115°F

6-5.  96.8°F

6-9.  55 ft

6-11.  133.5°F; 1.66 × 10⁶ Btu/hr

6-13.  4020 Btu/hr; 115°F

6-15.  0.777 lb_f/ft

6-17.  1290 Btu/hr-ft

6-25.  136 Btu/hr

6-29.  3.08 watts

7-4.  79.5 Btu/hr free convection
      73 Btu/hr forced convection

7-5.  126.3 Btu/hr

7-7.  497 Btu/hr-ft

7-9.  552 Btu/hr

7-16.  169 Btu/hr

7-18.  111,000 Btu/hr-ft²

7-23.  83.5 Btu/hr

7-25.  9.45 Btu/hr-ft²

8-2.  (a)  1.23 × 10⁴ Btu/hr-ft²
      (b)  2.29 × 10³ Btu/hr-ft²
      (c)  1.16 × 10² Btu/hr-ft²
      (d)  1.31 Btu/hr-ft²

8-5.  274°F

8-7.  8.20 Btu/hr

8-11.  2.54 × 10⁴ Btu/hr-ft with reflector

8-13.  77,000 Btu/hr

8-15.  259,000 Btu/hr lost by hot plate

8-17.  548°F; 8980 Btu/hr

8-19.  (a)  5.13 × 10³ Btu/hr-ft²
       (b)  5.13 × 10² Btu/hr-ft²
       (c)  1230°F

8-21.  555 Btu/hr

8-23.  280°F

8-25.  1140°F

8-29.  540°F

8-33.  1178°F

8-35.  242 Btu/hr; 9.75 Btu/hr

8-37.  476.4°F

9-4.  812 lb_m/hr-ft

9-5.  181.5 lb_m/hr

9-9.  34 percent

9-11.  0.00226 in.

9-13.  3.0 × 10⁵ Btu/hr

9-19.  26.3 lb_m/hr-ft

9-21.  2100 Btu/hr-ft

10-5.   Exchanger 2
10-7.   41.1 ft²; 0.294 percent
10-9.   80 percent
10-14.  2560 lb$_m$/hr
10-16.  67 ft²

11-2.   0.397 ft²/hr
11-5.   0.00124 lb$_m$/ft³
11-7.   2.57 × 10⁻⁵ lb$_m$/hr

12-1.   1519°R
12-2.   17.2 percent reduction
12-6.   304°F

10-21.  179°F
10-25.  50 ft²; 53.3 percent
10-27.  205.6°F
10.29.  215 ft²
10-31.  1.7 ft²

11-11.  58°F
11-12.  0.216 lb$_m$/hr

# INDEX